The Neurobiology - Psychotherapy - Pharmacology intervention triangle

The need for common sense in 21st century mental health

Edited by

João G. Pereira
Romão de Sousa Foundation, Casa de Alba, Portugal

Jorge Gonçalves
Universidade Nova de Lisboa, Portugal

Valeria Bizzari
University Clinic Heidelberg, Germany

Cognitive Science and Psychology

Vernon Press

www.vernonpress.com

In the Americas:
Vernon Press
1000 N West Street,
Suite 1200, Wilmington,
Delaware 19801
United States

In the rest of the world:
Vernon Press
C/Sancti Espiritu 17,
Malaga, 29006
Spain

Cognitive Science and Psychology

Library of Congress Control Number: 2018948069

ISBN: 978-1-62273-433-7

Table of Contents

Contributors

Birgitta Alakare, MD, Psychiatrist, Health District of Länsi-Pohja

João Albuquerque, MD, Clinical Director, Centro de Recuperação de Menores D. Manuel Trindade Salgueiro, Portugal

Sílvia Almeida, MSc, Clinical Psychologist, Champalimaud Research and Clinical Centre, Lisbon, Portugal

Pedro Amaro, Nurse, Centro de Recuperação de Menores D. Manuel Trindade Salgueiro, Portugal

Natália Andrade, Physiotherapist, Centro de Recuperação de Menores D. Manuel Trindade Salgueiro, Portugal

Claudinei Eduardo Biazoli Junior, MD, PhD, Professor, Universidade Federal do ABC, Brazil

Valeria Bizzari, PhD, Postdoctoral Researcher, Clinic for General Psychiatry, Centre of Psychosocial Medicine, Clinic University of Heidelberg, Heidelberg, Germany

Liberata Borralho, PhD Student, Department of Educational Sciences, University of Évora, Portugal

Inês Pinto Calisto, Research Fellow, Center for Research in Education and Psychology, University of Évora, Portugal

Chloe Campbell, PhD, Research Department of Clinical, Educational and Health Psychology, University College London, UK

Adelinda Araújo Candeias, PhD, Auxiliary Professor, Department of Psychology, University of Évora, Portugal

Giulio Caselli, MD, Psychiatrist, Policlinico Gemelli, Rome; Research Fellow, Catholic University of Sacro Cuore, Rome, Italy

Mário David, MD, Consultant Psychiatrist in CINTRA and Groupanalyst, Centro Hospitalar Psiquiátrico de Lisboa and Sociedade Portuguesa da Grupanálise, Portugal

Martin Debbané, Associate Professor and Director of the Developmental Clinical Psychology Research Unit, Faculty of Psychology and Educational Sciences, University of Geneva, Switzerland; Senior Lecturer, University College London, UK

Ana Carolina Florence, PhD Student, State University São Paulo, Brasil, Fellow with the Psychiatry Department at Yale University

Peter Fonagy, FBA, OBE, FAcSS, FMedSci, PhD, Anna Freud National Center for Children and Families, London, UCL; Research Department of Clinical, Educational and Health Psychology, University College London, UK

Alexandre Gomes, MD, Psychiatrist, Faro District Hospital, Portugal

Inês Gomes, Speech Therapist, Centro de Recuperação de Menores D. Manuel Trindade Salgueiro, Portugal

Jaime Grácio, PhD, Clinical Psychologist, Champalimaud Research and Clinical Centre, Lisbon, Portugal; Assistant Professor of Behavioural Medicine, NOVA Medical School, Universidade Nova de Lisboa, Portugal

Daniel Guerra, MSc, Psychologist in Training, Metalentejo, Portugal

Rex Haigh, MD, Consultant Medical Psychotherapist, Berkshire NHS, Honorary Professor of Therapeutic Environments and Relational Health, Nottingham University's School of Sociology and Social Policy

Robert Hinshelwood, Emeritus Professor, Center for Psychoanalyitic Studies, University of Essex; Fellow of the British Psychoanalytical Society and of the Royal College of Psychiatrists

Anabela Janeco, Social Worker, Centro de Recuperação de Menores D. Manuel Trindade Salgueiro, Portugal

Silvio Yasui, PhD, Full Professor, State University of São Paulo, Brazil

Margarida Morais, Nursing Director, Centro de Recuperação de Menores D. Manuel Trindade Salgueiro, Portugal

Tobias Nolte, MD, MSc, Wellcome Trust Center for Neuroimaging, University College London, United Kingdom; Anna Freud National Centre for Children and Families, London, UK

Albino J. Oliveira-Maia, MD, MPH, PhD, Head of the Neuropsychiatry Unit, Champalimaud Research and Clinical Centre, Lisbon, Portugal; Psychiatrist, Centro Hospitalar de Lisboa Ocidental, Portugal; Assistant Professor of Neuroscience and Psychiatry, NOVA Medical School, Universidade Nova de Lisboa, Portugal

Liliana Pacheco, Social Worker, Centro de Recuperação de Menores D. Manuel Trindade Salgueiro, Portugal

Dora Pereira, PhD, Auxiliary Professor, University of Madeira, Department of Psychology, Portugal

João G. Pereira, PhD, Clinical Director, Romão de Sousa Foundation, Estremoz, Portugal

António Portelada, PhD Student, Department of Educational Sciences, University of Évora, Portugal

Victor Amorim Rodrigues, MD, PhD, Clinical and Forensic Psychiatrist, Lisbon, Senior Lecturer, ISPA University Institute, Lisbon, Portugal

Célia Sales, PhD, Psychologist, Family Therapist, Researcher at the University of Porto, Portugal

Patrícia Santos, Psychologist, Centro de Recuperação de Menores D. Manuel Trindade Salgueiro, Portugal

Jaakko Seikkula, PhD, Professor of Psychotherapy, Department of Psychology, University of Jyväskylä

Elodie Toffel, Clinical Psychologist, University of Geneva, Switzerland

Pedro Valente, Psychomotor Therapist, Centro de Recuperação de Menores D. Manuel Trindade Salgueiro, Portugal

Josefa Ros Velasco, PhD, Postdoctoral Research Fellow and Teaching Assistant, Department of Romance Languages and Literatures, Harvard University

Foreword

Dr. Daniel J. Siegel
Clinical Professor, UCLA School of Medicine,
Founding Co-Director of the Mindful Awareness Research Center
at UCLA and Executive Director, Mindsight Institute

A plea for common sense in our modern age, *The Neurobiology - Psychotherapy – Pharmacology Intervention Triangle: The Need for Common Sense in 21st century Mental Health* is a book that provides an inspiring set of essays and summaries of cutting-edge science exploring the nature of our mental well-being and how we can embrace a complex array of frameworks as modern psychotherapists.

In the field of Interpersonal Neurobiology in which I work, we combine the broad range of scientific disciplines into one framework for understanding the mind and mental health. By examining the common findings across usually separate approaches, we are able to articulate "consilient" (Wilson, 1998) principles with which to view fundamental properties of nature that underlie our human journey. Within that view we see the mind as having at least four facets: Subjective Experience, Consciousness, Information Processing, and Self-Organization (Siegel, 2017). With the fourth facet, this aspect of the mind can be proposed to be defined as:

"The emergent, self-organizing embodied and relational process that regulates the flow of energy and information."

This definition then enables a mental health practitioner, an educator, a parent, or others helping the mind develop to address the next, natural question: What might a healthy mind be? From the mathematics of complex systems, the question then becomes how the mind might optimize its fundamental facet of self-organization. The study of complex systems—clusters of entities with the characteristics of being open, chaos-capable, and non-linear—reveal an answer with surprising application to the field of mental health. Optimal self-organization emerges with the linkage of differentiated parts of the system—something simply called integration—creating a state of criticality which "is the constantly shifting battle zone between stagnation and anarchy, the one place where a complex system can be spontaneous, adaptive, and alive." (Waldrop, 1992, page 12). Criticality is the mathematical space of a metaphoric river of integration that has a central flow of being flexible, adaptive, coherent, energized, and stable. That river of well-being is bounded on one bank by chaos, the other by the bank of rigidity.

Seen from this perspective, mental health arises from integration. Impairments to integration lead to chaos and rigidity. Integration then is the basis of health. Well-being, resilience, and flourishing arise from this common ground of integration. If differentiation or linkage, or both, are blocked, the flow of the system moves toward chaos or rigidity. An examination of descriptions of "mental dysfunction" and their symptoms can inspire a re-interpretation of deviations from well-being as consisting of chaos or rigidity or their combination. Health comes from integration; impairments to health result from impediments to integration and result in chaos and rigidity.

In this consilient approach, then, we can view the exciting discussions of this book as examples of this view even if these terms are not specifically utilized. For example, the brain is part of a larger neurobiological system in which energy and information flows. That is the neurobiological mechanism of energy and information flow in these discussions. Psychotherapy is a relational process in which energy and information are shared within the closely attuned interpersonal connections between therapist and client/patient. And psychopharmacology is sometimes needed to stabilize the neural flow of energy and information to stay within an integrative range of flexibility—that state of criticality—to enable the relational processes of integration to unfold. Seen this way, this book offers a natural unity across too commonly divided approaches to the individual within a social world.

The common finding among various studies of mental un-health or suffering and neuroscience studies can be seen as impaired integration in the brain. In bipolar illness, for example, linking inhibitory fibers between subcortical areas such as the amygdala are insufficient from the higher prefrontal regions (Chepinik et al., 2010). In the case of severe violations of attachment, such as with abuse and neglect, impaired integration is also found in a range of regions, including the prefrontal cortex, hippocampus, corpus callosum, and the interconnections of the "connectome" (See Teicher and colleagues).

The connectome is the term for the wide array of differentiated areas of the brain and how they are then linked to one another. The best predictor of well-being discovered in the Human Connectome Project was how interconnected the connectome is—how integrated the brain is predicts well-being (Smith et al., 2015).

From this line of empirical findings and this consilient thinking, the ways in which our inner energy and information flow (the "neurobiological" and the "psychopharmacological") is shaped interfaces with the interpersonal ways in which energy and information flow is shared within relationships, both with other people individually, and the environment of the social and natural worlds (see Siegel, 2017). Psychotherapy would entail a relational way in which the chaos and rigidity of mental life within and between were detected in the assessment and diagnostic phase of evaluation, and then therapeutic interventions would aim to take the identified impediments to integration and promote the necessary differentiation and linkage within the body and within the relational world (See Siegel, 2010a, and 2010b).

Taken from this Interpersonal Neurobiology perspective, this book offers exciting new discussions of how the field of mental health can blend the direct neurobiological and psychopharmacological interventions with psychotherapy assessment and treatment planning to create a consilient approach to the clinical field of mental health. For example, harnessing the power of practices that "train the mind" has been shown to enable a brain state to be repeatedly activated so that the neuroplastic changes in the brain can become traits of health in a person's life (Goleman and Davidson, 2017). Nothing excludes a therapist from utilizing this important discovery with medications used to stabilize self-regulation in dysregulated individuals to the sufficient state of equilibrium so that they can utilize mind-training practices that cultivate the research-established skills of focused attention, open awareness, and kind intention (Siegel, 2018). Medications alone may not cultivate integration in the brain, but they may at times be essential to stabilize a person's inner functioning so that they can then take part in therapeutic interactions and mind-training practices that can stimulate the growth of integration in the ever-growing brain.

The important discussions you are about to explore here offer an excellent way to consider these exciting times of integration in the field of mental health. Take in these new ideas and see how they fit with a new framework for understanding the mind and mental health in your own work and life.

References

Goleman, D and Davidson, RJ (2017). *Altered Traits.* New York: Penguin Random House.

Chepenik, L. G., Raffo, M., Hampson, M., Lacadie, C., Wang, F., Jones, M. M., et al. (2010). Functional connectivity between ventral prefrontal cortex and amygdala at low frequency in the resting state in bipolar disorder. *Psychiatry Research: Neuroimaging, 182,* 207–210.

McGreenery, C. E., Polcari, A., Samson, J. A., & Teicher, M. H. (2006). Sticks, stones, and hurtful words: Relative effects of various forms of childhood maltreatment. *American Journal of Psychiatry, 163,* 993–1000.

Siegel, DJ (1999/2012). *The Developing Mind.* New York: Guilford Press.

Siegel, DJ (2010a). *Mindsight.* New York: Bantam Random House.

Siegel, DJ (2010b). *The Mindful Therapist.* New York: WW Norton.

Siegel, DJ (2017). *Mind.* New York: WW Norton.

Siegel, DJ (2018). *Aware.* New York: Tarcher/Perigee Penguin Random House.

Smith, S. M., T. Nichols, D. Vidaurre, A. Winkler, T. Behrens, M. Glasser, K. Ugurbil, D. Barch, D. Van Essen, and L. Miller. (2015). A Positive-Negative Mode of Population Co-Variation Links Brain Connectivity, Demographics, and Behavior. *Nature Neuroscience* 18, no. 11: 1567–71

Teicher, M. H., & Andersen, S. L. (2009). Delayed effects of early stress on hippocampal development. *Journal of Child and Adolescent Psychopharmacology, 13*(1), 41–51.

Teicher, M. H., Andersen, S. L., Polcari, A., Anderson, C. M., & Navalta, C. P. (2002). Developmental neurobiology of childhood stress and trauma. *Psychiatric Clinics of North America, 25,* 397–426.

Teicher, M. H., Andersen, S. L., Polcari, A., Anderson, C. M., Navalta, C. P., & Kim, D. M. (2003). The neurobiological consequences of early stress and childhood maltreatment. *Neuroscience and Biobehavioral Reviews, 27*(1–2), 33–44.

Teicher, M. H., Dumont, N. L., Ito, Y., Vaituzis, C., Giedd, J. N., & Andersen, S. L. (2004). Childhood neglect is associated with reduced corpus callosum area. *Biological Psychiatry, 56*(2), 80–85.

Teicher, M. H., Samson, J. A., Polcari, A., & Andersen, S. L. (2009). Length of time between onset of childhood sexual abuse and emergence of depression in a young adult sample: A retrospective clinical report. *Journal of Clinical Psychiatry, 70*(5), 684–691.

Teicher, M. H., Samson, J. A., Sheu, Y., Polcari, A., & McGreenery, C. E. (2010). Hurtful words: Association of exposure to peer verbal abuse with elevated psychiatric symptom scores and corpus callosum abnormalities. *American Journal of Psychiatry, 167,* 1464–1471.

Teicher, M. H., Tomoda, A., & Andersen, S. L. (2006). Neurobiological consequences of early stress and childhood maltreatment: Are results from human and animal studies comparable? *Annals of the New York Academy of Sciences, 1071,* 313–323.

Wilson, E. O. (1998). *Consilience: The unity of knowledge.* New York: Vintage.

Chapter 1

The roots and Seeds of Humanistic Psychiatry

João G. Pereira
Romão de Sousa Foundation; Casa de Alba, Portugal

Jorge Gonçalves
Universidade Nova de Lisboa, Portugal

Valeria Bizzari
Centre for Psychosocial Medicine, Clinic for General Psychiatry University of Heidelberg, Germany

In the 18th and 19th century, the Western World went through dramatic changes in the understanding of mental health and in the provision of services. William Tuke (1732 - 1822) in the UK and Philippe Pinel (1745 - 1826) in France were in the forefront of a humanizing movement known as "moral treatment".

This humanization of services lost ground through the 19th and 20th centuries, giving rise to the maintenance and spreading of large mental health institutions where people suffering mental distress were severely separated/segregated from external reality and not given the rights of "normal" (adapted) people. A positive idea of asylum (with certain downsides – see below under "Critical Movements as base for a Humanistic Psychiatry") was replaced with segregation.

During the Second World War, a number of British doctors started the Therapeutic Community movement in the UK, in many ways returning to the ideas of Tuke and Pinel. They observed that transforming the environment of the "mentally ill" would also dramatically change their condition. A number of similar movements spread through Europe and the US giving rise to a new understanding of mental illness, sometimes even contesting the term illness itself, and often attempting treatment without medication.

Alongside the growing therapeutic community movement and the humanization of services another dominant trend was rising: the "technologization" and commodification of interventions based on value-free science. Whilst not the aim of that trend, it often dehumanized relationships. There is an implicit tension between "being-with" (humanistic values) and "doing-to" (technical expertise).

A "doing-to" stance has often been rooted in psychiatric manuals like DSM (Diagnostic and Statistical Manual of Mental Disorders) or ICD (International Statistical Classification of Diseases and Related Health Problems) which suggest the existence of objective states which are value-free and where concepts such as "autistic" or "schizophrenic" entered daily discourse and gained legitimacy. Increasingly we find ourselves in a "quick fix" culture dominated by a technical-rationality model of science, a change nicely termed by some authors as the "McDonaldization of Society" (Ritzer, 1993).

How can Being and Doing coexist in the service of patients and families? The relational paradigm and the scientific postmodern era arose at the same time that positivism and empiricism are growing. There are disparate movements of integration and sectarianism; important differences between affective and cognitive neuroscience; large gaps between theory and practice; contradictory evidence for and against "broken-brain" models. Is it possible for science to go back to "the ordinary" and start being human again, acknowledging the impossibility of separating figure from ground?

Neurobiology can be significantly modified through medication and psychotherapy, but also through play and occupational therapy, and by diet and lifestyle. The right weighting of the components of mental health, and the right measures of it can only be known through secure and trustworthy therapeutic relationships, helping to give meaning to interventions. The establishment of epistemic trust within psychologically enabling relationships is perhaps the only non-controversial ingredient of change, as research and practice consistently confirm (Pereira and Debbané, 2018; Norcross, 2002).

With this volume, we propose to open the debate between three main themes: psychotherapy (including psychological and philosophical influences), neurobiology (including cognitive and affective neuroscience) and psychopharmacology. The three main themes are clinically applied in what we call the "Intervention Triangle". The book is first focused on epistemologically distinct frameworks and gradually attempts to consider the integration of these three fundamental vertices of practice.

The volume will be particularly relevant to practitioners working towards integrative frameworks. Although unidisciplinary integration has been a theme in several research and theoretical publications, this book offers an interdisciplinary, comprehensive and reflexive view of mental health problems and approaches, avoiding developing into mere eclecticism.

Following a number of congresses, international meetings and the publication of Schizophrenia and Common Sense (Hipólito, Gonçalves and Pereira, eds. 2018) the first editor has decided to put together a second volume, grounded on the discussions undertaken during the Second International Mental Health Congress of Romão de Sousa Foundation. Several authors from this congress, and several others with an interest in the subject of integration and common sense mental health, agreed to publish and contribute towards the humanization and democratization of mental health services. This volume results from the honest effort of all the authors, editors, reviewers and assistants. We hope it can contribute towards common sense in 21st century mental health.

1.1 The "illnification" of mental distress

Biomedical views and the "illnification" of mental distress (see, for example, Szasz 1960) have dominated the field for some decades, many times tied up with socio-economic pressures, arising from a variety of sources, such as the pharmaceutical industry, health insurance companies, academia, the maintenance of professional status and power, and so on (see Florence et al. in this volume).

The biomedical model of mental health, alongside the "psychiatrization" and "psychologization" of mental distress, are parts of a wider sociocultural paradigm, the commodification of life in general and of mental health services in particular (see Hinshelwood in this volume). Mental distress, in this way, is viewed as a burden, an "illness" to get rid of, as quickly as possible and without pain. Medication fits very well into this "culture, providing hopes of "quick fixes". One possible problem arising from this view is that the "function" of mental distress is not fulfilled, the "message" it carries is ignored, and therefore the desired "quick fix" turns, in many cases, into chronicity and long term (dis)ability.

As mentioned further below in this chapter, opposition to reductionist biomedical and empirical views are not a new phenomenon, with the anti-psychiatry movement of the 60's and 70's being an example. A more balanced and integrative approach could be found in contemporary treatment systems such as the Finnish Open Dialogue (see Seikkula and Alakare in this volume), Mentalization-Based Therapies (see Nolte, Campbell and Fonagy, in this volume), or other rehabilitative programs, following WHO guidelines (see Gomes, in this volume). These models are strongly based in the therapeutic relationship as the central element, taking lessons from the vast amount of studies in the therapeutic alliance as the single most important factor for promoting change (Horvath and Bedi, 2002; Norcross, 2002). Medication is used but cautiously and reflexively, most notably in a selective and needs adapted way (Bergström et al., 2017). They also have in common the strong engagement with families, communities and other social networks, as well as the attention paid to attachment security, ruptures and repair in the working alliance (Safran, Murran and Eubanks-Carter, 2011). The different treatment systems are supported by research, from experimental to naturalistic studies (see this volume), and the results are not just promising but, as many practitioners have mentioned, rather obvious. The importance of consistency within services and teams is paramount, as demonstrated by the Open Dialogue system which, to operate in its full potential, should be adopted by the entire geographical catchment area (see Seikkula and Alakare, in this volume).

So, if the knowledge and success exist, what is needed in 21st Century Mental Health to make this humanistic culture dominant within mental health treatment systems? We claim, as the title of the book suggests, that it is Common Sense. The "manipulation" of macro-economic forces and the “illnification” of mental distress resemble, somehow, a collective psychosis, turning the common citizen apart from the compassion, tolerance and understanding that mental disturbance needs.

If societies express their wounds and disturbances through "symptoms" (rise in the number of suicides or terrorist attacks, persecutory culture, economic collapses or the rise in authoritarianism) so thus the individual. This essential form of communication must be heard and not silenced; it must be understood so that transformation is possible.

1.2 The Subjectivity of the Mind

The analysis of mental disorders necessarily requires a careful and multilayered reflection: in fact, we can affirm that psychiatry owns a “subject-object”, since it is focused on a complex entity which we cannot reduce to a mere empirical, spatio-temporal one. As Basaglia underlined (Basaglia, 2000), consciousness or subjectivity is the real core of psychiatry, its horizon of meaning. Human Beings are not objective entities: they are social animals, individuals, and subjective entities. In order to account for such complexity, Basaglia argued for an integrative approach, meaning to incorporate both psychiatry and philosophy, the only perspective able to understand mankind in its deepest core, to face life’s struggles, to account for the peculiarity of each person and of their authentic being.

In this view, pharmacotherapy is not entirely forgotten but is evaluated as a nonexhaustive perspective which cannot be considered the only source of recovery. The clinical gaze which is aimed at finding biological diagnosis and general treatments, following only causal-effect ways of thinking and empirical evidences, does not grasp the specificity of pain, which is always different, expressing itself in several ways.

On the other hand, neither a neurobiological approach can account for the complexity of sufferance: as Schneider (1919) has clearly stated psychiatry is not just a ramification of scientific-natural medicine, since pathological histology will never manage to understand psychic manifestations. Focusing only on brain disruption inevitably cuts off the experiential domain from the clinical setting. Pio Abreu, Fradique and Freire Lucas (2010) go even further, stating that psychiatry is a branch of medicine dealing with "information systems", rather than a medical specialty. Medical Specialties deal with sub-systems of the organism, whilst psychiatry, or to say better, psychic events, are, one way or another, involved in all body systems.

It seems that, in the analysis and treatment of mental disorders, one should choose between a "biological psychiatry" (according to which psychic events are neural events that we should treat through a pharmacological therapy) and a psychiatry which considers every psycho(patho)logical experience in its complexity.

In this volume, we argue for a shift from the predominant biomedical, empirical and pharmacological paradigm towards an integrative, interdisciplinary approach where pharmacological treatments are not the only source of healing, since the object is not a malfunctioning machine, but a suffering person. Jaspers (1959) emphasized the epistemological gulf between a biological perspective which tries to explain mental symptoms in terms of neurobiological malfunctions, and a psychological approach meant to understand mental disorders in terms of other mental phenomena. What one should recognize is that it is possible to embrace both perspectives, since a unilateral approach will never manage to account for the complexity of the human mind. This multiperspectival approach is called "methodological pluralism". In this sense, the study of mental disorders must stop over-relying on medication and focus on a more human and person-centered approach to treatment, that puts emphasis on the lived experiences of the patients and on the need for community-based interventions. In fact, a biological, reductionist approach is based upon a positivist epistemology which claims that empirical observations and rules of logic are the only sources of knowledge.

For this reason, psychopathology should not be a kind of biology only, but also part of the Humanities (Jaspers, 1959), because "any serious study of the mind, or 'psyche' must involve the consideration of consciousness, subjectivity, or the first-person point of view". (Parnas et al., 2012, p. 5). In this process, phenomenology plays a key role. This philosophical method of analysis is primarily directed towards the study of the facts of consciousness. Accordingly, it is also applicable to psychopathological facts, allowing to consider the patient as a consciousness, and taking into account the complexity of psychic life, the real core of psychological disorders. Phenomenology could be efficacious in understanding the patient without forgetting biological constraints: a vision of person as a psycho-physical entity is very useful for not underestimating either the psyche or the natural organism. This method has been used in the field of psychopathology: Jaspers's work (1959), in particular, represented a real revolution in psychiatry, since he used descriptive phenomenology in finding the implicit structures of subjectivity. Then, Binswanger (1956) applied existential analysis to the life history of patients: his work was primarily directed to the study of schizophrenia, but we can easily claim that it could be helpful in the study of other psychological conditions.

The advantages of adopting a phenomenological approach are both methodological and practical. From a methodological perspective, phenomenology allows one to focus on subjective experiences and not only on symptoms. Furthermore, the real object of psychopathology is the person and her/his subjective experiences instead of biological symptoms. From a practical point of view, phenomenology is helpful in hypothesizing therapy (Doerr-Zeggers & Stanghellini, 2013) and in modifying the relationship between the clinician and the patient, which becomes an empathic attitude where clinicians' primary purpose is understanding what it means to be in a certain psychopathological state, not merely in terms of brain disruptions, but in a more existential manner. The neurobiological causality, which belongs to a descriptive account, works together with the explanation of the first person subjective experiences (Sass, 2010). The result is a person-centered approach (Stanghellini & Aragona, 2016) which prioritizes patients' experiences and allows the clinician to consider them as individual consciousness.

1.3 Critical Movements as base for a Humanistic Psychiatry

1.3.1 Antipsychiatry

Although antipsychiatry in the general sense of the term is as old as psychiatry itself (Quétel, 2012), which emerges in the early nineteenth century, it is in the 1960s and 1970s that a movement known as "Antipsychiatry" emerges. Though some did not accept the label of "antipsychiatrist", the names of Michel Foucault, Thomas Szasz, Franco Basaglia, David Cooper and Ronald Laing among others were associated with this movement. We will not give a detailed description of the movement here, but only its general characterization insofar as it poses problems that would significantly influence the evolution of psychiatry.

Although there are several antipsychiatries, the common denominator to all of them is the struggle against the psychiatric institution synthesized in the figure of the physician and his power (Foucault, 1967; Rose, 2006). In this, anti-psychiatry met the anti-establishment movements of the time. We think that this critique of repressive psychiatric power is the most essential of the movement and its criticism of the concept of mental illness and of pharmacological therapies derived from this fundamental position.

"It seems to me that we could situate the different forms of anti-psychiatry according to their strategy in relation to these games of institutional power: to escape them in the form of a dual contract, freely consented to by both parties (Szasz); establishment of a privileged location where they should be suspended or refused if they are reconstituted (Kingsley Hall); blasts them one by one and destroy them progressively, inside a classic type institution (Cooper in Hall 21); to link them to other relations of power which, from outside the asylum, could already determine the segregation of an individual as mentally ill (Gorizia). Power relations constituted the a priori of psychiatric practice. "(Foucault, 1975, our translation, p. 102)

We will briefly mention here two philosophers, Michel Foucault and Thomas Szasz, who have clearly defined the ideas of the movement in theoretical terms. We will not present here the practices performed by the antipsychiatrists. *Madness and Civilization* (Foucault, 1967) was a book that influenced anti-psychiatrists, although Foucault's concerns were initially more epistemological. He argued in this book that the interpretation of madness as "mental illness" only occurs with the emergence of psychiatry in the early nineteenth century. The mentally ill persons are separated from other transgressors of the established order and locked up in asylums. Foucault does not see, however, in this movement any liberation but rather an attempt to adapt the patients to bourgeois society by psychological methods, the so-called "moral treatment." There appears a new figure, the doctor, a mixture of Judge and Father who is selected more by his moral capacities than medical. Later, under the influence of the positivist philosophy, a psychiatric knowledge will be elaborated, which serves essentially to justify the doctor's power over the sick. Anti-psychiatrists aimed to challenge the power of the institution, incarnated in the doctor's figure. They wanted to give the patient the right to take their "madness" by removing the doctor from the role of the one who knows what it is to be mad and to be normal.

Seeing that his book was having a lot of influence, Foucault took up the subject again, but in the meantime, he turned away from his initial conception of power, which he came to call the "repressive hypothesis." According to this conception, the psychiatric institution would be a microcosm of the society, being the power centered on the doctors. In *Psychiatric Power* (2006), Foucault reformulated this idea and maintained that power is not centralized in a subject (in this case, the medical class) but it is something diffuse and anonymous. Psychiatry is one of the devices of power that shapes individuals by "normalizing" them. Power is not only externally exercised over subjects because the subject is already a production of power (for example, the woman who "wishes" to submit to the patriarchal order). With this new version of power, however, becomes more difficult to understand the scope of Foucauldian criticism. Its first version was more linked to the anti-psychiatric movement giving a more concrete aim of action. In any case, in all of his critiques, Foucault does not deny the existence of madness, it only circumscribes it to a specific and contingent historical configuration.

Szasz (1960) directly attacked the concept of "mental illness", considering that it is a myth and a categorial error. There is psychological distress, living problems and moral conflicts, but none of these are "illnesses." "Illness" is a physiological concept that is described in a causal-mechanistic language while the so-called "mental illness" belongs to the domain of rationality and intentionality (Creswell, 2008). Even if the correlations between brain and mental illness are to be identified one day, it is defined in terms of behaviors that only make sense in a social value context. For example, a person only suffers from depression because she lives in a society where one thinks that life is something good and joyful. It is a value, not a natural fact. Thus, Szasz thinks that any power of the psychiatric institution on the individual is unjustifiable, defending contractual psychotherapy. Here, two individuals can come to an agreement on a job to do of psychological transformation without enforcing values.

We think that the antipsychiatric movement exerted considerable influence on psychiatric thinking. On the one hand, it elicited a response from neurobiological psychiatry that sought to reduce psychiatry to neuronal deficit factors by neglecting motivational and social factors.

However, we believe that it has also had an influence in the centralization of psychiatry in the medical value of harm for an individual and not so much in the deviation of social norms (Telles-Correia, Saraiva & Gonçalves, 2018). An example would have been the elimination of homosexuality from the catalog of mental illnesses, as it may be a deviation from dominant social values, but not an illness. By revealing how the so-called "mental illness" depends to a large extent on social values, anti-psychiatry will have inspired the creation of movements that aim to increase the power and freedom of patients, against a psychiatric institution that aimed primarily at social control.

1.3.2 Critical Psychiatry

Mainstream Psychiatry continues to be criticized nowadays, under various degrees of severity. A similar movement to antipsychiatry but also with important differences is called "Critical Psychiatry". This movement arose from a response against the proposals of the British Government to amend the 1983 Mental Health Act. A group of British psychiatrists met for the first time in 1999 expressing concerns about the implications of the proposed changes for human rights and the civil liberties of people with mental health problems. The Critical Psychiatry Network arose from here, and since then hundreds of peer-reviewed articles and dozens of books have been written, accumulating evidence on the following themes:

- The problems of diagnosis in psychiatry;
- The problems of evidence-based medicine in psychiatry and, related to this, the relationship between the pharmaceutical industry and psychiatry;
- The central role of contexts and meanings in the theory and practice of psychiatry, and the role of the contexts in which psychiatrists' work;
- The problems of coercion in psychiatry;
- The historical and philosophical basis of psychiatric knowledge and the practice of psychiatry.

Linked to critical psychiatry is the more theoretical "wing" called Postpsychiatry (Thomas and Brecken, 2004), who use similar principles to Critical Psychiatry whilst grounding their work more specifically in Heiddeger, Merleau-Ponty, Foucault and Wittgenstein.

1.3.3 Therapeutic Communities

Arising from an amalgamation of critical movements, Therapeutic Communities (TC´s) have been, and continue to be, a counterculture for mainstream mental health services. TC´s developed from models of treatment that broke from traditional practice and posed profound challenges to dominant thinking at various times: at the end of the eighteenth century (Tuke, 1813), during the Second World War (Bion, 1961; Harrison, 1999), and in the 1960s (Foucault, 1965; Goffman, 1968; Laing, 1967; Rapoport, 1960 cited in Haigh, this volume). More recent descriptions encourage medication reduction and use of other radical ideas (Haigh, 2007). Pearse and Haigh (2017) have recently published an entire volume dedicated to democratic therapeutic communities, including past references, present status quo and ideas for the future. This book also includes a recent RCT comparing personality disorder treatment in TC´s and treatment-as-usual, showing superior results for TC treatment.

Contrasting to many mainstream services, the clinical work of Therapeutic Communities takes account of the interface between the individual and their social and cultural setting (Jones, 1956). Democratic Therapeutic Communities use the concept of fluid hierarchy (Clarke, 2015) to sustain shared decision making and shared tasks in the day to day running of the community. Problems are believed to be of relational nature (Pereira, 2015; Pereira and Debbané, 2018) and symptoms are not exempt of meaning. Peer therapy, rather than only therapist-patients traditional therapy, ends up occurring as a consequence of the flattened hierarchy structure (See Haigh, in this volume). An example of user involvement in research in the context of a mental health therapeutic community is also given in this volume (Guerra, Pereira and Sales).

1.3.4. Other Critical Movements

In the US, Robert Withaker´s book *Mad in America: Bad Science, Bad Medicine, and the Enduring Mistreatment of the Mentally Ill* (Withaker, 2010) set the scene for another counter-movement that mainly criticizes the current drug-based paradigm of care. In their "Mad in America" website mission they claim that this model has failed society and that scientific research, as well as the lived experience of those who have been diagnosed with a psychiatric disorder, calls for profound change.

Also concerned with the dubious evidence about the efficacy of (long-term) pharmacological treatment, the Norwegian minister of health has recently advised that all psychiatric services should offer drug free wards, according to patient choice. The nordic city of Trömso has been the first to attempt this challenge, opening a 6-bed drug free ward for severe psychiatric disturbances. Patients and staff from this ward have recently reported on the 22nd International Meeting for the Treatment of Psychosis on what seems to be, thus far, a positive experience.

Close to Norway, the Finnish Open Dialogue system (see Seikkula and Alakare, in this volume) has, for long, been critical of mainstream psychiatry and abusive psychopharmacological treatment. They do not advocate against medication but instead promote a robust psychosocial structure in the psychiatric services and an organized, network way of approaching problems. By working in this way, they dramatically observe the reduction of coercive treatments and hospitalizations and well as the need for pharmacological treatment.

1.4 General Overview of the Book

In this volume, we bring together several contributions that address mental disorders from different but not divergent perspectives. In fact, our aim is to show that a more humanized, person-centered approach does not exclude neurobiological and pharmacological studies, but opens up the possibility for new integrative and transdisciplinary perspectives, where biological data and lived, first person experiences are equally useful for the treatment of mental disorders.

This emerges already in the first part, “Neurobiology and Pharmacology”, where it seems to be evident that we need a Gestaltic vision of subjectivity and mental functions. In *Neuroscientific Questions about Intrapsychic Phenomena and Interpersonal Processes,* M. David emphasizes the existence of interesting correlations between events that occur in the brain (conceived as a neurobiological organ) and intrapsychic phenomena and interpersonal processes of the mind. In his view, non-conscious communication phenomena are interfaces between neurophysiological processes and interpersonal dynamics. In fact, the brain is a plastic, dynamic organ that is shaped by the everyday functions of the human body, to such an extent that we can define it as a “mediating organ” (Fuchs, 2018). Arguing for an intertwining between neural processes and phenomenal ones, David claims that psychotherapeutic interventions can actually influence the neuroplasticity of the brain. In other words, psychotherapy has impact on patients’ brains, following a circular mechanism according to which the brain, the mind and the body holistically influence each other.

The importance of a multilayered therapy is also at the center of A. Florence et al.’s chapter, *Towards recovery-based practices in mental health: reframing long-term effects of neuroleptics and presenting alternatives,* where the authors critically analyze the outcomes of neuroleptic use in the treatment of individuals with psychosis and schizophrenia. They observe that there is limited evidence for the effectiveness of similar treatments in the long-term, and they argue for recovery-oriented strategies (such as the Open Dialogue approach) that make a selective use of pharmacological treatment, which involves network aspects and dialogue as the primary therapeutic components.

Among these components, as J. Grácio et al. remember us in the article *Embracing the Placebo effects in the treatment of depression: from neuropsychiatry to psychotherapy,* we should not undervalue placebo effects and the power of its healing-promoting properties. Through a careful review, the authors note that in pharmacotherapy and psychotherapy there are small differences between real treatments and placebo. Accordingly, they suggest considering placebo as a promotor of clinical improvement which should be used together with other therapeutic approaches. The efficacy of placebo is shown through the case of depression, a mental disorder where placebo seems to have a powerful, healing effect.

The complexity of facing mental disorders and the necessity of an integrative approach is emphasized by A. Gomes, who describes *Psychiatric rehabilitation: the efforts for conceptualization and meaning construction - a review.* He provides a historical understanding of the concept "rehabilitation" through a non-systematic review based on national reports, finding out that the origins of psychiatric rehabilitation mainly lay in historical and sociocultural events. The multidisciplinary nature of this therapy is based on two axes: the person and the environment, elements which are conceived in a dialectic relationship. In other words, the psychiatrists should consider both of these spheres and involve the patient in the community life, creating a support system able to improve patient's life.

Pharmacological, neurological, social and person-centered approaches work in a non-exclusive manner for improving the quality of patient's life, a patient who is not a disrupted brain, but a person who lives in a culturally-socially shaped environment which influences his behavior and his neural structures.

The intertwining between social elements, neural structures and behavioral components is at the center of the second part of the volume, which hosts contributions that interestingly testify the possibility of approaching mental health in a more human manner, focusing not only on the subjectivity of the patient, but also on the importance of the social, collective dimension. This involves the arising of new theoretical research directions and of new, innovative services. In fact, the dimension of the cure has enlarged its horizons from the study of the individual brain to the development of therapeutic communities. R. Haigh's chapter on *Therapeutic communities for the future: surviving modernization and staying at the radical edge* explains how a treatment which was seen at the forefront of progressive and radical practice in the mid-20th Century has survived and developed in very different sociocultural times and kept its critical stance in regard to mainstream psychiatry. Therapeutic communities encourage medication reduction and take into account the interface between the individual and their social and cultural setting. The chapter focuses on how group processes have been harnessed across the network of therapeutic communities to influence quality, training, research and innovation – and how the underlying therapeutic philosophy is now being used in new settings.

A concrete example of a similar practice is furnished by J. Seikkula and B. Alakare in *Open Dialogue principles and dialogical meetings for psychosis,* where they illustrate this innovative, family and social network-centered treatment approach that was initiated in 1984 in Finland (but only named as "Open Dialogue" in 1995). Open Dialogue involves two elements: first, open meetings in which all relevant members in the actual situation – including both the social network of the client and the professionals working as a team - participate from the outset to generate new understanding by dialogue; second, the guiding principles for the entire system of psychiatric practice in one geographic catchment area. The main forum for dialogues is the open treatment meeting where all management plans and decisions are made with everyone present. This kind of treatment seems to guarantee—among other benefits—psychological continuity, immediate response and long-term effectiveness for severe mental health problems. The positive outcomes presented may indicate that psychosis no longer needs to be seen as a sign of illness but can be viewed as one way of dealing with a crisis and that after this crisis many or most people are capable of returning to their active social life.

The continuity between what can be considered as "pathological" and "non pathological" is also described by M. Debbané and E. Toffel in *Mentalizing the early stages of the Psychosis Continuum,* suggesting that psychotic phenomena are expressed along a continuum ranging from non-clinical to more serious clinically psychotic manifestations. They focus especially on trauma and attachment insecurity, conditions that influence a developmental trajectory that strays away from investing in mentalizing skills, and that promotes a dissociative coping style in individuals having experienced hostile and sometimes traumatic early environments. This perspective could be useful in the prevention of risk. The enlargement of the horizon by putting attention not only on pharmacological treatments but also on the prevention is underlined by Nolte, Campbell and Fonagy in the paper *Social communicative processes in severe personality disorder*. At the center of their analysis, there is the so-called "p-factor", a general vulnerability index that captures the tendency of some individuals of being more prone to persistent and more severe psychopathology. Combining computational neuroscience and contemporary models of psychopathology, the authors shed light on the fact that "p" is strongly associated with high levels of childhood maltreatment and low mentalising capacities and emphasize aspects of brain structure-function relationships underpinning resilience processes. In other words, they account for a multilayered approach where neurobiological data and social, developmental structures have the same importance for the understanding of the pathology: broader social networks and the wider social environments also have neural structures, in a circular mechanism where the brain, the mind and the environment constitute a whole.

But how can we face this complexity in the clinical practice? In the third part of the volume we can find some examples: in *Multidimensional study of the infant population with severe and profound intellectual developmental disorder*, A. Janeco et al. take into account the case of mental deficiency, referenced as Intellectual Developmental Disorder (IDD). They present their work at the Center de Recuperação de Menores D. Manuel Trindade Salgueiro, which hosts 120 female residents aged from 3 to 17 years old. Their approach tries to assess the global needs of IDD population, evidencing the relevance of an integrated model of intervention, as well as illustrating how to organize a similar intervention in all technical areas in a structured and coordinated way. While the areas of intervention include psychiatry, nursing, social services and psychology— which are used especially in the diagnostic phase— the therapy areas include physical therapy, psychomotricity, speech therapy and occupational therapy, without forgetting patient's interests and needs.

A patient-centered approach is presented by D. Guerra et al. in *Practice based research at Casa de Alba: the perspective of service users*. Casa de Alba is a residential therapeutic community which adopts a participatory research approach, involving residents in planning and conducting research; and carries on therapies with patients and their families. In other words, it is a relational environment where everything is aimed at giving voice to patients in order to allow their opinions to influence future research and therapeutic priorities. The same happens to evaluation tools: in fact, the paper presents a participatory research study where, by carrying out a focus group, they explore the residents' views on the outcome measures used in the routine assessment of patient progress. Participants were asked to give their opinions on the advantages and disadvantages of two nomothetic measures and two client-generated outcome measures (CGOMS). Results provided valuable insights into the complementary nature of nomothetic measures and CGOMs, as well as into the effect that weekly progress monitoring can have on patients in residential psychiatric care.

A more human approach to mental disorders has the effects of changing treatments —as Casa de Alba and Centro of Recuperação de Menores D. Manuel Trindade Salgueiro testify— but also of modifying our view of mental illness. The changes involve the shift of attention from the patient conceived as an individual to the suffering person as being part of a community, and from the pathology as a specific, severe situation described by manuals that follow strict definitions, to a continuum state which can vary in its intensity and expressions. In this sense, mental health research and practice should be committed to creating new research paths, and a renovate social responsibility towards every citizen. A. Candeias et al. describe *Burnout in teaching: the importance of personal and social variables,* where they take into account a phenomenon often undervalued: the level of burnout in teachers. They present a study made on teachers of Portuguese public schools, using the Maslach Burnout Inventory (MBI; Maslach, Jackson & Leiter, 1996), which allows a three-dimensional characterization of burnout: personal accomplishment, emotional exhaustion, and depersonalization. The result leads to conclude that there is a strong need for an intervention promoting the health and wellbeing of teachers, oriented towards the training and professional valorization and the prevention of the emotional exhaustion that happens with the career advancement and years of experience.

The centrality of social dimensions in mental health is underlined by D. Pereira in *Parenting and/or Mental Health?* where she notes the difficulties in assessing parenting capacity evaluation. In this chapter, she argues for a flexible and dynamic approach, able to find a pattern through the difference gathered from multiples sources of information - parents, family members, neighbors, friends, professionals – and methodologies – home visits, interviews, psychological assessment, observation of interactions and contact with other professionals. Then, she presents the Parenting Capacity Assessment Guide as a tool that could help professionals in clinical judgements about parenting behavior.

The need for a new approach is also felt in research, where what is discussed is not only the right kind of treatment and the links among the different approaches, but also the nature of mental disorders themselves.

In fact, we cannot trace a line between madness and healthy: once again Basaglia's (2000) thought seems to be helpful in emphasizing that we cannot know what is "normal", since there are very thin boundaries between mental disorders and healthy conditions.

According to this perspective, even boredom seems to be problematic: on the one hand, there are specialists in psychology and psychiatry who understand boredom as a pathological personality trait and propose a wide variety of therapies to cope with it. Especially after the publication of the DSM-V, boredom is now being treated with medication. On the other hand, as J. R. Velasco argues in *Boredom: humanizing or dehumanizing treatment,* there is also an almost unknown understanding of boredom as an adaptive emotion through which it is possible to refocus attention on the context from which boredom comes instead of the subject itself.

All of these elements lead us to insist, in the final part of the volume, on the necessity of a human-centered and transdisciplinary approach: this is the reason why we found useful to give space to subjects not always taken into account within the debate on mental health. These subjects are phenomenology, existentialism and psychoanalysis. Restoring these perspectives seems to be necessary if we want to overcome the operationalized criteria of the Diagnostic Manual Disorder, which reduced mental disorders to a sum of signs and symptoms. The lived experience of the patient has been somehow ignored, reducing the clinical encounter itself to a crystalized, mechanistic relationship. Focusing on the essences of things and on an eidetic, intuitive method, phenomenology allows for the analysis of the substructures of consciousness (such as the formation of meaning, action planning, temporal continuity) that are disrupted in psychiatric pathologies. In fact, by describing and highlighting the pre-reflective aspects of experience, the phenomenological method emphasizes the importance of the analysis of this specific dimension as the core of psychiatric disorders. More specifically, in the analysis of psychiatric pathologies, the descriptive aim could be helpful also in the explanation of such pathologies. In other words, the eidetic description could contribute to the explanation of both the genesis and the structure of human experience: genetic and static phenomenologies can work together in the analysis of pathologies (cf. Sass, 2014). Furthermore, considering the patient as a Leib, a body which follows biological laws but is also inevitably linked to a psyche, this approach could help in developing a qualitative analysis of mental illness, explaining psychiatric disorders as subjective and material-bodily states.

In this way, it is possible to go beyond diagnostic classification: in fact, a phenomenological approach allows to adopt a first person perspective on the disturbance; and it can be combined with developmental psychology and other approaches, such as existential psychotherapy, opening up the possibilities for new therapeutic directions.

These themes are faced by V. A. Rodrigues' and V. Bizzari's papers. In the first—*Dialectic of the first and third person accounts in mental health-therapeutic implications*— V. A. Rodrigues, drawing on phenomenology and existential psychotherapy, describes clinical practice as a space where the therapist goes from a third person discourse to a first person case and back again to a third person analytic discourse. He claims that the clinical encounter could not be either a pure subjective space nor a space of pure reflection. In other words, the clinician owns "a position" that "is not that of a neutral anthropologist; it is rather one of a coach or a midwife. His/her trade is grounded on a sensitivity to the subtle indices of his interlocutor's phrasing, bodily language and expressiveness, seeking for indices (more or less explicit) which are inroads into the common experiential ground. Such encounters would not be possible without the mediator being steeped in the domain of experiences under examination, as nothing can replace that first-hand knowledge" (Varela & Shear, 1999, 10). Lived and first-person data are privileged with respect to scientific analysis: following this direction, Rodrigues contrasts third-person psychiatric and psychological reports with first-person reports of lived experience recorded of clinical sessions and proposes a hermeneutical critical method relevant for clinical practice.

On the other hand, in *From D.I.R. to D.I.R.E: the role of Embodiment in the treatment of self-disorders*, V. Bizzari shows the usefulness of phenomenological insights in the development of therapeutic paths: she combines Greenspan's "affective diathesis hypothesis" with a phenomenological account of subjectivity as a corporeal, kinesthetic entity, and points out the role of embodiment in the development of selfhood and intersubjectivity. Then, she proposes a special kind of therapy integrating Greenspan and Wider's "D.I.R." model—Developmental, Individual Difference and Relationship-based— with phenomenological insights, emphasizing the fact that, especially in the treatment of autism, we should take into account therapies aimed at strengthening the sense of the corporeal self.

Despite its many critics, another discipline that has played, and continues to play, a major role in the understanding of social and individual dynamics is psychoanalysis.

This is illustrated by R. Hinshelwood, who, in the chapter *Psychoanalytic roots of the therapeutic community*, explains how the therapeutic community movement in psychiatry (particularly, in Britain) was influenced by psychoanalysis. Considering the psychoanalytic encounter as a social field, we can thus claim that it allowed the theory to be applied in the social setting of mental hospitals, and thus emerged the practice of therapeutic communities. According to the author, the convergence of social field theory with object-relations psychoanalysis in the 1940s represented the innovative foundation for what we must think of as the group of relational therapies (psychoanalytic psychotherapy, group therapy, therapeutic community).

Psychoanalysis and its possible influences on other fields are also at the center of the last chapter, G. Caselli's *Ipsas Aquas Urere Consuevit. Psychiatric praxis, analytic thought, transference love* . This chapter has a twofold aim. On the one hand—like in V. A. Rodrigues' chapter—it reflects on the clinical encounter, arguing for the centrality of caring to such an extent that the therapist herself can be conceived as a relational instrument; on the other hand the author argues for an integrative approach where psychoanalysis (with a special attention to the notion of transference love) can have a role in the psychiatric practice, helping in building a renewed epistemology of psychiatry where emotions and love are primary tools for understanding the patient.

Psychiatry seems to resolve itself in a border science, which should be able to incorporate other disciplines, with the final aim of looking for connections within the fragmented experience of the patient, which should be considered as a continuous oscillation between subjectivity and objectivity, personhood and society, mind and body, individual needs and social expectations. In other words, psychiatry needs to become "ethic and kind" (Borgna, 2003).

Acknowledgments

The editors would like to especially thank the authors for their commitment. We also express our gratitude to Inês Hipólito, Cláudia Pedro and Joana Grave for their invaluable help to the editorial team. We want to thank Christoph Durt, Madalena Serra, Constança Biscaia, Sofia Tavares, João Gama Marques, Adrian Spremberg and Philipp Schmidt for their useful suggestions to improve the quality of the book. This volume would not have been possible without the help of Fundação Romão de Sousa, IFILNOVA from Universidade Nova de Lisboa and the Psychology Department of Universidade de Évora who jointly organized the 2nd International Mental Health Congress, where some of the chapters started to come alive.

1.5 References

Basaglia, F. (2000). *Conferenze brasiliane*, Cortina, Milano.

Bergström, T., Alakare, B. Aaltonen, J., Mäki, P., Köngäs-Saviaro, P. & Seikkula, J. (2017). The long-term use of psychiatric services within the Open Dialogue treatment system after first-episode psychosis. *Psychosis* Vol. 9, Iss. 4.

Binswanger, L. (1956). *Drei formen missglückten daseins. Verstiegenheit, verschrobenheit, manieriertheit.* Tübingen, Germany: Max Niemeyer.

Borgna, E. (2003). *Le intermittenze del cuore*, Feltrinelli, Milano.

Clarke, J.M. (2015). *Where the Change Is: everyday interaction rituals of therapeutic communities.* Unpublished PhD Thesis, University of Nottingham.

Creswell, M. (2008) Szasz and his interlocutors: reconsidering Thomas Szasz's 'myth of mental illness' thesis. *Journal Theory Social Behavior* 38:23–44. doi:10.1111/j.1468-5914.2008.00359.x.

Doerr-Zegers O. & Stanghellini, G. (2013). Clinical phenomenology and its psychotherapeutic consequences. *Journal of Psychopathology* 19, 228-233.

Foucault M. (1975) La maison des Fous. In: Basaglia F, Basaglia-Ongaro F, editors. *Criminini di Pace.* Turin: Eunadi. p. 151–69.

Foucault M. (1967) Chapter IX (The Birth of Asylum). In: Howard R, editor. *Madness and Civilization: A History of Insanity in the Age of Reason* (Abridged Version). New York: Random House. p. 241–78.

Foucault, M. (2006) *Psychiatric Power-Lectures at the College de France 1973-1974.* Palgrave Macmillan.

Fuchs, T. (2018). *Ecology of the Brain: the Phenomenology and Biology of the Embodied Mind,* Oxford University Press, Oxford;

Hipólito, I, Gonçalves, J. & Pereira, J. G. (Eds.), *Schizophrenia and Common Sense: Explainig the Relation between Madness and Social Values,* Springer International Publishing, Studies in Mind and Brain, Vol. 12.

Horvath A. O. & Bedi R. P. (2002) The alliance. In *Psychotherapy Relationships That Work: Therapist Contributions and Responsiveness to Patients,* ed. Norcross J. C., editor. (New York: Oxford University Press;), 37–69.

Jaspers, K. (1959) *Allgemeine Psychopatologie,* Springer, Berlin-Gottingen-Heidelberg; engl. transl. *General Psychopathology* (1963), Chicago: Il, University of Chicago Press.

Norcross J. C. (2002). *Psychotherapy Relationships That Work: Therapist Contributions and Responsiveness to Patients.* New York: Oxford University Press.

Parnas, J., Sass, L. & Zahavi, D. (2012). Rediscovering Psychopathology: the Epistemology and Phenomenology of the Psychiatric Object. *Schizophrenia Bullettin,* doi:10.1093/schbul/sbs153.

Pereira, J. G. (2015), An Integrative Relational Framework of Intervention at Casa de Alba: therapeutic community for severe mental health problems. *British Journal of Psychotherapy Integration,* Vol. 11 (2), 37-52.

Pereira, J. G. & Debbané, M. (2018). An Integrative Relational Approach in Schizophrenia: from philosophical principles to mentalization based practice. In I. Hipólito, J. Gonçalves & J. G. Pereira (Eds.), *Schizophrenia and Common Sense: Explainig the Relation between Madness and Social Values*, Springer International Publishing, Studies in Mind and Brain, Vol. 12.

Quétel C. (2012) *Histoire de la Folie – De l'Antiquité à nos jours.* Paris: Éditions Tallandier.

Ritzer, G. (1993). *The McDonaldization of Society.* California, Pine Forge Press.

Rose N. (2006) Foucault, Laing et le pouvoir psychiatrique. *Sociologie et sociétés*, 38(2):113–31. doi:10.7202/016375ar.

Szasz T. (1960) The Myth of mental illness. *American Psychology* 15:113–8. doi:10.1037/h0046535.

Sass, L. (2014), Explanation and Description in Phenomenological Psychopathology. *Journal of Psychopathology* 20, 366-376.

Schneider, K. (1919), Reine Psychiatrie, symptomatische Psychiatrie und Neurologie. *Zeitschrift fur die gesamte Neurologie und Psychiatrie* 57, pp.159-166.

Stanghellini, G. & Aragona, M. (2016). *An experiential Approach to Psychopathology. What is like to Suffer from Mental Disorders?*, Springer Nature.

Telles-Correia, D., Saraiva, S. & Gonçalves, J. (2018) Mental disorder – The need for an accurate definition. *Frontiers of Psychiatry* | https://doi.org/10.3389/fpsyt.2018.00064.

Pearse & Haigh (2017). T*he Theory and Practice of Democratic Therapeutic Communitty Treatment.* Jessica Kingsley Publishers, London.

Varela, F. J. & Shear, J. (1999). First-person methodologies: what, why, how?. *Journal of Consciousness Studies*, 6 (2–3), pp. 1-14.

Withaker, R. (2010) *Mad in America: Bad Science, Bad Medicine, and the Enduring Mistreatment of the Mentally Ill*, New York, Basic Books.

Part One:
Neurobiology and Pharmacology

Chapter 2

Neuroscientific questions about intrapsychic phenomena and interpersonal processes

Mário David
Centro Hospitalar Psiquiátrico de Lisboa, Portugal

This paper proposes some theoretical correlations between neurobiological processes that occur in the Brain (*Organ*) with certain intra-psychic phenomena and interpersonal processes that occur within the Mind (*Mental Apparatus*). The chapter focuses on recent neuroscientific information which points to new working hypotheses relevant within psychotherapeutic processes and psychotherapy.

2.1 Preamble

Over the last two decades, several lines of scientific investigation centered on intra-psychic phenomena and interpersonal processes have provided significant data and information that allow us to have a new scientific understanding and framework. They also suggest possible correlations between psychological and symbolical perspectives and also between neuro-dynamic and neurobiological levels.

This theoretical exercise constitutes an effort to discuss some hard questions about how we could relate the Brain's biological nature with its symbolic and mental functioning as emergence of a Mind (*Mental Apparatus*).

There are several lines of scientific investigation coming from different areas and perspectives, such as Social Neuroscience (Adolphs, 2003), Affective Neurosciences (Panksepp, 1998; Schore,1994; 2000; 2003; Panksepp & Lucy Biven, 2012) or the Attachment Model (Bowlby, 1969; 1988; Ainsworth, 1978; Main, 1993; 1995; Hoffer, 1995; 2014; Fonagy & Target, 2002; Fonagy, 1991; 2003; 2004; 2006; Stern, 1985/1998). These authors have done pioneering work that allows us to envision new ideas for future research about how psychotherapeutic processes could promote, under precise circumstances, significant cerebral changes in neuronal plasticity and neuronal networking levels towards the mental and symbolic levels.

These ideas have already been supported by recent propositions by Daniel J. Siegel (1999, 2007) and Louis Cozolino (2002/2010, 2006) who presented significant data and explanatory hypotheses about possible correlations between developmental and interpersonal psychological processes, such as connections, attunements and adjustments through proximal and significant affective relationships and under certain circumstances and conditions, which allows for some changes in neuronal plasticity and networking on certain areas of the Brain.

2.2. The neurobiological nature of the brain

The Brain, a highly specialized organ, generates, in a permanent way, flows of energy and information patterns by assembling neuronal networking processes with the information coming from either the *internal milieu* (body) through the spinal medulla, the autonomous nervous system (ANS) and different biochemical messengers (hormones, etc.), or from the *external milieu* (environment) through different pathways involving different sensorial systems which capture signs coming from the five senses (sight, hearing, taste, touch and smell). This information is generated by *firing* or *neuronal excitability* patterns which are organized and transformed into informational *representations* or *perceptions* through specialized neuronal circuits. Such information *representations* or *perceptions* will generate mental maps (Damásio, 1999, 2010) which give rise to higher cortical psychic products, such as feelings, thoughts and words. Simultaneously, a *self-awareness system* generates a variety of consciousness levels (*auto, self, enlarged, extended,* etc.) which are very important for managing enormous quantities of information generated at each moment, related with all these complex cognitive and affective capacities.

These phenomena support the emergence of one other kind of entity (Mind or *Mental Apparatus*) whose main function is to help the Brain (*Organ*) deal with and overcome automatic and non-conscious information through other levels of processing (pre-conscious and conscious) and also to handle higher levels of *mental representations,* such as non-linguistic symbols (ideas/thoughts) or linguistic symbols (words).

The correlations between neurobiological phenomena and the emergence of mental products are the object of endless scientific and philosophical discussions. Nowadays, deep intrinsic connections are deemed to exist between these two different aspects of the cerebral functioning. Some leading neuroscientists – Eric R. Kandel, Brenda Milner and Larry R. Squire – wrote: *the experience of models about mental processes through changing both activity and structure of connections among neurons, in which experience models directly responsible circuits for memoirs, emotionality and self-awareness* (Milner, Squire, & Kandel, 1998) and *excluding the influence of epigenetic factors and from disease situations or degenerative conditions, we know that the Brain is highly dependent on experiential and environmental factors* (Milner, Squire, & Kandel, 1998). More Recently, Daniel J. Siegel (1999, 2007) wrote: *the Mind is created within the interaction of internal neuro-physiological processes and interpersonal experiences and the structure and function of developmental Brain are determined by how experiences, especially, within interpersonal relationships, shape the genetically programmed maturation of the nervous system* (Siegel, 1999, p. 2).

The brain undergoes continuous reframing and lifelong adaptations, but it also accumulates errors and other operational difficulties due to enormous quantities of incoming information and while attempting to make adequate choices and decisions to accomplish its homeostatic needs. These errors or bad options will disturb the functioning of the Mind (*Mental Apparatus*), which is also highly susceptible to be reconfigured under the influence of intrapsychic and inter-relational processes.

2.3. Functioning principles for the brain/mind interface

The human brain is organized and structured under bio-psycho-social frameworks. These are organized by certain developmental and biological principles derived from bio-physical changes, natural selection and social forces which have worked through a long and slow hominization process until the Modern Man. This biological and social evolution has been followed by complexification over social interactions within larger and bigger social groups and societies.

In functional terms, at the time of birth the human brain is very immature and open to any external or internal stimuli, in particular, to prolonged parental care which influences it in very deep emotional and learning processes, generating significant information to be stored in different layers among implicit and explicit memories.

This functional immaturity is a space for potential specialization of neuronal systems during certain early life periods, so-called '*time windows*' or '*sensitive periods*', which are open by pre-programmed ways. During these '*sensitive periods*' certain networking and structural brain modifications occur in order to assure core capacities needed to perform basic tasks such as seeing, hearing, walking, talking, and so on.

This neuronal network reframing is widespread and results in neuroplasticity phenomena among different brain cells towards proliferation, differentiation, migration, apoptosis and even pruning phenomena, in particular, neurons and their terminal parts (axons, dendrites and receptors). This happens not only during early childhood and adolescence but also till very late in one's life. These neuronal phenomena are emphasized by the Rogers-Ramachandran's neuroscientist couple who have stated: "*the human brain is capable of more and faster functional reorganization than we had thought about*" (Ramachandran, Rogers-Ramachandran & Steward, 1992). They meant that the brain has very high capacities to retain, storage, reframe and reproduce huge quantities of information and it is able to store and reframe them from different and constant incoming new elements about affective/emotional and motor reactions, either from the '*inside milieu*' (Body) or from the '*physical and social environment*'.

A recent paradigm shift has occurred over how we understand circuitry systems information processing which functions at different levels (somatic/ emotional/ affective/ cognitive/ psychic) throughout cybernetic loops between lower neuronal structures (brainstem and other basal nuclei) and higher cortical structures (cerebral lobes and specialized areas). This new paradigm tells us that mental functioning is more determined by affective and emotional forces than by cognitive operations.

This new paradigm shift makes a huge difference in our understanding of the relationship between neuro-dynamic brain's functioning and mental phenomena. Thus, an emotional and affective overdetermination is proposed by two well-grounded neurobiological models:

- From the Jaak Panksepp (1998, 1999, 2011, 2012) neurobiological emotion model with its seven operative emotion systems: The emergence of four positive valence emotions: a '*SEEKING*' system, a '*LUST*' system, a '*CARE*' system and a 'PLAY' system which are located in deep subcortical circuitry and their information influences bottom-up memories, learning and other cognitive operations such as reasoning, decision making, etc. These emotional systems are modulated by secondary processes with information generated at upper limbic systems and higher circuitry levels influenced by mental activity, such as memories, ruminations, thoughts, etc. And another group of operative emotion systems where negative valence emotions emerge, represented by: a '*RAGE*' system, a '*FEAR*' system and a '*PANIC/GRIEF*' system which are influenced by a tertiary level of information processing, mainly at neocortical areas (Panksepp, 1998, 2011). For Jaak Panksepp, the Brain deals with neuronal information among these three processing levels: a primary level, a secondary level and a tertiary level. Higher levels of information (cognitive) are processed between top-down and bottom-up circular and cybernetic loops, in order to produce thoughts and actions (behaviors).

- Another neurobiological emotion model has been proposed by António R. Damásio and his colleagues (1994, 1999, 2000, 2010), where each '*basic emotion*' induces a multitude of significant bodily maps of higher complexity '*secondary emotions and sentiments*' and each fine sentiment is recognized by certain body maps representations that allow the brain to deal with these functional determinants over different cognitive operations.

2.4. The importance of emotional communication within inter-personal interaction dynamics

Several neuroscientists and neuro-psychoanalysts consider that non-conscious communication phenomena may occur within an interface between neurophysiological processes and interpersonal interaction dynamics. These non-conscious communication phenomena influence brain circuits responsible for social perception ('*social brain*'), which are deeply connected with neuronal circuits for basic control functions, such as attributive functions (creation of meaning), bodily states, modulation adjustment and maintenance of emotions, although some of these control functions have moments that also run around conscious communication levels (Panksepp, 1998, 2012; Schore, 1994, 2003; Siegel, 1999, 2007; Green, 2003). This interface phenomenon between neurophysiological processes and interpersonal interactions dynamics has been conceptualized by Louis Cozolino (2002/2010) as a '*social synapse*' for the brain´s functioning.

Several recent studies about interpersonal interactions ('*in physical proximity*' or 'how *can one mind can influence another mind*?') have clarified different ways how energy and information are being transferred between individuals. These information and energy transferences have certain direct effects, in separate ways, over modelling electrical activity in each one´s brain (McCraty et al., 1998). Also, the existence of '*hidden regulators*' was revealed. These are linked to tactile stimulation, human warmth and physical touch (Hoffer, 1995, 2014), which are extremely important in close relationships between parents and children (Sapolsky, 1997).

Other studies, *in higher physical distances*, have revealed some changes in brain waves activity and development through specific or unspecific signals, such as words and non-verbal prosodic components of speech, as voice tone, gestures and time response for both verbal and non-verbal behavior responses. These communication elements contain information that will create representational processes within the receiver's mind. They have a direct impact on social validation centers and emotional operation systems resulting in an expression of social signals. When the emissary's brain deals with attention guidance and appreciation of symbolic meanings, these activate the same circuits that mediate emotional responses in the receiver's brain.

In recent reviews done by neuroscientists, such as Jaak Panksepp & Lucy Biven (2012), Allan. N. Schore (1994, 2003), Daniel J. Siegel (1999, 2007) or Vivian Green (2003), it has been suggested that emotion activity and operative emotional circuitry emit basic information which overdetermines how incoming information is managed and stored as implicit and explicit memories and in other cortical areas. Also, interpersonal experiences affect the brain development directly and they are one of the most significant forces that model the brain structures and its functions. These interpersonal experiences can facilitate or inhibit this modelling when integrated in coherent and spontaneous communication exchanges, within interpersonal relationships and emotional interaction communications.

2.5. The neuronal structures and systems for a social brain

Jaak Panksepp & Lucy Biven (2012) consider three main levels for brain´s neuronal systems functioning, as states control and information processing which are conditioned by retroactive loop controls between all the levels:

- A primary processing level of instinctive or emotional nature from sub-neocortical areas, which generates: *"Basic-Primordial Affects: I) Emotional Affects; II) Homeostatic Affects; III) Sensory Affects"* (Panksepp & Biven, 2012, p. 10).
- A secondary learning process functioning essentially at limbic areas, which generates, via basal ganglia: "*Secondary Emotions through: I) Classical Conditioning; II) Instrumental & Operant*

Conditioning; III) Behavioral & Emotional Habits" (Panksepp & Biven, 2012, p. 10)

- A tertiary processing level which is largely constituted by neocortex areas, which generates: *"Tertiary Affects and Neo-cortical 'Awareness' Functions: I) Cognitive Executive functions; II) Emotional Ruminations & Regulations; III) 'Free-Will'"* (Panksepp & Biven, 2012, p. 10).

For Louis Cozolino (2002/2010, 2006), we can find several cerebral circuits responsible for social perception functions (Social Brain) and these functions are deeply connected with certain subcortical structures (amygdala, hippocampus and hypothalamus) and also with certain cortical structures, such as medio-orbital prefrontal cortex, somatosensory cortices, insula and cingulate (Cozolino, 2006). This is thought to be central for any social information processing (Adolphs, 2003) which is located at the limbic system apex, the medio-orbital prefrontal cortex, a convergence area for multi-sensorial, somatic and emotional information processing. This cortical area produces synthesis over incoming information from both internal and external '*Worlds*'. This cortical area plays a crucial role in control and inhibition autonomic functions and it also contributes enormously for behavior organization and affective adjustments throughout translation proceedings from positive reinforcement or punishment values over complex social information. This means that the control of facial expressions, postures and visual contact reveals a significant portion of information associated with our emotions and organizes attachment schemes (O'Doherty et al., 2001) (Tremblay & Schultz, 1999) (Hariri et al. 2000) (Zald & Kim, 2001).

The somatosensory cortices, which are located at central gyrus and draped in the parietal lobes around sylvan sulcus, process information over body experiences according to multiple body representations. They organize our experience from tactile, temperature, pain, and articulation position terms and also from our visceral states. These different information flows are combined and they create our subjective experiences about our '*Body-Self*'. According to António Damasio (1994, 2000), they participate in: "*intuition or gut feelings through an activation of implicit memories related to our experiences and they help us to take decisions which are guided by feelings*" (Damásio, 1994) or within neuronal maps: "*our bodies become models for our understanding and how we empathize with Others*" (Damásio et al., 2000.).

Other important areas for social information are the anterior frontal areas of superior temporal sulcus which contain the so-called '*mirror neurons'*. These specialized neurons, which are activated when we witness others engaging in functional behaviors or when we engage in what is known as *'resonance behaviors'* or *'reflexive imitation responses'*, when we are interacting with other people. According to Rizzolatti and colleagues (Rizzolatti et al. 1999), these *'resonance behaviors'* mechanisms may have evolved from group behavior's synchronization when we were hunters, gatherers or during long standing emigration movements. They have connections with visceral and emotional circuitry, allowing these same systems to support emotional resonance, attunement and empathy. It is hypothesized that these *'mirror neurons'* systems and *'resonance behaviors'* have evolved into our ability to attune other people's emotional states.

Also, we have sensory, motor, and affective neuronal systems which aren't yet very well defined and are mainly located at the temporal lobes, where our sensorial senses are integrated, organized and combined with deep primitive drives of emotional significance. They are responsive for complex social information input coming from eye gazes, facial expressions and body postures. Some of these *'regulatory systems'* have already been defined as:

1) Stress Regulation or Hypothalamic-Pituitary-Adrenal (HPA) System

Any stress signal is controlled and regulated by a hypothalamic-pituitary-adrenal axis (HPA) which regulates hormones secretion involved in body's responses to stress or any other threat, either facing short-term survival with a fear reaction activated by amygdala or throughout other brain centers that promote fight-flight responses or even when facing long-term effects, like parental deprivation, failures to attachment or early trauma.

2) Fear Regulation System

When amygdala and several other centers trigger a fight-flight response and activate the sympathetic branch of the autonomic nervous system (ANS), it results in anxiety, agitation and panic symptoms. At a higher cortical function level, fear has a regulatory circuit system placed at medio-orbital prefrontal cortex that inhibits conscious awareness from amygdala-based processes.

3) Social Engagement System

The 10th cranial nerve or vagal nerve system from the autonomic nervous system (ANS) is a complex system of communication between the Brain and multiple points of the Body (heart, lungs, throat and digestive system). This vagal nerve is involved with homeostatic regulation and maintenance in order to achieve biological balances which represent good physical health and emotional well-being (Porges, Doussard-Roosevelt & Maiti, 1994). The vagal nerve allows us to maintain continued social engagement by sympathetic arousal modulation and fine-tuning during emotional interpersonal exchanges. This vagal nerve enhances the social engagement system through an inhibition effect over self-regulation phenomena which seem to be dependent from attachment qualities acquired since early childhood relationships. This allows us to internalize what we learn from life experiences with our primary caregivers as a somatic regulation based on a '*moment to moment*' mode. In fact, the vagal nerve system controls innumerous muscles from the face, mouth and throat which are primarily involved in communication procedures, connects them to our self-awareness and self-control over our internal biological states and also coordinates emotional and cognitive processes required to handle human relationships.

4) Social Motivation System

E. Nelson and J. Panksepp (1998) propose a '*social motivation system*' modulated by oxytocin, vasopressin, endogenous endorphins and other neurochemical substances related to reward, decreased physical pain and the emergence of well-being feelings. These circuits and neurochemicals are thought to regulate attachment, pair bonding, empathy and altruistic behavior (Decety & Lamm, 2006; Seitz, Nickel & Azari, 2006).

Simultaneously, Helen E. Fisher (1998, 2000) suggested another '*motivation social system*' distributed in quite similar areas but subdivided into three subsystems or circuits:

A) '*Bonding and Attachment*' (regulated by peptides, vasopressin and oxytocin);
B) '*Attraction*' (regulated by dopamine and other catecholamine)
C) '*Sex Drive*' (regulated by androgens and estrogens).

These neuronal networks are very important for control and regulation functions of body states, emotions modulation, memories workout, meaning creation and decision making, and many do not work at conscious levels of communication (Fisher, 1998, 2000).

2.6. Face recognition and facial expression

Another investigation area about emotional and social interaction is related to face recognition and facial expression because these express huge amounts of emotional and social information at an interpersonal level. They participate in an interpersonal emotional regulation of our interactions with other people over intimacy or threat expressions and over the exercise of social control or facilitating coordination and cooperation between people (Freire et al., 2004; Kleinke, 1986). The increased pupil diameter can mean acceptance/assessment states, for example, "*capture of any pleasant image or the act of listening to music or even a state while it is solving problems mentally*" or "*even when we're feeling positive feelings, or we're interested in people*" (Hess, 1975; Siegle et al., 2004).

When we are looking, we are always interested in '*where*', '*what* 'and if the '*Other*' is watching us or not. There is a facial expression that has been well studied, '*blushing*', which is a unique human facial expression. Usually, the blushing facial expression reveals, '*that we are aware that others recognize us*' (Crozier, 2004) or an act of blushing always arises when one of these four situations occurs: 1) *"When our social identity is threatened"*; 2) *"When we are being scrutinized"*; 3) *"When we're being adulated"*; 4) *"When we say that we're blushing even if you don't"* (O´Leary et al., 1992).

In pathologic terms, we find '*eritrophobia*' as a blushing expression deriving from social phobia and other anxious disorders and it is also an expression of guilt and shame feelings or even of how '*sorry you are*' in order to obtain sympathy and positive approval from other people about something that you have transmitted. Finally, it is also a conciliatory behavior.

Face recognition involves both brain hemispheres and it functions close to visual-spatial mechanisms and holistic strategic information processes from the right hemisphere.

2.6.1 Attachment and attunement within interpersonal relationships

We presumed that emotional communications are one of the primary ways that promote neuronal circuitry and network modeling through attachment experiences and other early relationship dynamics. These are supported by mental representations of interpersonal experiences which directly influence the way we mentally build our '*view about World and Life*'.

Newborns establish certain binding processes with their primary caregivers through tuning and emotional resonance and also by reciprocity responses between each other which are revealed by certain phenomena, such as mutual recognition knowledge and turn-taking, etc. Early emotional adjustments are established by parent-child synchronization, contributing to neural network organization and integration and also to child's self-regulation: "*Mother-child emotional attunement during 1st year of life is predictive for child's self-control of his 2 years of age, controlling temperament, intelligence levels (QI) and maternal style*" (Feldman, Greenbaum & Yirimiya, 1999).

The combined experience of internal safe feelings, the liberation of anxieties and other excitement moments which are generated through attachment processes will provide an affective background experience for vital and spontaneous expressions of true self.

The early maternal deprivation derived from maternal separation or depression promotes in human babies a sufferance that: "*resulted in a decrease in brain function, in higher levels of anxiety and greater difficulties in a subsequent attachment*" (Brennan et al., 2008) (Brennan & Shaver, 1995) (Tyrka et al., 2008) or "*low levels of maternal care are correlated with more fearful, less positive, less positive attention and brain frontal lobe activation with a right side deviation and all these data were correlated with higher levels of stress and excitement*" (Hane & Fox, 2006).

The increased complexity for neuronal networking is associated with learning mechanisms, development of different memories and ever more with regulation and inhibition capacities over all sorts of impulses, in particular, sexual and aggressive impulses. This impulse inhibition will increase attention and mental focus capacities and maximize neuronal networking plasticity that will support higher cognition functions. Contemplation and empathy are promoted by basic and nuclear communication processes, attunement modelling and affective regulation that occur in human interactions between neural structures and affective and cognitive experiences in two continuous ways of changes between brains and social phenomena.

The existence of attachment or binding schemes in both human and non-human babies was discovered by biologist and psychoanalyst John Bowlby (1969, 1988) through his interest about links between mother-child interactions, the importance of exploratory behavior and the impact of separation and loss in the normal development of children. He noticed a significant influence of the mother concerning care and protection to assure survival and development of babies. These studies of attachment schemes revealed issues, such as attachment figures, pursuit of proximity or secure base (Bowlby, 1969). This line of research was continued by Mary Ainsworth (1978) and her collaborator Mary Main (1993, 1995) who developed research methodologies that allowed the scientific testing of John Bowlby´s theoretical ideas.

Attachment schemes are particularly apparent in stressful situations because they play a central role in affective regulation and other attachment issues. They are mediated by autonomic nervous system (ANS) adjustments and in a cascade of biochemical processes that create approach and avoidance reactions, as well as the emergence of positive emotions or negative emotions. These attachment schemes are activated by affective relationships and they will guide us, in subsequent searches or avoidance of closeness or intimacy. They also determine how we are able to use future intimate relationships in the sense of physiological and emotional homeostasis (Schore, 1994, 2000, 2003) (Schore & Schore, 2008) (Fonagy et al., 2003, 2004, 2006).

Louis Cozolino (2006) proposed a conceptualization about an '*internalized good mother*', as paradigm for successful attachment, which means that we should have good memories about our earliest interactions with our mother, organized in visceral, somatic and emotional memories networks. These different unconscious memories intend to enhance: "*1) The core of our own self-esteem; 2) Our ability to self-reassurance; 3) Self-determination over nature and quality of our relationships, when we are adults*" (Cozolino, 2006) This author wrote that we all needed to be handled adequately, to be tucked gently and securely, to be comforted, in order to experience homeostatic regulations of our basic biological needs, such as sleep, hunger and other sensory stimuli. In addition, we should have had repeated experiences about emotional transitions from distress states to reassuring moments and also have experienced sustained positive emotional states.

More recent studies were carried out by P. Fonagy, H. Steele, M. Steele (1991); P. Fonagy & M. Target (2002) and P. Fonagy, (2003) P. Fonagy, G. Gergely, E. Jurist & M. Target (2004) about attachment processes in a developmental perspective. They have conceptualized an '*interpersonal interpretative function*' (IIF) which is different from J. Bowlby's 'i*nternal working model*'. Within this '*interpersonal interpretative function*', they have separated cognitive-oriented processes (IIF-C) from a set of processes guided by emotions or affections (IIF-A). In this last set of emotions or affections processes, they have proposed an idea of '*Mentalized affection*', a mental process that refers to experiences and simultaneously acknowledges a feeling. For these authors, affect regulation or ability to modulate emotional states is closely related with a '*Mentalization or reflective function*'...."*which plays a key role in blossoming of Self-awareness and representation. In this perspective, affection regulation is a harbinger for Mentalization*" (Fonagy, 2003)

Peter Fonagy and Mary Target (2002) have proposed four control mechanisms which contribute to unfold developments of interpretative functions: affection recognition and labelling, excitement regulation, control and mentalization of specific capabilities, which are organized by three distinct interconnected and interactive brain intersections (nodes) related with social detection, affective regulation and cognitions regulation (Adolphs, 2003; Nelson et al. 2005).

The first interactive and interconnection node consists of a predetermined structure set that classifies stimuli as social and decodes or detects their final purposes. The brain areas which are involved in this social detection node are: anterior fusiform '*face area*', superior temporal sulcus and anterior temporal cortex (Gauthier et al., 2000) (Fonagy, 2006). The second interactive and interconnected node involved with affection covers brain areas involved with reward and punishment stimuli. Affection emergence provides social stimulus with emotional significance and adjusts or regulates emotional stimulation. This system plays a significant role in evaluating attachment experiences, and it is activated by stimuli related to these attachment experiences. The brain regions involved in this emotion regulation node are: amygdala, hypothalamus, nucleus accumbens and strial terminalis basal node. These regions interact with the social detection node, giving emotional significance to social stimuli (Fonagy, 2006).

The third node is involved in cognition regulation and its key-functions include: inhibition of pre-potent responses, mediation for objective-oriented behavior and mentalization. The regions participating in this cognition regulation node are the medio-dorsal prefrontal cortex and the ventral prefrontal cortex (Fonagy, 2006).

2.7 Interpersonal interaction in physical proximity situations, 'How can one mind influence another mind?'

Psychotherapeutic interventions create new knowledge and narratives about the patient's personal truths, life story and about each one's Self, as new ways to understand real life and mental fantasy contexts and how to deal with each one's personal past. Meanwhile, psychotherapeutic interventions promote several neuro-dynamic phenomena, such as awareness of implicit memories; acknowledge reframing; reorganization of explicit memories; network modelling for better affective and emotion regulation and creation of new personal meanings and memories.

In the group analytic model, the "*Mental Apparatus*" emerges intrinsically from Man´s social nature and it is susceptive to change and expand self-awareness and self-knowledge openness (Foulkes & Anthony, 1957/1984), (Foulkes, 1964/1982). These authors deeply believed that in the group situation the '*group therapeutic factors*' would emerge, induce and promote significant ego changes in each participant of the group. This was designated as '*Ego training in action*' (Foulkes, 1957/1984), which supposes '*corrective emotional experiences*' occurring in group space or '*self-development through subjective interaction*' according to Dennis Brown´s conceptualization (1994). More recently, a Hungarian group analyst named Tom Ormay (2012) proposed the existence of an autonomous psychic instance, '*NOS*', which urges people to engage in communication modes and emotional mental sharing, especially in group situations. Each group member feels urges to share thoughts and feelings and these are illustrated by a variety of verbal and non-verbal signals supported by more or less spontaneous verbalizations. For this author (Ormay, 2012), therapeutic groups have '*natural*' and '*intrinsic*' '*powers*' whose forces emerge from their deep social and sociological nature and also through intense interaction situations.

2.8 Final reflections and remarks

It is known that trauma and similar situations generate high-stress levels that induce several kinds of neuronal and networking disturbances and neuroplasticity transformations. It is suggested that the use of moderate and appropriate levels of stimulation allows a better access to cortical and sub-cortical mechanisms, promoting neuronal reintegration and networking reorganization: '*through activation of associative areas from frontal, parietal and temporal lobes*' (Cozolino, (2002/2010), '*which they coordinate, regulate and guide multiple neuronal circuits for memory and emotion*' (Cozolino, (2002/2010). Scientific evidence underlines that a simultaneous activation of emotions and cognitions supports: '*the idea that moderate levels of arousal optimize the production of neurotransmitters*' and '*neural growth hormones that enhance long-term memory (LTP), learning and cortical reorganization*' (Cowan & Kandel, 2001; Zhu & Waite, 1998).

Also, current scientific data indicates that psychotherapies can have significant impacts on patients' brains, improving neuroplasticity, synchronization and networking reframing. The brain´s areas more susceptible to be sculpted by psychotherapeutic processes are the prefrontal lobe with the premotor and motor sensitive areas, the anterior and posterior cingulate areas; the parietal lobe and the hippocampus with the memory retrieval and memory consolidation (McCarthy, 1995), (Price et al., 1996) (Cozolino, 2002/2010), (Geuze et al., 2005). All these areas play important roles in physical, social and emotional information syntheses and so they could become target areas for all kinds of psychotherapies.

Any unconscious or conscious communication phenomena between the patient and his psychotherapist are facilitated by affective dispositions, cognitive abilities and also through his motivation and desire for the search of personal "*truths*" and "*knowledge*". The psychotherapist´s personal qualities and capabilities, his mental dispositions and emotional attitudes play a role towards a needed safe, confident and intimate human setting in which all kinds of psychotherapies should find place!

When any psychotherapist proposes to help someone, he must have a certain internal disposition besides his accumulated clinical experience in order to promote friendly, peaceful, empathic, supportive and stimulating attitudes towards his patient. He should help his patient in his quest for deeper and truthful emotional, as well as, cognitive issues. The psychotherapist should create a holding and nurturing environment under confidential conditions to allow an internal (psychological) space which enhances his patient for new and liberated personal understandings and narratives about himself.

Concerning the patient, he should be able to endure certain levels of psychological stress, anxieties and fears, in order to re-actualize his mental defenses which, prevent him from having access to his self-awareness over previous and possibly painful memories. Usually, patients show significant improvements from the moment they start to be able to liberate their emotions and sentiments associated with their most painful memories or when they start to face their most personal fearful situations.

Every creative psychotherapeutic process should support any effort from the patient in order to gain new ways of reframing his memories, affective self-regulation, new symbolic meanings, in order to reach new personal psychological understandings about his mental and life circumstances, new self-awareness and personal possibilities to become a more balanced psychological creative person.

The above research review over this interface between the Brain and the Mind has provided significant and pertinent data from different investigation areas centered at intrapersonal and interpersonal phenomena; it opens important questions related to short and long-term neurobiological developments and pertinent questions could be raised, such as:

- *"What kind of neurobiological benefits are obtained with long-term or with short-term psychotherapeutic processes (individual or group), for example, emotion resilience reinforcement?)"*
- *"What kind of cognitive reserves enlargement can occur in patients?"*
- *"What are the most significant outcomes in neurobiological and neuro-dynamic networking after a long-term psychotherapeutic process (individual or group)?"*

Also, this review stands for further developments with update current scientific data envisioning an ongoing dialogue over possible correlations between mental and neurobiological phenomena which emerge between a still mysterious and unknown Brain and its Mind within a deep relation with one Body.

2.9 References

Adolphs, R. (2003) Cognitive neuroscience of human social behavior. *Nature Reviews Neuroscience.* 4, 165-178.

Ainsworth, M.D.S. et al. (1978) *Patterns of Attachment: A psychological study of a strange situation.* Hillsdale - New Jersey, Erlbaum.

Bowlby, J. (1969) Attachment and Loss. In: *Attachment. (Vol.1),* New York: Basic Books.

Bowlby, J. (1988) *A secure base, Clinical applications of attachment theory.* London, Routledge.

Brennan, K.A. & Shaver, P.R. (1995) Dimensions of adult attachment: affect regulation and romantic relationship functioning. *Personality and Social Psychology Bulletin.* 21, 267-283.

Brennan, P.A. et al. (2008) Maternal depression and infant cortisol, Influences of timing, co-morbidity and treatment. *Journal of Child Psychology and Psychiatry.* 49, 1099-1107.

Brown, D. (1994) Self-development through subjective interaction. In: Brown & L. Zinkin (eds.) *The Psyche and the Social World.* London, Routledge.

Cozolino, L. (2002/2010) *The Neuroscience of Psychotherapy – Healing the Social Brain.* (2nd Edition), New York, W. W. Norton & Company Inc.

Cozolino, L. (2006) *Neuroscience of Human Relationships – Attachment and the Developing of the Social Brain.* New York, W. W. Norton & Company Inc.

Cowan, W. M. & Kandel, E. R. (2001). A brief history of synapses and synaptic transmission. In: W. M. Cowan, T.C. Sudhof, & C. F. Stevens (eds.), *Synapses,* pp. 1-88. Baltimore, John Hopkins University Press.

Crozier, W.R. (2004) Self-Consciousness, exposure, and the blush. *Journal for Theory of Social Behavior.* 34, 1-17.

Damásio, A.R. (1994) *Descartes' Error.* New-York, Putnam and Sons.

Damásio, A.R. (1999) *The Feeling of what happens: Body and emotion in the making of consciousness.* New York, Harcourt Brace.

Damásio, A., et al. (2000) Sub-cortical and cortical brain activity during the feeling Self-generated emotions. *Nature Neuroscience.* 3, 1049-1056.

Damásio, A.R. (2010) *Self comes to mind, constructing the conscious brain.* New-York, Pantheon Books.

Decety, J. & Lamm, C. (2006) Human empathy through the lens of social neuroscience. *Scientific World Journal.* 6, 1146-1163.

Feldman, R., Greenbaum, C.W. & Yirimiya, N. (1999) Mother-infant affect synchrony as an antecedent of emergence of self-control. *Development Psychology. 35,* 223-231.

Fisher, H.E. (1998) Lust, attraction, and attachment in mammalian reproduction. *Human Nature.* 9, 23-52.

Fisher, H.E. (2004) *Why we love. The nature and chemistry of romantic love.* New-York, Holt Paperbacks.

Fonagy, P. Steele, H., & Steele, M. (1991) Maternal representations of attachment during pregnancy predict the organization of infant-mother attachment at one year of age. *Child Development.* 62, 891-905.

Fonagy, P. & Target, M. (2002) Early Intervention and the development of the self-regulation. *Psychoanalytic Inquiry.* 22, 307-335.

Fonagy, P. (2003) The interpersonal interpretative mechanism: the confluence of genetics and attachment theory in development. In: Viviane Green (ed.) *Emotional Development in Psychoanalysis, Attachment Theory and Neuroscience – Creating Connections.* USA & Canada, Brunner-Routledge, pp.107-126.

Fonagy, P. et al. (2004) *Affection Regulation, Mentalization and the development of the Self.* London, Karnac Books.

Fonagy, P. (2006) The Mentalization-Focused Approach to Social Development. In: Jon G. Allen & Peter Fonagy (eds.), *Handbook of Mentalization-based Treatment.* USA, John Wiley & Sons, pp. 53-99.

Foulkes, S.H. & Anthony, E.J. (1957/1984) *Group Psychotherapy – The Psychoanalytic Approach.* (2nd Fac-Simile Edition), London, Karnac Books.

Foulkes, S.H. (1948/1982) *The Therapeutic Group Analysis.* London, Karnac Books.

Freire, R. Eskritt M. & Lee, K. (2004) Are eyes windows to deceived soul? Children's use of another's eye gaze cues in a deceptive situation. *Development Psychology.* 40, 1093-1104.

Gauthier I. et al. (2000) The fusiform 'face area' is a part of a network that processes faces at individual level. *Journal of Cognitive Neuroscience.* 12, 495-504.

Geuze, E, Vermette, E., & Bremner, J.D. (2005) MR-based in vivo hippocampal volumetrics, 2. Findings in neuropsychiatric disorders. *Molecular Psychiatry.* 10, 160-184.

Green, V. (2003) Emotional development – biological and clinical approaches – towards integration. In: *Emotional Development in*

Psychoanalysis, Attachment Theory and Neuroscience – Creating Connections. USA & Canada, Brunner-Routledge, pp. 1-19.

Hane, A. & Fox, N. (2006) Ordinary variations in maternal care-giving influence human infants' stress reactivity. *Psychological Science.* 17, 690-694.

Hariri, A.R., Bookheimer, S.Y. & Mazziotta, J.C. (2000) Modulating emotional responses, Effects of a neocortical network on the limbic system. *NeuroReport.* 11, 43-48.

Hess, E.H. (1975) The Role of pupil size in communication. *Scientific American.* 233, 110-119.

Hoffer, M.A. (1995) Hidden regulators: Implications for a new understanding of attachment, separation, and loss. In: S. Goldberg, R. Muir & J. Kerr (eds.), *Attachment theory, Social development and clinical perspectives.* Hillsdale - New Jersey, Analytic Press, pp. 203-217.

Hoffer, M.A. (2014) The emerging synthesis of development and evolution: a new biology for psychoanalysis. *Neuropsychoanalysis.* 16 (1) 3-22. Available from: "http://dx.doi.org/10.1080/15294145.2014.901022"

Kleinke, C.L. (1986) Gaze and eye contact: a research review. *Psychological Bulletin.* 100, 77-100.

Main, M. (1993) Discourse, prediction and the recent studies in attachment: Implications for Psychoanalysis. *Journal of the American Psychoanalytic Association.* 41, 209-244.

Main, M. (1995) Attachment, Overview, with implications for clinical work. In: S. Goldberg, R. Muir, & J. Kerr (eds.) *Attachment Theory, Social, Development and Clinical Perspectives.* Hillsdale – New Jersey, Analytic Press, pp. 407-474.

McCarthy, G. (1995) Functional neuroimaging of memory. *The Neuroscientist.* 1, 155-163.

McCraty, R., Atkinson, M, Tomassion, D. & Tiller, W.A. (1998). The electricity of touch: Detection and measurement of cardiac energy exchanges between people. In: K. H. Pibram & J. King (eds.), *Brain and values: Is a biological science of values possible?,* pp. 359-379. Hillsdale, New Jersey: Erlbaum.

Milner, B., Squire, L.R. & Kandel, E.R. (1998). Cognitive neuroscience and the study of memory. *Neuron,* 20, 445-468.

Nelson, E.E. & Panksepp, J. (1998) Brain substrates of infant-mother attachment: Contributions of opioids, oxytocin and norepinephrine. *Neuroscience and Biobehavioral Reviews.* 22, 437-452.

Nelson, E. E. Leibenluft, E., McClure, E. B. & Pine, D. S. (2005) The social re-orientation of adolescence, a neuroscience perspective on the process and its relation to psychopathology. *Psychological Medicine.* 35, 163-174.

O'Doherty et al. (2001) Abstract reward and punishment representations in the human orbitofrontal cortex. *Nature Neuroscience.* 4, 95-102.

O'Leary, M.R. et al. (1992) Social blushing. *Psychological Bulletin.* 112, 446-460.

Ormay, T. (2012) *The Social Nature of Persons - One Person is No Person.* New International Library of Group Analysis, London, Karnac Books.

Panksepp, J. (1998) *Affective Neuroscience. The Foundations of Human and Animal Emotions.* In: Series, Affective Science, Oxford & New-York, Oxford University Press.

Panksepp, J. (1999) Emotions as viewed by psychoanalysis and neuroscience, an exercise in consilience. *Neuropsychoanalysis.* 1, 15-38.

Panksepp, J. (2011) Cross-species affective neuroscience decoding of the primal affective experiences of humans and related animals. *PLoS ONE.* 6. Available from: www.http://doi,10.131371/journal.pone.0021236

Panksepp, J, & Biven L. (2012) *Archeology of the Mind - Neuroevolutionary Origins of Human Emotions.* New-York, W.W. Norton & Company.

Porges, S.W., Doussard-Roosevelt, J.A. & Maiti, A.K. (1994) Vagal tone and the physiological regulation of emotion. *Monographs of the Society for Research in Child Development.* 59 (2-3, Serial nº 240), 167-186.

Price, J.L., Carmichael, S.T. & Drevets, W.C. (1996) Networks related to the orbital and medial prefrontal cortex, a substrate for emotional behavior? *Progress in Brain Research.* 107, 523-536.

Ramachandran, V.S., Rogers-Ramachandran, D. & Stewart, M. (1992) Perceptual correlates of massive cortical reorganization. *Science.* 258, 1159-1160.

Rizzolatti, G. et al. (1999) Resonance behaviors and mirror neurons. *Archives Italiennes de Biologie.* 137, 85-100.

Sapolsky, R.M. (1997) The importance of well-groomed child. *Science,* 277, 1620-1621.

Schore, A.N. (1994) *Affect regulation and origin of the self, the neurobiology of emotional development.* Hillsdale - New Jersey, Erlbaum.

Schore, A.N. (2000) Attachment and the regulation of the right brain. *Attachment and Human Development.* 2, 23-47.

Schore, A.N. (2003) *Affect Dysregulation & Disorders of the Self.* New-York & London, W.W. Norton & Company.

Schore, J. & Schore, A. (2008) Modern attachment theory, the central role of affect regulation in development and treatment. *Clinical Social Work Journal.* 36, 9-20.

Seitz, R.J., Nickel, J. & Azari N.P. (2006) Functional modularity of medial prefrontal cortex, Involvement in human empathy. *Neuropsychology.* 20, 743-751.

Siegel, D. J. (1999) *The Developing Mind, Toward a Neurobiology of Interpersonal Experience.* New-York, Guilford Press.

Siegel, D.J. (2007) *The Mindful Mind: Reflection and attunement in the cultivation of well-being.* New York, W. W. Norton & Company Ltd.

Siegle, G.J., Steinhauser, S.R., & Thase, M.E. (2004) Pupillary assessment and computational modeling of Stroop task in depression. *International Journal of Psychophysiology.* 52, 63-76.

Stern, D. N. (1985/1998) *The interpersonal world of the infant.* New-York, Basic Books.

Tremblay, L. & Schultz, W. (1999) Relative reward preference in primate orbitofrontal cortex. *Nature,* 398, 704-708.

Tyrka, A.R. et al. (2008) Childhood parental loss and adult hypothalamic-pituitary-adrenal function. *Biological Psychiatry.* 63, 1147-1154.

Zald, D.H. & Kim, S.W. (2001) The orbitofrontal cortex. In: S. P. Salloway, P. F. Malloy & J. D. Duffy (Eds.). *The frontal lobes and neuropsychiatric illness.* Washington: DC.

Zhu, X. O. & Waite, P. M. E. (1998). Cholinergic deplction reduces plasticity of barrel field cortex. *Cerebral Cortex,* 8, 63-72.

Chapter 3

Towards recovery-based practices in mental health: reframing long-term effects of neuroleptics and presenting alternatives

Ana Carolina Florence
State University of São Paulo, Brazil

Silvio Yasui
State University of São Paulo, Brazil

Claudinei Eduardo Biazoli Junior
Universidade Federal do ABC, Brazil

The use of neuroleptics as a first-line treatment for individuals diagnosed with schizophrenia or psychosis is recommended in technical literature and has become widespread in western society. However, long-term outcomes associated with the continuous use of neuroleptics over many years have not been comprehensively appreciated. Building upon the concepts of specificity, effectiveness and risk evaluation, we critically reviewed some of the current evidences on long-term outcomes of neuroleptics use, particularly concerning recovery rates. For this critical evaluation, the political and theoretical implications of the current hegemonic model for the mental health problems and the role of pharmaceutical industry were taken into account. Giving the limited evidence for the effectiveness of neuroleptics in the long term and well-established safety issues, we argued in favour of recovery-oriented strategies for mental health policies, clinical practice and research. Finally, the principles and main outcome results of the Open Dialogue approach for the treatment of severe mental health problems were presented. Of particular interest, this approach offers a set of practices actively avoiding the early introduction of pharmacological treatment.

3.1. Introduction

The main question addressed in the first part of this paper can be articulated as follows: what is currently established, from the existing literature, about the long-term effects of neuroleptics use in severe mental health problems? This broad question led us to the definition of three key concepts of the evidence-based approach in medical research (Sackett et al., 1996), namely specificity, effectiveness and risk assessment. Such concepts guided the critical scope narrative literature review here attempted. The concept of specificity applies to both diagnostic procedures and therapeutic efficacy. Effectiveness is related to the prognosis and long-term outcomes of a given intervention in concrete, "real-world" conditions. Finally, risk assessment is critical to evaluate therapy safety. The choice of specificity, effectiveness and risk assessment as key dimensions to evaluate currently hegemonic psychiatric treatments is arguably strategic in identifying the gaps in evidence-based knowledge and to determine what steps should be taken in order to advance evidence-based practices, ultimately benefiting patients, practitioners and public policy planners.

Moncrieff (2008) and Whitaker (2010) have extensively investigated the role of these dimensions in psychiatric research. In fact, both authors pointed out to a central gap in psychiatric evidence-based knowledge regarding long-term outcomes, treatment safety and effectiveness. This gap was attributed to two aspects that have been guiding and influencing research in psychiatry. The first one is the prevailing psychiatric paradigm that defines mental disorders as chemical imbalances in the brain (Moncrieff, 2008); and the second is the pharmaceutical industry influential role in the scientific and medical communities (Angell, 2005; Moncrieff, 2008, Whitaker, 2010).

The aforementioned aspects pose certain constraints to a properly unbiased literature review. Most studies are designed within the framework of the prevailing model of mental disorders and therefore measure particular effects of a specific intervention in controlled settings, comparing them with a group that receives either an inert placebo or a gold-standard treatment. This double-blind and randomized clinical trial (RCT) design is suitable to assess the short-term relative *efficacy* of a given treatment. However, RCTs results may not be readily generalizable to more extended timescales or contexts that sensitively differs from the controlled setting (Seikkula, 2014). In other words, demonstrating the *efficacy* of a treatment does not assure its *effectiveness*. Pharmaceutical industry funding is another crucial aspect as it adds to well-recognized literature biases: negative results tend to remain unpublished, risks posed by drugs might be overlooked and conclusions generally tend to favour pharmacological interventions (Angell, 2005; Breggin, 1993; Moncrieff, 2008; Whitaker, 2010). Considering all these caveats and their political and theoretical implications, we aimed to critically review the literature on the long-term effects of neuroleptics from the advanced standpoint of clinicians, mental health workers and policy planners directly implicated in effectively managing serious mental health problems. For this purpose, we adopted scoping review (Arksey & O'Malley, 2005) as a methodological framework. A scope review consists in proposing a narrative that includes a wide variety of study designs, critically prioritizing crucial aspects of the literature. Many systematic reviews on neuroleptic use for the treatment of individuals with schizophrenia have indeed been conducted, and we did not intend to duplicate these efforts. Rather, we have based upon those previous efforts in order to further support the statement that hegemonic psychiatric rationale regarding drug treatment is far from consensual and flawless, justifying the need for recovery-oriented approaches.

Accordingly, the second part of this paper outlined the main principles of the Open Dialogue approach which has been developed in Finland and subsequently adapted and implemented in a smaller scale in the US. We argued that the dialogical perspective of the Finnish set of practices might provide an alternative to the medical model of severe mental health problems. Moreover, the relevance of the Open Dialogue approach relied on proposing a way of working with patients and families that have systematically shown high recovery rates and better long-term outcomes (Seikkula, 2002; Seikkula, 2008; Seikkula & Arnkil, 2006; Seikkula & Arnkil, 2014). Of great importance to our analysis, the Open Dialogues approach have actively avoided the use of neuroleptics by introducing an effective and concrete set of practices for clinicians and health planners.

It is worth mentioning some of the terminologies we have adopted. Hereby, the term "mental health problems" will be applied whenever possible, as it is less stigmatizing than mental disorders terminology. The term schizophrenic, commonly used in psychiatric literature, was thoroughly avoided. When referring to studies specifically employing this terminology, we rather used the expression "individuals with schizophrenia or psychosis". "Mental illness, disease and disorder" were also actively avoided to refer to mental health problems. Finally, the term "antipsychotic" was not employed since this nomenclature implies that neuroleptic drugs treat specific illnesses (Moncrieff, 2008), and is not based in a pharmacological definition as usually applied for drugs in other fields.

3.2 Long-term effects of neuroleptics: a scope review

3.2.1 Neuroleptics and the disease model of psychosis

Chlorpromazine was first synthesized in France in December 1950 and was reported as an anesthetic agent in 1952 by Laborit (Dundee, 1954). This French surgeon has also noticed that intravenously administering certain doses of the drug produced an effect of disinterest without affecting the level of consciousness. The medical community readily recognised the potential of chlorpromazine for psychiatric applications due to this tranquilizing effect. Afterwards, the work of Delay and Deniker (1952) has set the basis for the use of chlorpromazine by psychiatrists as an effective way to control the behaviour of agitated patients in hospital wards.

The first studies of the effects of chlorpromazine, when its pharmacological mechanisms were not yet known, offered detailed descriptions of patients' behaviour following the drug administration. Those observations were indeed described as a syndrome characterised by indifference to stimuli, pathological tranquility, emotional neutrality, but with minimal effects on the consciousness level (Moncrieff, 2008). A lack of interest in the surroundings and an appearance of being heavily drugged contrasted with preserved higher psychological functions (Dundee, 1954). One of the most interesting among these early accounts on chlorpromazine's effects can be found on Elkes and Elkes (1954): "The schizophrenic and paraphrenic patients continued to be subject to delusions and hallucinations, though they appeared to be less disturbed by them" (p. 562).

The introduction of neuroleptics in hospital wards had a great impact, since the new drugs calmed down patients and rendered them docile in institutions. In parallel, chlorpromazine effects were often described as an awakening of patients from years of psychosis (Healy, 2002). At the time, physical treatments such as electroconvulsive therapy, insulin shock therapy and frontal lobotomy were accompanied by dramatic effects on personality and consciousness level, and therefore, could not be considered symptom-specific (Breggin, 1993). The concept of specificity actually became important after the "bacterial revolution" in the late 19th century. The success in identifying pathogens for several diseases promoted the idea that most medical conditions should be caused by specific agents. The generalization of the infectious diseases paradigm led to the emergence of the quest for "magic bullets", which should be designed to target specific aetiologies and cure the also specific diseases

(Healy, 2002; Moncrieff, 2008; Whitaker, 2010). According to Healy (2002), in the early experiences with chlorpromazine, no theoretical rationale was behind its use. Posteriorly, the dopamine hypothesis for schizophrenia was put forward to support neuroleptics-based treatment in hospital wards, and not the other way around. In psychiatry, the rise of neuroscience from the 1970's led to several biological hypotheses of mental illnesses and new neural mechanisms were proposed. Nevertheless, the clinical relevance of neuroscientific findings and advances in psychiatry and mental health are yet to be established, as quite a little impact on diagnosis and therapeutics has been observed so far (Healy, 2002).

The need for a biological rationale to support drug therapy gave rise to the disease-centred model in psychiatry, described by Moncrieff (2008) as the idea that psychotropic drugs treat specific biological diseases, postulated as chemical imbalances in the brain. This rising of the biological psychiatry model was concomitant to a paradigmatic void in psychopathology during the 1980's. The third version of the Diagnostic and Statistical Manual of Mental Disorders (3rd ed, Reviewed, American Psychiatric Association, 1987) marked a shift away from phenomenology and psychoanalysis towards a tentatively a-theoretical classification system in psychopathology. The descriptive and a-theoretical approach to psychiatric syndromes was constructed intending to support empirical-experimental research (Dunker & Kyrillos Neto, 2011), and helped pave the way towards a hegemonic biological approach to mental health problems. In the particular case of neuroleptics and psychosis, the discovery that clorpromazine and halopedirol act as dopaminergic receptor antagonist has led to the hypothesis that psychotic symptoms should be caused by an excess of dopamine in the brain. Hence, the classical dopaminergic hypothesis of schizophrenia has been posed as follows:

> "*The classical dopamine hypothesis of schizophrenia postulates that an excess of dopamine subcortically is associated with positive symptoms of schizophrenia. This belief was based on two observations: firstly, that sustained exposure to D2 receptor agonists induces schizophrenia-like positive symptoms and secondly, that all drugs with proven antipsychotic effects block D2 receptors to some degree.*" (Abi-Dargan, 2004, p. S2)

As this excerpt illustrates, the main evidence to support the dopaminergic hypothesis was based on drug actions. Underlying this conception was the idea that drug induced-psychotic states constitute a valid model of schizophrenia. Particularly problematic, in attempting to confirm that schizophrenia is caused by excessive dopamine in the brain using this model, a rather circular argument is advanced (Moncrieff, 2008). Furthermore, recent researches suggest that stimulant-induced psychotic states are better characterized by a distinct symptomatology from that observed in individuals with schizophrenia (Murray et al., 2013). Some worth-noting differences between drug-induced psychosis and schizophrenia include: the absence of thought disorganisation (Bell, 1965), of delusions of control, and of emotional flattening (Snyder, 1972 as cited by Moncrieff, 2008). In a recent systematic review on drug-induced psychotic states, Murray et al. (2013) argued that "the various drug models mimic different aspects of the syndrome" rather than account for a unifying aetiology (p. 666).

Early neuroimaging studies in the 1990's that have described significant brain structural differences between subjects with schizophrenia and subjects without symptoms failed to take into account that the vast majority of subjects diagnosed with schizophrenia had been submitted to drug therapy with neuroleptics for most of their lives (Moncrieff, 2008; Whitaker, 2010). The findings of D2 cortical density increases in individuals with schizophrenia, for instance, were promptly connected to the illness (Breggin, 1993), regardless the fact that this brain alteration had been previously described as a compensation mechanism due to the dopamine blockade promoted by neuroleptics (Muller & Seeman, 1977). More recent findings have shown that both first and second-generation neuroleptics cause a compensatory increase of around 34% in the density of D2 receptors in humans (Silvestri et al., 2000), in line with previous findings in non-human animals (Muller & Seeman, 1977). Along with this compensatory up-regulation mechanism of D2-receptor, several other dopaminergic system changes due to neuroleptics exposure have been reported, such as dopamine receptors hypersensitivity and increased dopamine synthesis and release (Aderhold & Stastny, 2015). Critically, at least some of these changes are dose-dependent (Samaha, 2008). These findings provide a basis for the withdrawal symptoms presented by patients following abrupt neuroleptics suspension (Moncrieff, 2008).

Neuroleptic-induced brain alterations in dopaminergic receptors might have important and still largely unappreciated consequences, as proposed by Aderhold and Stastny (2015). For instance, 30% to 40% of patients treated with neuroleptics will endure a loss of drug effectiveness, probably due to these dopaminergic compensation mechanisms. Also, when faced with low responses to drug therapy, physicians tend to increase the dosage seeking to obtain a therapeutic response. Finally, patients might experience rebound effect in the short-term from dose reductions or abrupt withdrawal. Aderhold and Stastny (2015) also propose that prolonged drug exposure might lead to a “supersensitive psychosis” phenomenon, characterized by an exacerbation of psychotic symptoms upon dose reduction, or even when the dose is maintained. Such patients might also present abnormal involuntary movements unrelated to neuroleptic-induced Parkinsonism and an increase in reactivity to stressors, leading to more frequent relapses.

Generally, hypersensitivity to a drug receptor should lead to withdrawal symptoms after drug therapy interruption (Healy, 2002). Considering that the current dopaminergic hypothesis states that imbalances in the dopaminergic system are the very cause of schizophrenia, the emergence of withdrawal symptoms might be obscured and interpreted as additional evidence supporting the dopaminergic hypothesis. Since withdrawal symptoms are thought to be very similar to the original disturbance, the difficulty in diagnosing such symptoms has great practical implications, beyond the logical circularity (Aderhold & Stastny, 2015; Breggin, 2012)

Regardless the limited evidence supporting the disease-centred model of mental health problems (Aderhold & Stastny, 2015; Breggin, 2012; Healy, 2002; Moncrieff, 2008; Whitaker, 2010), the long-term use of neuroleptics for psychotic symptoms management is yet the most widespread strategy in patient care worldwide.

3.2.2 Current evidence for neuroleptics effectiveness

A current consensus on both clinical practice and public policy-making was constructed around evidence-based medicine. In this context, clinical practice and policies aiming at planning and delivering health services should be based on the best evidence available (Rose, Thornicroft & Slade, 2006). While properly conducted double-blind Randomized Clinical Trials (RCTs) are considered the gold-standard for producing high-quality evidence in medicine in general, mental health issues pose some further obstacles to this approach as some of psychiatry's "basic premises of diagnosis and treatment are strongly contested" (Rose, 2014, p. 150). The gap between research and practice in mental health demands a critical approach to the suitability of RCTs alone to inform practice, services delivery and management. Instead, other stakeholders should be included in research strategies, as more perspectives are needed to produce valid and relevant knowledge (Rose, Thornicroft & Slade, 2006). In fact, service users, practitioners, families and policy makers often disagree with standard academic knowledge and including their views is crucial if the production of recovery-oriented evidence is desired, to the benefit of individuals at the receiving end of mental health systems. Furthermore, mental health problems are complex experiences, determined by multiple factors and research strategies should be suited to grasp more accurately the nuances of recovering (Davidson, 2003). Taking these specificities of an evidence-based approach in mental health into account, we critically review the topics of safety and outcomes of neuroleptics use.

High-quality systematic reviews on long-term outcomes and safety of neuroleptic use have already been conducted (Aderhold & Stastny, 2015; Breggin, 1993; Moncrieff, 2008; Whitaker, 2010), supporting a critical approach to drug therapy. It is important to emphasize some of the methodological issues associated with summarizing the current evidence for neuroleptic use. One of the main obstacles is the well-documented and extensive influence of the pharmaceutical industry in psychiatric research, which greatly difficult evaluating the quality of the evidence. For instance, pharmaceutical industry funding of studies was shown to introduce important biases on results presentation and interpretation (Angell, 2005; Gøtzsche, 2014; Moncrieff, 2008; Whitaker, 2010), eventually ending up in massively harmful strategies such as disease mongering (Moynihan & Henry, 2006). A particularly important interpretation bias consists in systematically attributing good outcomes to the drug therapy while the negative ones are ascribed to a supposedly degenerative evolution of schizophrenia, when untreated (Breggin, 1993). Furthermore, reporting biases are difficult to detect as some trials remain unpublished for years after completion and biased descriptions of trial designs, conducts and results may lead to invalid conclusions with harmful consequences to healthcare practices (Doshi et al., 2013)

In a Cochrane systematic review attempting to compare neuroleptics introduction to either placebo or psychosocial treatment on early episodes of schizophrenia-spectrum disorder, only 6 out of 681 studies have met quality-based inclusion criteria (Bola, Kao & Soydan, 2011). The authors concluded that the data then available were too limited to assess whether the drug introduction improves outcomes, contrasting their findings with part of the current technical literature recommending drug therapy for first-episode psychosis for a period of 6 to 24 months.

Since most RCTs compare the effects of a drug intervention against inert placebo over relatively short periods, they fail to provide evidence for long-term outcomes, including recovery. Aiming to assess the clinical relevance of findings regarding the use of neuroleptics, a systematic review was conducted translating the usually applied symptom scores evaluations into clinical relevance scores (Lepping et al., 2011). When symptom scores assessed through the Positive and Negative Syndrome Scale (PANSS) and the Brief Psychiatric Rating Scale (BPRS) were translated to Clinical Global Impression scales of severity and change

(CGI–S and CGI–C), both first and second-generation neuroleptics showed a quite limited clinical significance:

> "*Comparison of first- and second-generation antipsychotics as categories shows neither class of drug achieving a mean CGI–C score of great clinical significance*" (Lepping et al., 2011, p. 344).

When interpreting the review findings, despite presenting a standard account based on the assumption that neuroleptics are more effective than other treatments, the authors noted that: "other explanations for our findings include the possibility that the clinical relevance of antipsychotics is in fact limited" (Lepping et al., 2011, p. 344).

A systematic review and meta-analysis on the recovery of individuals with schizophrenia have shown that recovery rates did not increase over time, in spite of the emergence of new technologies such as drug treatment, psychosocial approaches and deinstitutionalisation (Jääskeläinen et al., 2013). In fact, studies tracking long-term outcomes of individuals with schizophrenia have suggested that those who have not used psychotropic drugs continually fared better (Harrow, Jobe & Faull, 2014).

The Soteria project took place in the United States of America between 1979 and 1983 and consisted of a psychossocial approach to individuals newly diagnosed with schizophrenia. The project aimed to provide intense treatment focusing mainly on peoples' relationships. Care was provided in a homelike environment with a non-professional staff actively avoiding the use of neuroleptics during the first 6 weeks (Bola & Mosher, 2003). Individuals diagnosed with schizophrenia receiving care in Soteria houses were compared with subjects treated in conventional hospitals (Bola & Mosher, 2003). A drug prescription was provided to 24% of the individuals in the Soteria group compared to 94% in the hospitalized group. In a 2-year follow up, the number of hospital readmissions and days in readmissions, ability to live independently, level of global psychopathology, improvement in psychopathology and working status were assessed. Overall, Soteria-treated individuals had better results in all measured outcomes, including lower psychopathology scores and fewer hospital readmissions (Bola & Mosher, 2003). Soteria Houses can still be found in Germany and some Scandinavian countries but were never adopted as a larger-scale Mental Health strategy for catchment areas or as part of public policy initiatives.

In 1978 Rappaport et al. compared four groups of patients. The first one received placebo during hospitalisation and no medication after discharge; the second received placebo during hospitalisation and medication after discharge; the third received medication during hospitalisation and no medication was prescribed after discharge, and the last group was prescribed with psychotropics both in the hospital and afterwards. The groups were longitudinally assessed in several follow-ups over the course of three years. Only 1 out of the 24 patients who had never received medication was re-hospitalised during this period. All the other groups showed worse outcomes with re-hospitalisation rates ranging from 47% to 73% (Rappaport et al., 1978).

A longitudinal study conducted by Harrow, Jobe and Faull (2014) have shown similar results. Authors assessed 136 individuals diagnosed with schizophrenia and mood disorder at hospitalization and in 6 follow-ups over a period of 20 years. Aiming to assess whether medicated subjects had less psychotic symptoms over time, authors found that over 70% of continually medicated subjects presented psychotic symptoms in at least 4 of the 6 follow-ups (Harrow, Jobe & Faull, 2014). On the other hand, patients who had not taken medication continuously presented less psychotic symptoms, fewer hospitalisations and better outcomes when compared to those using medication continuously (Harrow, Jobe & Faull, 2014). The study design was naturalistic and, therefore, patients were not randomised into drug therapy and placebo groups. Nevertheless, an important finding was that differences between continuously and not-continuously medicated groups only surfaced after 2.5 years, reinforcing that conventional 4 to 6 weeks trials are not suitable to infer long-term recovery.

A Scandinavian study compared two groups of subjects, one receiving treatment according to the Open Dialogue approach in Finland and the second receiving traditional western psychiatric treatment in Sweden (Seikkula & Arnkil, 2014). Differences in the outcomes over a five year-span between the groups included: (1) patients receiving care in Finland (n=72) had an average of 31 days of hospitalisations, compared to an average of 110 in Sweden (n=75); (2) among the Finnish, 33% used some neuroleptic and 17% continued long-term use, whereas in the Swedish group, 93% used some neuroleptic and 75% continued long-term use; and (3) in Stockholm, 62% of patients were incapable of working and were on government aid after five years, compared to 19% of patients treated with the Open Dialogue approach and minimal use of neuroleptics. This study compared different treatment methods in countries with otherwise similar social and economic structures, showing that patients using less medication and hospitalised for shorter periods of time have had lower rates of work incapacitation (Seikkula & Arnkil, 2014). Furthermore, in a 10-year period, 84% of the patients treated with the Open Dialogue approach in Western Lapland, Finland, were reported to have returned to an active social life, with full employment or schooling after a psychotic episode (Seikkula, Alakare & Aaltonen, 2011).

Wunderink et al. (2015) conducted a 7-year follow up study comparing the outcome of early neuroleptic dose-reduction or discontinuation with maintenance drug therapy in remitted first episode psychosis. 103 individuals symptomatically remitted enrolled an 18-month trial comparing dose reduction to dose maintenance and were followed up at 7 years. Recovery was assessed in two dimensions: symptom remission and functionality. Individuals assigned to drug reduction showed twice the recovery rate of drug maintenance individuals (Wunderink et al., 2015). No significant differences were found in symptomatic remission rates. However, dose reduction individuals had higher functional remission-rates which accounted for an overall better recovery-rate after 7 years. Authors concluded that neuroleptic dose reduction performed in the early stages of remitted first episode psychosis improved outcome in recovery and functionality when compared to neuroleptic drug maintenance.

Nordentoft et al. (2016) conducted a cohort study including 496 individuals with schizophrenia spectrum disorders diagnosis and assessed long-term outcome of individuals with no current use of neuroleptic medication. Authors report that 30% of the individuals assessed had symptom remission in the 10-year follow up while not being on medication. In line with Bola, Kao and Soydan (2011), Nordentoft et al. (2016) also point to a contrast in long-term outcome findings of unmedicated individuals and technical recommendations that advise against neuroleptic suspension.

Wils et al. (2016) found similar results in a cohort longitudinal study of long-term outcomes in individuals with first episode psychosis concluding that after 10 years, 30% of participants achieved a remission of psychotic symptoms while unmedicated at the 10-year follow up. Authors suggest that great heterogeneity can be found regarding the course of severe mental health problems and that good outcomes in non-medicated individuals are not accounted for in the technical literature (Wils et al., 2016)

In a 2013 meta-analysis of 37 studies on social and clinical dimensions of recovery in schizophrenia (Jääskeläinen et al., 2013), a recovery rate of about 13% over 2 years was found. Remarkably, this study has shown that, in spite of the development of new technologies in care, including psychosocial approaches and drug therapy, recovery rates have not improved over time (Jääskeläinen et al., 2013). Of great importance for public policy making and clinical research, these results reinforce the need for recovery-oriented investigations, focusing on long-term outcomes rather than on symptoms-assessment over short periods. Guidelines seem to neglect a large group of patients who recover without the use of neuroleptics (Harrow, Jobe & Faull, 2014; Norderntoft et al., 2016; Wils et al., 2016; Wunderink et al., 2015).

3.2.3. Risk assessment: neuroleptics safety

Distinguishing symptoms associated with psychosis from those caused by neuroleptics is crucial to evaluate the potential benefits and risks of drug therapy (Aderhold & Stastny, 2015). When randomising patients in clinical trials, for instance, those allocated in the placebo group might have been taking neuroleptics before starting the trial (Gøtzsche, 2014; Moncrieff, 2008). If some of these individuals present withdrawal symptoms, the comparison between placebo and treatment arms would be compromised, biasing the results to better outcomes for drug treatment (Aderhold & Stastny, 2015; Breggin, 1993; Moncrieff, 2008; Whitaker, 2010). The well-established cognitive decline and brain morphology alterations associated with the diagnosis of schizophrenia were once considered part of the natural evolution of a degenerative brain disease (Breggin, 2012, Moncrieff, 2008). However, recent findings in epidemiological and neuroimaging studies have shown that both cognitive impairments and brain structure alterations are strongly correlated to neuroleptic use, especially with use over longer periods (Ho et al., 2011).

Among the widely described side effects of neuroleptic drugs are akathisia, tardive dyskinesia, flattened mood, cognitive decline and secondary Parkinsonism (Moncrieff, 2008). Furthermore, structural brain alterations, obesity, metabolic syndrome, sudden cardiac death, cardiovascular diseases, diabetes and shortened life expectancy are all effects correlated with neuroleptic use (Aderhold & Stastny, 2015). Next, we will present evidence for the aforementioned side effects highlighting the risks posed by prolonged neuroleptic exposure.

Neuroleptics are the leading cause of drug-induced akathisia, a sense of psychomotor restlessness, both subjective and objective, with a prevalence of 18% to 25% in patients treated with neuroleptics, according to Juncal-Ruiz et al. (2017). Despite being related to worst outcomes, dropping out of treatment and possibly a greater risk of suicide, akathisia is frequently ill-recognized and under-diagnosed (Juncal-Ruiz et al., 2017)

Tardive dyskinesia seems to affect 15% to 30% of individuals in the long-term use of neuroleptics (Cornett et al., 2017), the syndrome is characterised by abnormal repetitive involuntary movements of face and mouth and seems to be accompanied by cognitive decline (Moncrieff,

2008). The condition was first described in 1964, raising concern among psychiatrists (Healy, 2002). Unsure about the exact cause of the symptoms, a debate surfaced with two expressed positions:

> "*First, blame the disease not the treatment. Second, even if there are some problems with the treatment, there will be an even greater number of problems if patients are scared away from treatment.*" (Healy, 2002, p. 247)

Nowadays, the connection between the onset of involuntary movements and D2 receptor occupancy is well established: the higher the occupancy rates, the greater the chance of developing symptoms (Yoshida et al., 2014). It is important to highlight that tardive dyskinesia is chronic and non-reversible in most cases. As Aderhold and Stastny (2015) noticed, almost 20 years were necessary to connect tardive dyskinesia to neuroleptic exposure. This condition is currently established as a direct effect of neuroleptics treatment.

Beyond the severe and highly prevalent side effects mentioned above, increased incidence of metabolic syndrome is associated with second-generation neuroleptics use (Aderhold & Stastny, 2015). A Swedish national population study including data from over 10 million subjects showed that individuals diagnosed with schizophrenia have a higher mortality rate compared to the general population (Westman et al., 2017). The diagnosis of schizophrenia is associated with a decrease of 10 to 25 years in relation to general population life expectancy, with cardiovascular diseases as the major cause (Olfson et al., 2015; Westman et al., 2017). There is a 3-fold increase in coronary heart disease (CHD) mortality for people with severe mental health problems aged 18 through 49 years (Osborn et al., 2007) and the increased risk of CHD and stroke associated with neuroleptics in a dose-dependent fashion (Osborn et al., 2007). However, whether second-generation neuroleptic use is causally linked to increased mortality is still controversial (Westman et al., 2017).

Among the putative causal links between neuroleptics use and increased cardiovascular mortality is metabolic syndrome, defined by the presence of three of the following findings: weight gain, hypertension, lipid dysregulation, glucose tolerance or insulin resistance (Aderhold & Stastny, 2015). However, other factors also account for the excess of cardiovascular mortality as high rates of smoking, poor diet and other lifestyle risk factors.

Extensive data also have linked second-generation neuroleptics use to an increased incidence of sudden cardiac death. These drugs indeed prolong QT intervals, blocking repolarizing potassium currents (Ray et al., 2009). For example, a 2009 retrospective cohort study evaluating the risk of sudden cardiac death associated with first and second-generation neuroleptics have found that:

> "*both current users of typical antipsychotic drugs and current users of atypical antipsychotic drugs had an increased risk of sudden cardiac death as compared with nonusers, with incidence-rate ratios of 1.84 (95% CI, 1.50 to 2.26) and 1.99 (95% CI, 1.61 to 2.46), respectively*" (Ray et al., 2009, p. 230).

Another important conclusion of this study was that the risk increase was dose-related. Moreover, the increased risk was observed for both short and long-term neuroleptic users, suggesting that acute drug-related events might also be involved (Ray et al., 2009).

Overall, the evidence presented does not support the assumption that second generation neuroleptics are safer than their predecessors, as was promoted by the time of introduction of these drugs. Actually, several meta-analytic results found no differences among older and newer drugs regarding safety (Moncrieff, 2008; Ray et al., 2009)

Non-human animal studies have demonstrated volumetric brain reduction and brain morphology alterations due to Olanzapine and Haloperidol (Dorph-Petersen et al., 2005). Lieberman et al. (2005) assessed 161 individuals diagnosed with schizophrenia randomised for olanzapine or haloperidol groups and compared with a control group of 58 individuals without psychiatric diagnosis. Magnetic Resonance Imaging (MRIs) was performed at baseline and periodically for a year, showing a progressive decrease in grey matter in the haloperidol-treated group. Olanzapine-treated subjects also had volumetric reductions, affecting mostly frontal, occipital, and parietal areas, but to a lesser extent (Lieberman et al., 2005). Volumetric changes, in general, were interpreted as part of schizophrenia's degenerative progression, allowing olanzapine, but not haloperidol, to be framed as neuroprotective. Needless to say, Eli Lilly, the pharmaceutical company selling olanzapine, funded the study.

In 2011, a neuroimaging study assessed 211 patients diagnosed with schizophrenia over 7.2 years on average, showing that volumetric brain reduction was directly correlated to the dose and duration of neuroleptics use (Ho et al., 2011). The quote below was extracted from the paper conclusions, illustrating with great clarity how unfavourable results can be transformed into arguments supporting drug use if the dubious premise of effectiveness is accepted:

> *"Antipsychotics are effective medications for reducing some of the target clinical symptoms of schizophrenia: psychotic symptoms. In medicine we are aware of many instances in which improving target symptoms worsen other symptoms. Hormone therapy relieves menopausal symptoms but increases stroke risk. Non-steroidal anti-inflammatory drugs relieve pain but increase the likelihood of duodenal ulcers and gastrointestinal tract bleeding. It is possible that, although antipsychotics relieve psychosis and its attendant suffering, these drugs may not arrest the pathophysiologic processes underlying schizophrenia and may even aggravate progressive brain tissue volume reductions."* (Ho et al., 2011, p.135)

In the light of the review of evidence on effectiveness and safety of neuroleptics, we favour the idea that the disease-centred model has greatly influenced the interpretation of research findings and introduced several biases in study designs and results reports. Specifically, this model introduces a bias towards valuing drug treatment over other approaches to severe mental health problems (Moncrieff, 2008). Furthermore, long-term outcome studies show that a significant group of individuals manage to recover without the use of neuroleptics, indicating that RCTs alone are insufficient to determine neuroleptic use effectiveness long-term.

As Aderhold and Stastny (2015) and Kapur et al. (2006) pointed out, neuroleptics mainly reduce subjects' preoccupation with hallucinations and delusions, rather than eliminate symptoms, in line with the early description of chlorpromazine's effects by Elkes and Elkes (1954). Detachment and less preoccupation with symptoms can be helpful in some cases and in specific clinical circumstances. However, treatment and research have to be framed in novel ways, evaluating when it would be desirable to introduce such drug effects on psychotic experiences, rather than assessing drug efficacy in improving scores of symptoms' scales. This reframing depends on abandoning for good the idea that mental health problems are caused by neurotransmitters imbalances. Overall, the disease-centred model led to ill-posed aetiological hypotheses that failed to translate into effective therapeutical approaches, widely disregarding long-term outcomes and safety issues. In summary, given the dubious effectiveness and the safety concerns regarding neuroleptics use, recovery-based practices avoiding drug treatment are warranted.

3.4 The recovery-oriented approach

One widespread argument favouring psychotropic drug therapy, in general, is that the psychopharmacological revolution was responsible for the deinstitutionalisation of thousands of patients. As Gronfein (1985) have demonstrated, the process of deinstitutionalisation was quite complex and multi-causal. Many studies were conducted to assess whether the introduction of psychotropic drugs was causally related to the decrease in hospital population. Despite some studies proposing the introduction of neuroleptics as a cause of deinstitutionalization, great variability in population discharge was found across the US even after the use of neuroleptics had become widespread (Gronfein, 1985). In addition, deinstitutionalisation began about 20 years after psychotropic drugs were introduced in France and, in the United Kingdom, the decline in hospital population began more than 40 years before chlorpromazine (Gronfein, 1985; Healy, 2002; Moncrieff, 2008; Whitaker, 2010).

In Brazil, the Psychiatric Reform Movement gained momentum in the 1970s and, although the somewhat delayed arrival of psychopharmacology in Brazil, the deinstitutionalisation can be better explained by a series of public policy changes, political pressures, and shifts in care delivery paralleling the implementation of the National Health System. The introduction of psychotropic drugs certainly changed day-to-day life inside the hospitals, calming down patients and rendering them more docile to staff intervention; however, that doesn't necessarily translate to actual changes in patients' lives outside the hospital.

Individual's ability to recover from or manage their mental health problems has been a widely disregarded question (Davidson, 2003). Alternatively, Moncrieff (2008) have proposed a drug-centred model to investigate psychotropic drugs' effects. Such a drug-centred account would shift research towards producing knowledge on the psychoactive or subjective effects of the drugs, helping to determine the circumstances in which their use actually benefit individuals going through mental health problems. Considering that some individuals may benefit from short-term use of psychotropic drugs, a safer and more need-adapted way of prescribing is desirable, taking into account individuals' experiences with the drug effects and carefully assessing risk.

Understanding Mental Health crisis beyond the medical model and reframing psychotropic drugs' role in treatment could improve the evidence-based medical consensus towards the development of recovery-oriented research. Such hegemonic model greatly impacts not only the way we do research, plan health services and provide care but also culture and society. On another level of analysis, diagnostic expansion, as seen in the recently published DSM-V, may lead us to a future when all forms of suffering are translated into psychiatric terms and paired with the correspondent pharmacological treatment.

Whitaker (2010) have proposed that most Americans think of mental health problems, such as depression, as biochemical imbalances amenable to correction by drugs. While a critical discussion of psychotropic drugs' effectiveness and risks is taking place within the scientific community, the cultural and social views regarding mental health remain stable. The power of the medical corporation in producing protocols and steering practices remains strong, as Gøtzsche (2014) pointed out. The pharmaceutical industry has heavily marketed the disease-centered model of mental health problems and drug companies have committed fraud, corrupted psychiatrists and promoted drugs illegally in order to do so (Gøtzsche, 2014). Providing services and care beyond the disease-centred model of mental health problems is a first-step towards the development of new recovery-oriented technologies. These experiences might be crucial to producing positive changes in Mental Health knowledge production and practices, ultimately benefiting those in suffering.

3.4.1The Open Dialogue experience

During the 1980s', a team in the Keropudas Hospital in Tornio, Finland, gradually developed a new way of planning services and delivering care to patients and families: the Open Dialogue approach (Seikkula, 2002). Open Dialogue is a family-oriented approach to severe mental health problems developed in the health district of Western Lapland, Finland. More than a clinical approach, Open Dialogue is the entire mental health system in that region. The impact of the Open Dialogue approach as a mental health system for all psychiatric care in Western Lapland was extensive. According to Aaltonen, Seikkula and Lehtinen (2011), 40% of patients referred to the psychiatric hospital in the first year of implementation of the Open Dialogue approach were not hospitalized. From 1992 onwards, no new long-stay hospital patients with a diagnosis of schizophrenia have emerged and the incidence of all schizophrenic disorders diagnosed individuals has significantly decreased from 1985 to 1994 (Aaltonen, Seikkula & Lehtinen, 2011).

Seven main principles emerged from research on the effectiveness of the model and treatment-process evaluations: (1) immediate help; (2) a network perspective; (3) the need for adaptedness; (4) group responsibility; (5) continuity of psychological treatment; (6) tolerating uncertainties; and (7) dialogical posture (Seikkula & Arnkil, 2014). Although research was mostly conducted on psychotic episodes, the approach is not diagnostic-specific and targets all psychiatric crisis in the catchment area.

According to Seikkula (1994), relevant events take place in the boundary between the hospital and the patient's family, and changes in that system could not be forced in an attempt to control patient's behaviours. The team has had to rethink their position and modify decision-making processes along the treatments. Network meetings including the person in crisis and their social network were conceptualized as a dialogical forum in which information about the problem could be gathered and therapeutical paths might be traced through jointly produced diagnosis. Hence, instead of planning interventions in staff meetings and informing patients and families of professional decisions, the very decision-making process became the treatment.

In the Open Dialogue approach, dialogue in itself is seen as therapeutic. A dialogical perspective develops from a view of meaning as a co-production in a process of exchange between people that generates new understandings and a shared language to address problems (Seikkula & Arnkil, 2006). In this scenario, psychotic experiences are understood as:

> "*attempts to make sense of one's experience and to cope with experiences so difficult that it has not been possible to construct a rational spoken narrative about them*" (Seikkula, 2002, p. 264).

The first principle of Open Dialogue is an immediate response to crisis situations, i.e., a group of professionals is responsible for scheduling a family meeting, with the presence of the patient's social network, within 24 hours of the first contact. The person who responds to the patient's first contact is responsible for gathering a team for the first meeting and help to invite the patient's network.

The implementation of this strategy helped preventing hospitalisation in most cases and reduced the untreated period of psychosis (Seikkula & Arnkil, 2014), furthermore, as a result of the development of the Open Dialogue treatment system in Western Lapland, a decrease in the number of hospital beds was reported from 4.2 to 0.8 beds per 1000 inhabitants from 1983 to 1992 (Aaltonen, Seikkula & Lehtinen, 2011). At the beginning of a crisis, many meetings may be necessary, while later into recovery, meetings can occur weekly and treatment duration should be flexible to address the specific needs of patients and families. The team should also be able to adjust to patients' and families' language and lifestyle specificities (Seikkula & Arnkil, 2006).

According to Seikkula and Arnkil (2014), most hallucinations and disorganised behaviour tend to disappear quickly, giving rise to new needs. Therefore, meetings should be arranged as soon as possible so that words and meanings for difficult experiences can be generated, allowing patient and family to overcome the crisis. It is crucial that the responsible team becomes involved as early as possible, guaranteeing a safe and trusting environment even when significant thought and behaviour disorganisation and psychotic content are present (Seikkula & Arnkil, 2014). In Western Lapland, the period of untreated psychosis is of around 3 weeks, and the average age of first psychotic episode is 20 years old (Seikkula, Alakare & Aaltonen, 2011). Early intervention, avoiding the use of neuroleptics and preventing hospitalisation, are thought to account for a decrease in schizophrenia incidence in Western Lapland.

Defining a catchment area is another important aspect of the approach. Defined areas allow families and other health services to know where treatment can be accessed. Implementing 24/7 mobile crisis teams helped to consolidate immediate response, assuring assistance to families, patients and the involved professionals (Seikkula & Arnkil, 2014).

Ideally, the first responding team assures treatment continuity, as the recovery-process greatly benefits from relationships between team members, patients and their networks (Seikkula & Arnkil, 2014). The principle of flexibility also assured that the Open Dialogue approach was not meant to compete with other therapeutical approaches. Other professionals eventually involved in the recovery process are encouraged to take part in the meetings in order to avoid neglect and care fragmentation.

Tolerating uncertainty is another treatment key-point in Open Dialogue approach. In the early moments of a crisis, major decisions are avoided and efforts are made to create a trusting environment and to develop therapeutic relationships (Seikkula & Arnkil, 2014). Avoidance of neuroleptics use in the early stages of the crisis is also actively attempted. Better outcomes have been reported for patients not using neuroleptics or starting medication later into the treatment (Seikkula & Arnkil, 2006). While the medical model, according to Seikkula and Arnkil (2006), considers recovery as a decrease or removal of psychotic symptoms, the Open Dialogue approach encourages dialogicity as a way to generate words and meanings to refer to difficult experiences. A decrease in symptomatology is seen as a result of a dialogical perspective in treatment meetings. In order to generate new meanings and to recover, patients must mobilize their psychological resources, according to Seikkula and Arnkil (2006) "neuroleptic medication has a sedative effect that calms down psychological activity, and thus it may be a hindrance to psychological work" (p. 59)", justifying their advice to wait at least three meetings before making a decision regarding the use of neuroleptic drugs.

The Open Dialogue approach pointed to the importance of a selective use of neuroleptics rather than complete avoidance. A retrospective cohort study that evaluated the Open Dialogue approach effectiveness in all first episode psychosis in the catchment area of Western Lapland showed that from 1992 to 2015 74% of the subjects at the initial contact did not receive any neuroleptics (Bergström et al., 2017). Considering the entire period studied, 45% of subjects did not receive neuroleptics at any point of the treatment. The data showed that individuals who received neuroleptic medication at the onset of the crisis had the highest hospital admission rates and received psychiatric care for a significantly longer period when compared to non-medicated subjects. Authors ponder that higher admission rates and longer treatment periods cannot be solely attributed to medication use. However, the results Bergström et al. (2017) found are in line with other longitudinal studies comparing outcomes in medicated versus non-medicated subjects, such as the Harrow, Jobe and Faull (2014) study.

To be feasibly part of a mental health care system, a clinical approach has to be evaluated regarding effectiveness, outcomes and recovery. Aware of the relevance of such evaluations, the group that developed the Open Dialogue approach has been systematically dedicating efforts towards researching their way of working, particularly emphasizing outcomes (Seikkula, 2002; Seikkula & Arnkil, 2006; Seikkula, 2008; Seikkula & Arnkil, 2014, Seikkula, et al., 2006), studies repeatedly showed recovery rates of roughly 80% of subjects treated in Western Lapland with the Open Dialogue approach, considering symptomatology and social functioning dimensions. Positive outcomes have remained stable for the past 20 years. This posture, based on the understanding that practical solutions are utterly important, helped to guide research into real-world scenarios and naturalistic research methodologies are preferred as this particular study design is more suited to outcome assessment. In that sense, the distance between academic knowledge and real world problems was reduced, and research and real world solutions might converge.

The Open Dialogue approach has advanced a way to overcome the deadlocks of medicalization by actively avoiding the use of psychotropics in crisis care through selective use of drug therapy, quickly responding to suffering by reducing illness untreated period, avoiding hospitalisation and rejecting traditional ways of diagnosing and planning treatments (Seikkula, 2002; Seikkula & Arnkil, 2006; Seikkula, 2008; Seikkula & Arnkil, 2014). The experience of Open Dialogue also demonstrated that investigative strategies in real world scenarios can produce relevant knowledge for service planning, emphasizing long-term outcome and recovery. But the extents to which the Open Dialogue approach can be successfully implemented in other contexts have remained largely untested until very recently. Effectiveness and feasibility of this approach might be due to context-specific factors such as social, economic and cultural vicissitudes of a small and developed country. But the emergence of recent experiences in the USA has been changing this scenario.

The implementation of Open Dialogue in the USA was described in a 12-month feasibility and effectiveness study conducted by seven clinicians and one psychiatrist, all with Open Dialogue training (Gordon et al., 2016). The service provided included a mobile team available 24/7 and an outpatient service. The meetings took place both at the clinic facility and at patients' homes. Comprehensive assessments were made at three, six and twelve months. Diverse scales and clinical records were analysed, including data on hospitalisation, school attendance and medication prescribed. Sixteen participants aged 14 to 35 years old and experiencing psychotic symptoms were enrolled in the study. Overall, the results have shown fairly good clinical and positive changes in symptoms and functionality were reported. Regarding neuroleptics introduction effects, there was no significant difference in the outcome between patients taking or not taking the medication. The first study, although quite small and preliminary, suggests that the Open Dialogue model can be adapted and implemented to a certain extent in the USA or other countries.

The Parachute program was established in 2013 by the New York Department of Health and Mental Hygiene in an effort to decrease the number of hospitalisations in the city. Similar to the Open Dialogue approach, the Parachute program is also based in peer-support. Currently, the program covers around 900 people in its respite centre in NY city, and another 700 are carried by their mobile team on house calls. The state of New York has funded the initiative that has already shown to be more cost-effectiveness then hospitalisations. With the emergence of recovery-based initiatives in different countries further studies on their feasibility and effectiveness are expected to improve evidence-based practices and health services planning and delivery.

3.5 Final thoughts

The currently hegemonic theoretical accounts of mental health problems, such as the disease-centred biomedical model, have not succeeded in translating their findings into practices ultimately helping people to recover. Investigations on the pathophysiological aspects of mental health problems are relevant to foster psychopathological processes understanding but reducing complex life-experiences to simplistic biological models actually decreases our ability to effectively tackle mental health issues. Many and diverse interests and factors seem to have been playing an important role in steering research, clinical practice and public opinion into a narrow comprehension of human suffering. Not a minor interest behind the widespread use of medication, the oversimplification of diagnostic rationality and the hegemony of biomedical model of mental health problems, the pharmaceutical industry's power to deceive people and their extraordinary marketing skills have been extensively demonstrated. But there are also many factors and interests at stake. As for the mental health field, an ethical issue seems to surface. Should we plan services and deliver care in light of the current hegemonic model, albeit the lack of evidence to support it, or should we question its ability to describe and change human experiences?

Acknowledgements

We would like to thank Aline Amazonas, Rafael Colaço and Robert Whitaker for their careful reading and support. Authors would also like to express their gratitude to the reviewers for their helpful input.

3.6 References

Aaltonen, J., Seikkula, J. & Lehtinen, K. (2011) The Comprehensive Open-Dialogue Approach in Western Lapland: I. The incidence of non-affective psychosis and prodromal states. *Psychosis: Psychological, Social and Integrative Approaches.* 3:3, 179-191. Available from: http://dx.doi.org/10.1080/17522439.2011.601750.

Abi-Dargham, A. (2004). Do we still believe in the dopamine hypothesis? New data bring new evidence. *The International Journal of Neuropsychopharmacology, 7*(S1), S1-S5. Available from: http://dx.doi.org/10.1017/S1461145704004110.

Aderhold, V., & Stastny, P. (2015). A Guide to Minimal Use of Neuroleptics: Why and How. Available from: http://psychrights.org/research/digest/nlps/Guide2MinimalUseOfNeurolepticsAde rhold-Stastny2015.pdf.

American Psychiatric Association. (1987). *Diagnostic and Statistical Manual of Mental Health Disorders (DSM-III-R).* American psychiatric association.

Angell M. (2010) *A verdade sobre os laboratórios farmacêuticos: como somos enganados e o que podemos fazer a respeito.* 5th ed. Rio de Janeiro: Record.

Arksey, H., & O'Malley, L. (2005). Scoping studies: towards a methodological framework. *International journal of social research methodology. 8*(1), 19-32. Available from: http://dx.doi.org/10.1080/1364557032000119616.

Bell, D. S. (1965). Comparison of amphetamine psychosis and schizophrenia. *The British Journal of Psychiatry, 111*(477), 701-707. Available from: http://dx.doi.org/10.1192/bjp.111.477.701.

Bergström, T., Alakare, B., Aaltonen, J., Mäki, P., Köngäs-Saviaro, P., Taskila, J. J., & Seikkula, J. (2017). The long-term use of psychiatric services within the Open Dialogue treatment system after first-episode psychosis. *Psychosis. 9*(4), 310-321. Available from: http://dx.doi.org/10.1080/17522439.2017.1344295.

Bola, J. R., & Mosher, L. R. (2003). Treatment of acute psychosis without neuroleptics: two-year outcomes from the Soteria project. *The Journal of nervous and mental disease. 191*(4), 219-229. Available from: http://dx.doi.org/10.1097/01.NMD.0000061148.84257.F9.

Bola, J., Kao, D., Soydan, H., & Adams, C. E. (2011). Antipsychotic medication for early episode schizophrenia. *Cochrane Database of Systematic Reviews 2011.* Issue 6. *Art. No.: CD006374.* Available from http://dx.doi.org/10.1002/14651858.CD006374.pub2.

Bralet, M. C., Yon, V., Loas, G., & Noisette, C. (2000). Cause of mortality in schizophrenic patients: prospective study of years of a cohort of 150 chronic schizophrenic patients. *L'Encephale, 26*(6), 32-41.

Breggin, P. R. (2012). *Psychiatric drug withdrawal: A guide for prescribers, therapists, patients and their families.* Springer Publishing Company.

Breggin, P. R. (1993). *Toxic psychiatry: Drugs and electroconvulsive therapy: The truth and the better alternatives.* HarperCollins UK.

Cornett, E. M., Novitch, M., Kaye, A. D., Kata, V., & Kaye, A. M. (2017). Medication-Induced Tardive Dyskinesia: A Review and Update. *The Ochsner Journal, 17*(2), 162-174. Available from: http://www.ochsnerjournal.org/doi/full/10.1043/TOJ-16-0108.

Davidson, L. (2003). *Living outside mental illness: Qualitative studies of recovery in schizophrenia.* NYU Press.

Delay J, Deniker P. 38 cas de psychoses traitèes par la cure prolongèe et continuè de 4560 RP. CR. *Congr Méd Alién Neuro*l (France). 50: 503-13.

Deniker, P. (1990). The neuroleptics: a historical survey. *Acta Psychiatrica Scandinavica. 82*(S358), 83-87. Available from: http://dx.doi.org/10.1111/j.1600-0447.1990.tb05295.x

Dorph-Petersen, K. A., Pierri, J. N., Perel, J. M., Sun, Z., Sampson, A. R., & Lewis, D. A. (2005). The influence of chronic exposure to antipsychotic medications on brain size before and after tissue fixation: a comparison of haloperidol and olanzapine in macaque monkeys. *Neuropsychopharmacology, 30*(9), 1649. Available from: http://dx.doi.org/10.1038/sj.npp.1300710.

Doshi, P., Dickersin, K., Healy, D., Vedula, S. S., & Jefferson, T. (2013). Restoring invisible and abandoned trials: a call for people to publish the findings. *British Medical Journal. 346,* f2865. Available from: https://doi.org/10.1136/bmj.f2865.

Dundee, J. W. (1954). A review of chlorpromazine hydrochloride. *British Journal of Anaesthesia. 26*(5), 357-379. Available from: https://doi.org/10.1093/bja/26.5.357.

Dunker, C. I. L. & Kyrillos Neto, F. (2011). A crítica psicanalítica do DSM-IV-breve história do casamento psicopatológico entre psicanálise e psiquiatria. *Revista Latinoamericana de Psicopatologia Fundamental, 14*(4). Available from: http://dx.doi.org/10.1590/S1415-47142011000400003.

Elkes, J., & Elkes, C. (1954). Effect of chlorpromazine on the behaviour of chronically overactive psychotic patients. *British Medical Journal, 2*(4887), 560. Available from: https://www.ncbi.nlm.nih.gov/pmc/articles/PMC2079576/ [Accessed: 23rd August 2017]

Gordon, C., Gidugu, V., Rogers, E. S., DeRonck, J., & Ziedonis, D. (2016). Adapting open dialogue for early-onset psychosis into the US health care environment: a feasibility study. *Psychiatric Services. 67*(11), 1166-1168. Available from: https://doi.org/10.1176/appi.ps.201600271.

Gøtzsche, P. C. (2014). Why I think antidepressants cause more harm than good. *The Lancet Psychiatry. 1*(2), 104-106. Available from: http://dx.doi.org/10.1016/S2215-0366(14)70280-9.

Gronfein, W. (1985). Psychotropic drugs and the origins of deinstitutionalization. *Social Problems, 32*(5), 437-454.

Harrow, M., Jobe, T. H., & Faull, R. N. (2014). Does treatment of schizophrenia with antipsychotic medications eliminate or reduce psychosis? A 20-year multi-follow-up study. *Psychological Medicine. 44*(14), 3007-3016. Available from: https://doi.org/10.1017/S0033291714000610.

Healy, D. (2002). *The creation of psychopharmacology*. Harvard University Press.

Ho, B. C., Andreasen, N. C., Ziebell, S., Pierson, R., & Magnotta, V. (2011). Long-term antipsychotic treatment and brain volumes: a longitudinal study of first-episode schizophrenia. *Archives of general psychiatry, 68*(2), 128-137. Available from: https://doi.org/10.1001/archgenpsychiatry.2010.199.

Jääskeläinen, E., Juola, P., Hirvonen, N., McGrath, J. J., Saha, S., Isohanni, M. & Miettunen, J. (2012). A systematic review and meta-analysis of recovery in schizophrenia. *Schizophrenia bulletin. 39*(6), 1296-1306. Available from: https://doi.org/10.1093/schbul/sbs130.

Juncal-Ruiz, M., Ramirez-Bonilla, M., Gomez-Arnau, J., de la Foz, V. O. G., Suarez-Pinilla, P., Martinez-Garcia, O., & Crespo-Facorro, B. (2017). Incidence and risk factors of acute akathisia in 493 individuals with first episode non-affective psychosis: a 6-week randomised study of antipsychotic treatment. *Psychopharmacology*. 1-8. Available from: https://doi.org/10.1007/s00213-017-4646-1.

Kapur, S., Agid, O., Mizrahi, R., & Li, M. (2006). How antipsychotics work—from receptors to reality. *NeuroRx. 3*(1), 10-21. Available from: https://doi.org/10.1016/j.nurx.2005.12.003

Lieberman, J. A., Stroup, T. S., McEvoy, J. P., Swartz, M. S., Rosenheck, R. A., Perkins, D. O., & Severe, J. (2005). Effectiveness of antipsychotic drugs in patients with chronic schizophrenia. *N Engl j Med, 2005*(353), 1209-1223. Available from: https://doi.org/10.1056/NEJMoa051688

Lepping, P., Sambhi, R. S., Whittington, R., Lane, S., & Poole, R. (2011). Clinical relevance of findings in trials of antipsychotics: systematic review. *The British Journal of Psychiatry. 198*(5), 341-345. Available from: How antipsychotics work—from receptors to reality. *NeuroRx. 3*(1), 10-21. Available from: https://doi.org/10.1192/bjp.bp.109.075366.

Moncrieff J. (2008) *The myth of the chemical cure: a critique of psychiatric drug treatment.* New York: Palgrave Macmillan.

Moynihan, R., & Henry, D. (2006). The fight against disease mongering: generating knowledge for action. *PLoS medicine. 3*(4), e191. Available from: https://doi.org/10.1371/journal.pmed.0030191

Muller, P., & Seeman, P. (1977). Brain neurotransmitter receptors after long-term haloperidol: dopamine, acetylcholine, serotonin, α-noradrenergic and naloxone receptors. *Life sciences. 21*(12), 1751-1758. Available from: https://doi.org/10.1016/0024-3205(77)90155-2.

Murray, R. M., Paparelli, A., Morrison, P. D., Marconi, A., & Di Forti, M. (2013). What can we learn about schizophrenia from studying the human model, drug-induced psychosis?. *American Journal of Medical Genetics Part B: Neuropsychiatric Genetics. 162*(7), 661-670. Available from: https://doi.org/10.1002/ajmg.b.32177.

Nordentoft, M., Wills, R., & Gotfredsen, D. (2016). Social cognition, functional capacity and symptoms in the longitudinal prediction of outcome in subjects with first-episode schizophrenia. *European Psychiatry. 33*, S67. Available from: https://doi.org/10.1016/j.eurpsy.2016.01.971.

Olfson, M., Gerhard, T., Huang, C., Crystal, S., Stroup, T.S. Premature Mortality Among Adults with Schizophrenia in the United States. *JAMA Psychiatry.* 2015;72(12):1172–1181. doi:10.1001/jamapsychiatry.2015.1737

Osborn, D. P., Levy, G., Nazareth, I., Petersen, I., Islam, A., & King, M. B. (2007). Relative risk of cardiovascular and cancer mortality in people with severe mental illness from the United Kingdom's General Practice Research Database. *Archives of general psychiatry. 64*(2), 242-249. Available from: https://doi.org/10.1001/archpsyc.64.2.242

Rappaport, M., Hopkins, H. K., Hall, K., Belleza, T., & Silverman, J. (1978). Are there schizophrenics for whom drugs may be unnecessary or contraindicated. *International Pharmacopsychiatry, 13*(2), 100-111. Available from: https://doi.org/10.1159/000468327

Ray, W. A., Chung, C. P., Murray, K. T., Hall, K., & Stein, C. M. (2009). Atypical antipsychotic drugs and the risk of sudden cardiac death. *New England Journal of Medicine. 360*(3), 225-235. Available from: https://doi.org/10.1056/NEJMoa0806994

Rose, D. (2014). Patient and public involvement in health research: Ethical imperative and/or radical challenge?. *Journal of health psychology. 19*(1), 149-158. Available from: https://doi.org/10.1177/1359105313500249

Rose, D., Thornicroft, G., & Slade, M. (2006). Who decides what evidence is? Developing a multiple perspectives paradigm in mental health. *Acta Psychiatrica Scandinavica. 113*(s429), 109-114. Available from: https://doi.org/10.1111/j.1600-0447.2005.00727.x

Sackett, D. L., Rosenberg, W. M., Gray, J. M., Haynes, R. B., & Richardson, W. S. (1996). Evidence based medicine: what it is and what it isn't. *British Medical Journal.* Jan 13;312(7023):71-2. Available from: https://www.ncbi.nlm.nih.gov/pmc/articles/PMC2349778/ [Accessed: 24th August 2017]

Samaha, A. N., Reckless, G. E., Seeman, P., Diwan, M., Nobrega, J. N., & Kapur, S. (2008). Less is more: antipsychotic drug effects are greater with transient rather than continuous delivery. *Biological psychiatry. 64*(2), 145-152. Available from: https://doi.org/10.1016/j.biopsych.2008.01.010.

Seikkula, J. (2008). Inner and outer voices in the present moment of family and network therapy. *Journal of Family Therapy, 30*(4), 478-491. Available from: https://doi.org/10.1111/j.1467-6427.2008.00439.x

Seikkula, J. (2002). Open dialogues with good and poor outcomes for psychotic crises: Examples from families with violence. *Journal of marital and family therapy. 28*(3), 263-274. Available from: https://doi.org/10.1111/j.1752-0606.2002.tb01183.x.

Seikkula, J. (1994). When the boundary opens: family and hospital in co-evolution. *Journal of Family Therapy. 16*(4), 401-414. Available from: http://dx.doi.org/10.1111/j.1467-6427.1994.00805.x

Seikkula, J., & Arnkil, T. E. (2014). *Open dialogues and anticipations- Respecting otherness in the present moment.* Thl.

Seikkula, J., & Arnkil, T. E. (2006). *Dialogical meetings in social networks.* Karnac books.

Seikkula, J., Aaltonen, J., Alakare, B., Haarakangas, K., Keränen, J., & Lehtinen, K. (2006). Five-year experience of first-episode nonaffective psychosis in open-dialogue approach: Treatment principles, follow-up outcomes, and two case studies. *Psychotherapy research. 16*(02), 214-228. Available from: http://dx.doi.org/10.1080/10503300500268490.

Seikkula , J., Alakare, B. & Aaltonen, J. (2011) The Comprehensive Open-Dialogue Approach in Western Lapland: II. Long-term stability of acute psychosis outcomes in advanced community care. *Psychosis: Psychological, Social and Integrative Approaches.* 3:3, 192-204. Available from: http://dx.doi.org/10.1080/17522439.2011.595819.

Silvestri, S., Seeman, M. V., Negrete, J. C., Houle, S., Shammi, C. M., Remington, G. J. & Seeman, P. (2000). Increased dopamine D2 receptor binding after long-term treatment with antipsychotics in humans: a clinical PET study. *Psychopharmacology. 152*(2), 174-180. Available from: https://doi.org/10.1007/s002130000532.

Whitaker, R. (2010). *Anatomy of an Epidemic: Magic Bullets, Psychiatric Drugs, and the Astonishing Rise of Mental Illness in America.* New York: Broadway Books.

Westman, J., Eriksson, S.V., Gissler, M., Hällgren, J., Prieto, M.L., Bobo, W.V., Frye, M.A., Erlinge, D., Alfredsson, L., Ösby, U. (2017). Increased cardiovascular mortality in people with schizophrenia: a 24-year national register study. *Epidemiology and Psychiatric Sciences,* 1-9. Available from: http://dx.doi.org/doi:10.1017/S2045796017000166.

Wils, R. S., Gotfredsen, D. R., Hjorthøj, C., Austin, S. F., Albert, N., Secher, R.G., Thorup, A.A.E., Mors, O., Nordentoft, M. (2017). Antipsychotic medication and remission of psychotic symptoms 10years after a first-episode psychosis. *Schizophrenia Research. 182,* 42-48. Available from: http://dx.doi.org/10.1016/j.schres.2016.10.030.

Wunderink, L., Nieboer, R., Nienhuis, F., Sytema, S., & Wiersma, D. (2015). Long Term Recovery Rates in Schizophrenia After Early Antipsychotic Dose Reduction. *European Psychiatry. 30,* 67. Available from: https://doi.org/10.1016/S0924-9338(15)30056-0.

Yoshida, K., Bies, R. R., Suzuki, T., Remington, G., Pollock, B. G., Mizuno, Y., ... & Uchida, H. (2014). Tardive dyskinesia in relation to estimated dopamine D2 receptor occupancy in patients with schizophrenia: analysis of the CATIE data. *Schizophrenia research. 153*(1), 184-188. Available from: http://dx.doi.org/10.1016/j.schres.2014.01.017.

Chapter 4

Embracing the placebo effect in the treatment of depression: from neuropsychiatry to psychotherapy

Jaime Grácio
Champalimaud Research and Clinical Centre, Champalimaud Centre for the Unknown, Lisbon, Portugal; NOVA Medical School | Faculdade de Ciências Médicas, Universidade Nova de Lisboa, Portugal

Sílvia Almeida
Champalimaud Research and Clinical Centre, Champalimaud Centre for the Unknown, Lisbon, Portugal

Albino J. Oliveira-Maia
Champalimaud Research and Clinical Centre, Champalimaud Centre for the Unknown, Lisbon, Portugal; Department of Psychiatry and Mental Health, Centro Hospitalar de Lisboa Ocidental, Portugal; NOVA Medical School | Faculdade de Ciências Médicas, Universidade Nova de Lisboa, Portugal

While the placebo effect has been a ruler to set the pace for research in clinical trials, its healing-promoting properties are frequently disregarded. The placebo effect is often labelled as an unspecific effect, frequently halting, rather than promoting, the development of new forms of treatment. However, a growing research interest in this field has increasingly demonstrated that this effect may be an important ingredient in many forms of treatment. In this chapter we will describe the placebo-controlled evidence supporting the most frequently used treatment options for depression, namely pharmacotherapy and psychotherapy, as well as some of the options for treatment-resistant depression, namely electroconvulsive therapy and transcranial magnetic stimulation. The most relevant meta-analysis and review studies in the field will be discussed, focusing on the benefits attributable to the placebo effect, compared across interventions to provide novel integrative insights on the relevance of this effect for the treatment of depression.

4.1. Introduction

The placebo effect is generally defined as a favorable outcome of treatment, attributable to the patient's positive expectancy of the results of treatment. It is commonly viewed as an unspecific effect, as opposed to the specific effects of biologically-based or psychologically-based interventions (Gold and Lichtenberg, 2014). What we currently call the placebo effect is thought to have been the basis of the healing process across cultures and over many centuries, namely in the so-called pre-scientific period of medicine. However, the current terminology and concepts related to placebo derive mostly from controlled therapeutic trials, the cornerstone of modern clinical research. These trials are designed to dissociate and compare the clinical improvement obtained from application of an active treatment being investigated (the active arm of the trial), to the induction of the belief that such treatment is being applied in a separate group of patients (the placebo arm of the trial). The paradigm of controlled clinical trials was undoubtedly a great achievement in health sciences, in which placebo effect has been a key element, as a baseline to measure the specific effects of a certain intervention. However, this has also equated to a vision of placebo as a concept akin to deception, rather than related to the more positive concepts of therapeutic relationship and expectation of therapeutic effect. This is particularly relevant in conditions with strong psychological determinants (Ayers and De Visser, 2011; Kirsch and Sapirstein, 1998), such as pain and depression (Hróbjartsson and Gøtzsche, 2010). Nonetheless, a growing interest in research on the healing-promoting properties of the placebo effect in several clinical conditions is generating a new vision of the concept (Colagiuri et al., 2015).

Depression is one of the most prevalent mental disorders (Wittchen et al., 2011; World Health Organization, 2016) and a leading cause of disability (Whiteford et al., 2015). Special attention must thus be paid to improve treatment outcomes for this clinical condition. The placebo effect is very relevant in the treatment of depression, and the administration of pill-placebos has well recognized biological effects, leading some to consider that these should no longer be considered inert substances (Shapiro and Shapiro, 1997). In fact, neuroimaging research of the placebo effect supports the existence of similarities between mechanisms that are activated by placebo and those that are activated by medication (Benedetti et al., 2011; Mayberg et al., 2002). In this chapter, we will focus on the relevance of the placebo effect in the treatment of depression, namely for first-line treatments such as pharmacotherapy and psychotherapy, as well as with brain stimulation alternatives typically used in treatment-resistant patients, namely electroconvulsive therapy and repetitive transcranial magnetic stimulation. We will describe and integrate results from the most relevant systematic review and meta-analyses for each of these treatment modalities, and thus attempt to infer on the global and treatment-specific role played by the placebo effect in antidepressant treatments.

4.2. Method

We identified adequate systematic review and meta-analyses through searches on electronic databases, namely PubMed/MEDLINE and EBSCO Host (Academic Search Complete; Education Research Complete; Education Source; ERIC; and PsycINFO). Non-systematic searches were performed using "depression", "placebo effect", "treatment", "pharmacotherapy", "psychotherapy", "transcranial magnetic stimulation", "electroconvulsive therapy" as key-words. Additional hand-searches were performed on the reference list of the selected papers. In order to be comparable, the included studies had to be performed on depression severity data collected using the Hamilton Rating Scale for Depression (HRSD) to assess treatment outcome. This scale was chosen because it is one of the gold standards to assess depression severity, and one of the most frequently used scales to assess treatment outcome in trials of depression. Cut-off points for assessing the severity of depression are: mild depression (8–16); moderate depression (17–23); and severe depression (≥24) (Zimmerman et al., 2013). When several meta-analyses were identified for a specific treatment method, the most relevant were selected based on the larger number of independent trials included in the analyses, the year of publication (most recent were preferred) and an assessment of the quality and relevance of the study (e.g. studies performed by groups of experts in the field and that informed clinical guidelines were preferred). A total of four studies were selected for review. Table 4.1. describes a summary of these studies.

Table 4.1. Descriptive summary of the studies selected for review

Treatment	Authors	Number of Included trials	Total sample size (n)
ADs	Kirsch et al. (2008)	35	5133
Psychotherapy	Cuijpers et al. (2010)	3	588
ECT	The UK ECT Review Group (2003)	63	256
rTMS	Lepping et. al (2014)	63	3236

4.3. Assessing the relevance of the placebo effect in the treatment of depression

4.3.1. Antidepressants

Antidepressants (ADs) are amongst the most frequently used treatments for depression. A population-based survey from 10 European countries demonstrated that approximately 6% of the general population had been prescribed ADs in the previous year (World Health Organization, 2009). Paradoxically, while the treatment gap for depression is well recognized, and evident even in developed countries with structured health systems (Saxena et al., 2007), there is also considerable concern regarding overtreatment, especially because of the high prescription levels for antidepressant pharmacotherapy (Gartlehner et al., 2015). Such concern is typically focussed on the use of ADs as first-step therapy by many clinicians, namely in cases of mild depression treated in primary healthcare centres (Johnson et al., 2017). For these cases, most guidelines recommend non-pharmacological alternatives, such as active monitoring and guided self-help programmes (National Institute for Health and Clinical Excellence, 2009).

The concerns of pharmacological overtreatment of depression are typically based, among other arguments, on the results of meta-analyses summarizing the results of the numerous randomized and placebo-controlled trials (RCTs) of AD efficacy. Indeed, when comparing the antidepressant efficacy of placebos and ADs, most research shows that there are only marginal differences between these two treatment conditions (Kirsch and Sapirstein, 1998). This has also sustained the positions expressed in most guidelines, recommending against antidepressant use as the first-step treatment in mild to moderately severe depression (NICE, 2009).

The placebo response rate when using pill placebos in RCTs of AD efficacy has been reported to be around 30% (Walsh et al., 2002), with some studies demonstrating higher values, from 35% to 40% (Furukawa et al., 2016). One of the best-known demonstrations of the magnitude and determinants of the placebo effect in the treatment of depression was the meta-analysis by Kirsch and colleagues (2008), including studies from the US Food and Drug Administration (FDA) dataset. This study used meta-analytic techniques to compare data from 35 RCTs, and thus assess improvement for medication and pill-placebo groups, according to the initial mean severity of depression. The authors found that, overall, ADs resulted in a mean improvement of the HRSD of 9.6 points. However, placebos also generated a clinically significant improvement, with a mean improvement of 7.8 points in the HRSD, relative to the initial severity. The authors of this meta-analysis thus argued that the mean efficacy differences between the two treatment conditions (1.8 points) were below the three-point difference considered to be the cut-off point for clinical significance in the HRSD. Importantly, this study also described that ADs had a larger advantage over placebos in the treatment of patients with severe depression. In fact, when considering trials including only patients scoring above certain HRSD scores (either 24 or 28) at baseline, differences between the mean improvements obtained with ADs and placebos were considerably larger than the 1.8 points aforementioned (3.5 and 4.8 points respectively). However, it is also important to note that the greater benefit of ADs in the most severely ill patients was due to a lower response to placebo in this patient group, rather than a greater benefit of the active drug. In fact, while placebo response was reduced amongst patients with greater initial depression severity, there were no differences in drug response according to initial depression severity (Kirsch et al., 2008).

The findings reported by Kirsch and colleagues raise an intriguing point. In fact, it is interesting to note that the recommendations to avoid ADs in patients with mild-to-moderate depression seem unrelated to the uncontrolled therapeutic potential of these substances in this patient group and may actually be a result of a better response to placebo. However, placebo is not an option for treatment outside of RCTs, raising interest on developing and testing open-label placebos for depression, as has been reported for other conditions (Carvalho et al., 2016; Kaptchuk et al., 2010). In studies testing open-placebo, patients are aware that they are taking a placebo, yet it is also explained that the pill can nonetheless have a positive impact in the treatment of their condition (Kaptchuk et al., 2010). The placebo effect in depression seems to occur also when patients are administered open-label pill-placebos, as demonstrated by Kelley and colleagues (2012). However, treatment effects were modest when compared with double-blind placebos, with a mean improvement in the HRSD of only 4.8 points, lower than the 7.8 points found by Kirsch and colleagues (2008). Additional research on this subject is thus definitely necessary.

Irrespective of the potential future use of placebo in routine clinical practice, it is clear from the available RCTs that the placebo effect is highly relevant for the current use of ADs in the treatment of depression. We believe this should not be a cause of scepticism regarding the efficacy of ADs in the treatment of depression, but rather reveal the placebo effect as a critical ingredient of treatment, reflecting the capability to harness hope and positive expectations. Research on the treatment of depression should thus devise methods that optimize recruitment of this effect, rather than try to avoid it.

4.3.2. Psychotherapy

Psychotherapy is a process of treatment for mental disorders based on a structured patient-therapist relationship, using verbal and nonverbal communication in order to convey specific therapeutic techniques and relational-contextual therapeutic interactions that stem from (or lead to) many different theoretical backgrounds (Ayers and De Visser, 2011). Understanding psychotherapy thus entails understanding the idiosyncrasies of each approach and the respective theoretical orientation, as well as the common or unspecific factors underlying the efficacy of the several different approaches (Zarbo et al., 2016). Over the last decades, psychotherapy has been increasingly recognized as an effective treatment approach on a wide variety of disorders, namely in depression (NICE, 2009). Cognitive therapy, developed by Aaron T. Beck and colleagues, was probably the first structured psychotherapeutic treatment approach to produce robust evidence in the treatment of depression (Beck and Rush, 1979). However, a wide variety of treatment approaches subsequently also proved its efficacy, namely interpersonal therapy, psychodynamic therapy, behavioural activation and problem-solving (Cuijpers et al., 2013a; Patel et al., 2009).

After several decades of research on the efficacy of psychotherapy, the specific components that most contribute to the efficacy of this treatment approach are still elusive (Lambert and Barley, 2001; Norcross, 2002; Norcross and Lambert, 2011). In fact, surprisingly, while most research efforts have been dedicated to understanding the importance of factors that are specific to each psychotherapeutic approach, such factors are thought to explain only 15% of variance of its efficacy, while the remaining 85% is thought to result from unspecific factors, namely spontaneous remission (40%), common therapeutic factors (30%) and the placebo effect (15%) (Norcross, 2002). While some authors argue that placebo should be conceived stringently as a distinct component of psychotherapy, that increases its effectiveness operating through psychological mechanisms (Patterson, 1985), it is important to note that, at a conceptual level, the several unspecific elements of psychotherapeutic efficacy mentioned above correspond to what is globally referred to as placebo in the AD RCT literature. It is also curious to verify that in the meta-analysis by Kirsch and colleagues (2008), the mean placebo effect (HRSD reduction of 7.8 points) is 81.25% of the mean effect of ADs (HRSD reduction of 9.6 points), which is very similar to the 85% of psychotherapeutic efficacy that is attributed to unspecific factors (Kirsch et al., 2008; Norcross, 2002). Indeed, when considered globally as the unspecific component of treatment, the placebo effect seems to be a critical component of psychotherapeutic efficacy.

Nevertheless, the empirical challenge for placebo-related research in the context of psychotherapy is to find a placebo intervention, as adequate as what is available for AD RCTs (Shapiro and Shapiro, 1997). Some argue that it is hard, or even impossible, to separate placebo, defined stringently, from psychotherapy, considering that it is an active part of the process, that can be maximized or minimized by the therapist (Patterson, 1985). In fact, it is difficult to determine what should be considered placebo in psychotherapy, since elements of placebo can be easily confused with active elements of psychotherapy (Vasco, 2018). As a result of that difficulty, the methodology for most psychotherapy placebo-related research is inconsistent (Patterson, 1985). Authors have suggested that befriending is the ideal control intervention for clinical trials of psychotherapy. Befriending is a non-directive intervention, whereby the patient is introduced to a trained individual, who offers the patient social support (Mead et al., 2010). Furthermore, the intervention must be limited so as to not add any other elements that could be considered interventions of any kind of psychotherapy (Bendall et al., 2003). Befriending was originally used as a control intervention in cognitive-behavioural therapy (CBT) trials for the treatment of resistant psychosis

(Sensky et al., 2000), but it has also been used in other disorders, namely depression (Bendall et al., 2003). While there is evidence that befriending has a modest effect on depressive symptoms (Mead et al., 2010), the most recent empirical research shows that the effect of befriending on these symptoms is not significant, even though it can facilitate social integration and recovery for some individuals (Siette et al., 2017). Nonetheless, while befriending is probably the best placebo to assess psychotherapy-specific efficacy in a RCT scenario, there is a clear lack of high-quality research involving this study design (Cuijpers et al., 2013b).

There is, nevertheless, some controlled research regarding the efficacy of psychotherapy for depression. While there is no clear evidence to support superiority of one form of psychotherapy over others, most studies of psychotherapy for depression in adults have focussed on CBT (Cuijpers et al., 2013a). One meta-analysis by Cuijpers and colleagues (2010) assessed the quality of the studies and included three high-quality RCTs comparing CBT with pill-placebos for treatment of depression using HRSD as treatment outcome measure (Dimidjian et al., 2006; Elkin et al., 1989; Jarrett et al., 1999). We focused our analysis solely on these studies because they were the only that provided suitable information for our comparisons. Overall, 291 patients were included in these comparisons (140 in the CBT arm and 151 in the pill placebo arm). Not surprisingly, the HRSD score means change in the placebo groups (6.9 points) was similar to that found in the RCTs of ADs. On the other hand, patients offered CBT improved 9.9 points, rendering a mean difference between CBT and pill-placebo groups of three points.

Importantly, ADs and CBT have equivalent efficacy in the short-term (Cuijpers et al., 2013a). Nevertheless, CBT might generate more enduring results in the treatment of depression by reducing the risk for subsequent symptom relapses (Hollon et al., 2006). Furthermore, there are suggestions that the difference between pill-placebo and combined psychotherapy and pharmacotherapy is twice the difference between pharmacotherapy alone and pill-placebo (Cuijpers et al., 2014).

4.3.3. Electroconvulsive therapy

Electroconvulsive therapy (ECT) is one of the oldest, and probably the most controversial, treatment option for depression (Rauschenberg and Crawford, 2003). The concern that ECT could cause permanent brain damage is possibly the main contribution to nurture this controversy and is sustained by findings of memory impairment after treatment (Sackeim, 2014; Squire, 1977; Sterling, 2000). However, there is also abundant research demonstrating that ECT is effective and safe (Devanand et al., 1994; Fink and Bailine, 1998), with cognitive side-effects limited mostly to the period during and immediately after application of ECT (Semkovska and McLoughlin, 2010). Most regulatory bodies, such as the National Institute for Health and Care Excellence (NICE) consider ECT as an effective treatment option for depression, promoting at least short-term improvements in symptoms (NICE, 2003). Notwithstanding, in recognition of the risks associated with this treatment, NICE strictly recommends ECT as a treatment option only for severe depression, typically when there is a risk to life and/or severe self-neglect. While the combination of ECT with pharmacotherapy does not seem to promote better short-term outcomes (NICE, 2009), maintenance treatment with tricyclic antidepressants and/or lithium reduces relapse rates in patients that respond favourably to ECT. CBT is another alternative to overcome the limitation of the short-term nature of treatment benefits from ECT, and it has been shown to sustain ECT treatment effects for as long as 12 months (Brakemeier et al., 2014).

The role of the placebo effect in ECT can be assessed through the results of RCTs that used simulated ECT (also known as sham ECT) as a control. In the simulated ECT group, patients receive all of the components of ECT, including general anaesthesia, but the electric stimulus is omitted, and the associated convulsion does not occur. One systematic review and meta-analysis summarized data from six clinical trials, involving 256 patients and using simulated ECT as control (The UK ECT Review Group, 2003). The authors found that real ECT was significantly more effective than sham ECT. It is worth to note that, in this study, specific data describing the mean improvement on HRSD separately for real and simulated ECT was not presented. Thus, we were not able to assess the individual impact of these interventions, as described above for pharmacotherapy and psychotherapy. Nonetheless, authors presented the difference of mean HRSD improvements between these two conditions, which was found to be 9.7 points, favouring real ECT.

The large difference between real and simulated ECT could be accounted for by a lesser relevance of the placebo effect. In fact, ECT is typically offered to the most severely ill patients that were shown to have a smaller placebo effect in AD trials (Kirsch et al., 2008). Additionally, the six trials in this meta-analysis were conducted between 1963 and 1985, a period when ECT was an unpopular procedure (Abrams, 2002), possibly lowering positive expectations related to treatment, and thus also the placebo effect. On the other hand, enhanced treatment differences between active and control groups could reflect the efficacy of ECT as a very powerful method for short-term alleviation of depression (American Psychiatric Association, 1990; NICE, 2009). For example, when compared with pharmacotherapy, ECT was superior by an impressive mean difference of a 5.2 points reduction in the HRSD (The UK ECT Review Group, 2003). In any case, given the significant ethical and practical challenges associated with the use of simulated ECT in severely ill patients, data regarding placebo effects in ECT is still lacking.

4.3.4. Repetitive transcranial magnetic stimulation

Amongst the treatment options for depression mentioned here, repetitive transcranial magnetic stimulation (rTMS) is the most recent. It involves non-invasive stimulation of the dorsolateral prefrontal cortex, using a magnetic coil held against the scalp at specific locations (Oliveira-Maia et al., 2017; Pascual-Leone et al., 1996). rTMS is a safe procedure that can be delivered on an outpatient basis, without the need for anaesthesia (NICE, 2015). While there is some heterogeneity of procedures for rTMS, namely with regards to stimulation frequency and stimulation target, a number of relevant reviews and meta-analyses on rTMS for depression were conducted between 2007 and 2014 (Lepping et al., 2014), demonstrating the interest on this new treatment approach. In fact, evidence shows that rTMS promotes a reduction in depressive symptoms, especially in severe depression (NICE, 2015).

Similarly to ECT research, controlled research on rTMS relies on sham rTMS procedures for comparisons regarding efficacy in clinical trials. Sham rTMS is conducted identically to real rTMS, but procedures are in place to eliminate the magnetic field or its effect on the cortex. Lepping and colleagues (2014) conducted a systematic review of 63 studies, including RCTs, comparing real vs. sham rTMS. In the real rTMS arms, the authors found a reduction in depression severity of 37% across all patients (mean 9.3 points improvement on the HRSD), and 48% for treatment-resistant depression (TRD; 11.4 points). In patients receiving sham rTMS, the reduction in depression was only 22% across all patients (5.4 points), and 23% for TRD (5.7 points). Another meta-analysis (Liu et al., 2014) found similar results when assessing the potential of rTMS to enhance AD efficacy in patients with TRD. The six studies included in this meta-analysis demonstrated a mean improvement on HRSD of 10.9 in the real rTMS arm, as opposed to the 6.3 in the sham groups. Overall, the evidence for the placebo effect on rTMS is overlapping to that described for other treatment modalities, but there is insufficient evidence to support that it varies according to baseline depression severity.

Figure 4.1. Differences in HRSD response between treatment and placebo groups across different treatment modalities.

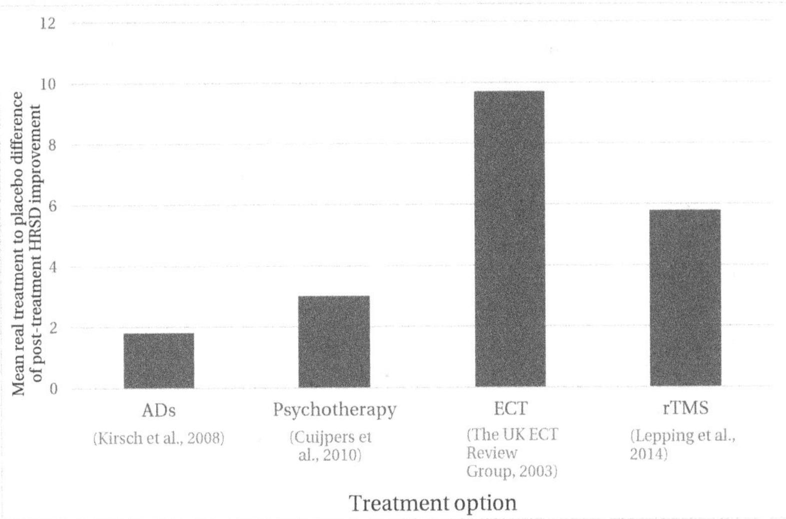

4.4. Discussion

In this chapter, we described relevant meta-analyses and review studies of the most established treatment options for depression. Our intention was to focus on the relevance of the placebo effect for each treatment, inferred by comparisons of clinical improvements, as measured by the HRSD, in RCTs including a placebo arm, or placebo-like arm (simulated ECT and sham rTMS). According to the typical method of assessing treatment efficacy, we compared the several treatment options according to the difference in HRSD mean improvement after treatment with placebo or with the real intervention. Using the studies included in our review, we found that ECT resulted in the larger difference between real treatment and placebo, followed by rTMS. CBT and pharmacotherapy resulted in the smaller differences between the two conditions (Figure 4.1.).

The variability in differences between real treatment and placebo could essentially result from a lower placebo effect and/or from the higher efficacy of the real-treatment. Thus, the large difference between ECT and placebo might suggest, for example, that placebo is less relevant in this treatment approach. In our review, it was impossible to retrieve raw data of the mean HRSD changes in either of the arms of the ECT trials, which does not allow for direct comparisons of the changes promoted by the placebo condition relative to other treatment approaches. There are, however, direct comparisons between active treatment with ADs and with

ECT, demonstrating an additional mean reduction of 5.2 points on the HRSD, for patients treated with ECT (The UK ECT Review Group, 2003). In the data used for our review, the difference of post-treatment HRSD reduction between real treatment and placebo was 9.7 for ECT (The UK ECT Review Group, 2003) and 1.8 for ADs (Kirsch et al., 2008). Since the difference between these two values (7.9) is substantially larger than the difference in efficacy between ECT and ADs (5.2), it is possible that, in fact, the placebo effect might be reduced in ECT trials. This would be consistent with the fact that guidelines for ECT use in depression recommend its use for some of the most severely ill patients (NICE, 2003) that could be less responsive to placebo (Kirsh et al., 2008). ECT is also a somewhat unpopular treatment, and as result patients may feel less enthusiasm and less positive expectancy when receiving ECT (Reisner, 2003). Nevertheless, this rationale is speculative, and further analysis of previously collected data, or conduction of additional trials, would be necessary to have a clear picture of the placebo effect in ECT, and thus to consider potential strategies to harness its use towards enhancing treatment efficacy.

When comparing with ECT, in the opposite extreme of the magnitude of the difference between real treatment and placebo, we found pharmacotherapy (ADs) and psychotherapy (CBT). In these treatment approaches, differences between real treatment and placebo were much smaller. In fact, placebo could explain around 80% of the efficacy of ADs and CBT in cases of mild to moderate depression, demonstrating that the placebo effect is highly relevant in both cases (Figure 4.2). Naturally, this does not mean that these treatments are not efficacious or less useful. On the contrary, the placebo effect should be considered as a relevant promoter of clinical improvement, and clinicians should be aware of this fact when treating depression. However, the small difference between placebo and real treatment in AD trials has stimulated criticism regarding the efficacy of ADs (Kirsch et al., 2008; Kirsch and Sapirstein, 1998), with suggestions that the demonstration of AD efficacy in clinical trials actually results from imperfect blinding associated with the perception of drug side effects. However, recent publications have demonstrated that ADs are superior to placebo even in the absence of adverse side effects (Hieronymus et al., 2017). On the other hand, while high-quality RCTs for psychotherapy in depression, using psychological placebos (e.g. befriending therapy), are not available, it seems likely that psychological interventions for depression could also be effective due to the healing-promoting properties of common therapeutic factors, as has been demonstrated in treatment of other conditions (e.g. Grácio et al., 2016). Consequently, it seems that one of the major criticism that has been pointed to pharmacotherapy, which is the magnitude of the placebo effect, could also be applicable to psychotherapeutic inventions.

Figure 4.2. HRSD mean improvements across different treatment options for depression

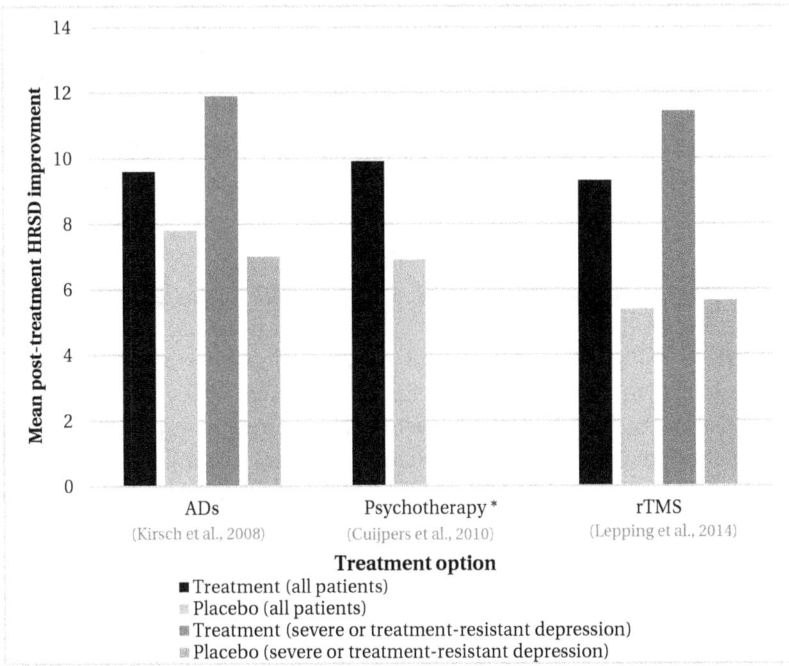

*** It was not possible to obtain data for only severe depression in psychotherapy from studies included in our review**

Finally, rTMS seems to be a very promising treatment option for depression (Aleman et al., 2014). It is less invasive and safer than ECT, but is nevertheless associated with higher absolute improvements in depression, and higher differences between control and real treatment conditions, when compared to ADs and psychotherapy. Furthermore, while the placebo effect is very similar for ADs and CBT, it seems to be less relevant in rTMS, as compared with ADs and CBT (Figure 4.2). Similarly, to what has been discussed for ECTs, this could reflect the inclusion of more severe patients in rTMS trials, given that this treatment has been approved for treatment-resistant depression. However, data presented in our comparisons does not support this idea. As opposite to what happen with ADs, in which placebo effect decreases from 7.8 (all levels of severity) to 7 (severe depression) points mean improvement in HRSD, in rTMS the placebo effect is similar between all patients and TRD (5.4 and 5.7 points, respectively; Figure 4.2). Alternate possibilities are that sham conditions could be inadequate in some rTMS trials and/or that rTMS does not promote the positive expectations that underlie the placebo effect. Further research is needed to assess these hypotheses. Importantly, it is also fundamental to note that the comparisons between treatment approaches described here suffer from a number of limitations, mostly regarding a lack of homogeneity between the different RCTs and the several meta-analyses. For example, as noted previously, baseline characteristics of patients in the different RCTs were not homogeneous and, additionally, the follow-up times varied between studies, ranging from immediately post-treatment to 8 weeks or more after treatment.

4.5. Conclusion

By considering the relevance of the placebo effect in the treatment of depression, as judged from controlled treatment trials, we aimed to raise concern over the possibility of misjudgment about the relevance of this concept in non-experimental clinical practice. We describe data demonstrating that the placebo effect is a strong component in the treatment of depression with ADs, namely in cases with mild to moderate severity. Other treatment approaches like psychotherapy seem to be equally sensible to the placebo effect but for treatments like rTMS and ECT the placebo effect may be less important. In accordance with these findings, we conceive placebo as a healing-promoter ingredient in the treatment of depression, rather than a form of treatment in itself. We thus propose that clinicians should not deny this very useful tool to improve the efficacy of their approaches, adapting treatment options not only to the patient's diagnosis, but also to the patient's characteristics, including their expectations about different forms of treatment. Research on the contribution of the placebo effect as a healing-promoter ingredient in real-treatment conditions remains to be undertaken. Such research could underlie the development of novel techniques, possibly based on approaches such as modelling and classical conditioning, which could be used across treatment options to enhance positive expectancy towards treatment, and possibly improve anti-depressive efficacy.

4.6. References

Abrams, R., (2002). *Electroconvulsive therapy*, 4th ed. ed. Oxford University Press, Oxford ; New York.

Aleman, A., Slotema, C.W. & Sommer, I.E., (2014). rTMS deserves a fair chance as a novel treatment for depression. *Acta Psychiatrica Scandinavica* 130, 324–325. doi:10.1111/acps.12277

American Psychiatric Association (Ed.), (1990). *The practice of electroconvulsive therapy: recommendations for treatment, training, and privileging: a task force report of the American Psychiatric Association.* American Psychiatric Association, Washington, DC.

Ayers, S., De Visser, R., (2011). *Psychology for medicine.* SAGE, Los Angeles.

Beck, A.T. & Rush, A.J. (Eds.), (1979). *Cognitive therapy of depression,* 13. print. ed, The Guilford Clinical psychology and psychotherapy series. Guilford Press, New York.

Bendall, S., Killackey, E., Jackson, H. & Gleeson, J., (2003). *Befriending Manual,* 1st ed. ORYGEN Research Centre.

Benedetti, F., Carlino, E. & Pollo, A., (2011). How Placebos Change the Patient's Brain. *Neuropsychopharmacology* 36, 339–354. doi:10.1038/npp.2010.81

Brakemeier, E.L., Merkl, A., Wilbertz, G., Quante, A., Regen, F., Bührsch, N., van Hall, F., Kischkel, E., Danker-Hopfe, H., Anghelescu, I. & Heuser, I., (2014). Cognitive-behavioral therapy as continuation treatment to sustain response after electroconvulsive therapy in depression: a randomized controlled trial. *Biological Psychiatry* 76, 194–202.

Carvalho, C., Caetano, J.M., Cunha, L., Rebouta, P., Kaptchuk, T.J. & Kirsch, I., (2016). Open-label placebo treatment in chronic low back pain: a randomized controlled trial. *PAIN* 157, 2766–2772. doi:10.1097/j.pain.0000000000000700

Colagiuri, B., Schenk, L.A., Kessler, M.D., Dorsey, S.G. & Colloca, L., (2015). The placebo effect: From concepts to genes. *Neuroscience* 307, 171–190. doi:10.1016/j.neuroscience.2015.08.017

Cuijpers, P., Berking, M., Andersson, G., Quigley, L., Kleiboer, A. & Dobson, K.S., (2013a). A meta-analysis of cognitive-behavioural therapy for adult depression, alone and in comparison with other treatments. *Canadian Journal of Psychiatry* 58, 376–385.

Cuijpers, P., Sijbrandij, M., Koole, S.L., Andersson, G., Beekman, A.T. & Reynolds, C.F., (2014). Adding psychotherapy to antidepressant medication in depression and anxiety disorders: a meta-analysis. *World Psychiatry* 13, 56–67.

Cuijpers, P., Sijbrandij, M., Koole, S.L., Andersson, G., Beekman, A.T. & Reynolds, C.F., (2013b). The efficacy of psychotherapy and pharmacotherapy in treating depressive and anxiety disorders: A meta-analysis of direct comparisons. *World Psychiatry* 12, 137–148.

Cuijpers, P., van Straten, A., Bohlmeijer, E., Hollon, S.D. & Andersson, G., (2010). The effects of psychotherapy for adult depression are overestimated: a meta-analysis of study quality and effect size. *Psychological Medicine* 40, 211. doi:10.1017/S0033291709006114

Devanand, D., Dwork, A., Hutchinson, E., Bolwig, T. & Sackeim, H., (1994). Does ECT alter brain structure? *American Journal of Psychiatry* 151, 957–970. doi:10.1176/ajp.151.7.957

Dimidjian, S., Hollon, S.D., Dobson, K.S., Schmaling, K.B., Kohlenberg, R.J., Addis, M.E., Gallop, R., McGlinchey, J.B., Markley, D.K., Gollan, J.K., Atkins, D.C., Dunner, D.L. & Jacobson, N.S., (2006). Randomized trial of behavioral activation, cognitive therapy, and antidepressant medication in the acute treatment of adults with major depression. *Journal of Consulting and Clinical Psychology* 74, 658–670. doi:10.1037/0022-006X.74.4.658

Elkin, I., Shea, M.T., Watkins, J.T., Imber, S.D., Sotsky, S.M., Collins, J.F., Glass, D.R., Pilkonis, P.A., Leber, W.R. & Docherty, J.P., (1989). National Institute of Mental Health Treatment of Depression Collaborative Research Program. General effectiveness of treatments. *Archives of General Psychiatry* 46, 971–982; discussion 983.

Fink, M., Bailine, S., (1998). Electroconvulsive therapy and managed care. *American Journal of Managed Care* 4, 107-12; quiz 113-114.

Furukawa, T.A., Cipriani, A., Atkinson, L.Z., Leucht, S., Ogawa, Y., Takeshima, N., Hayasaka, Y., Chaimani, A. & Salanti, G., (2016). Placebo response rates in antidepressant trials: a systematic review of published and unpublished double-blind randomised controlled studies. *Lancet Psychiatry* 3, 1059–1066. doi:10.1016/S2215-0366(16)30307-8

Gartlehner G., Gaynes B.N., Amick H.R., Asher G., Morgan L.C., Coker-Schwimmer E., Forneris C., Boland E., Lux L.J., Gaylord S., Bann C., Pierl C.B. & Lohr K.N. Nonpharmacological Versus Pharmacological Treatments for Adult Patients with Major Depressive Disorder: Current State of the Evidence, (2016). *Comparative Effectiveness Review* No. 161. (Prepared by the RTI International–University of North Carolina Evidence-based Practice Center under Contract No. 290-2012-00008-I.) AHRQ Publication No. 15(16)-EHC031-EF. Rockville, MD: Agency for Healthcare Research and Quality; December 2015. www.effectivehealthcare.ahrq.gov/reports/final.cfm.

Gold, A. & Lichtenberg, P., (2014). The moral case for the clinical placebo. *Journal of Medical Ethics* 40, 219–224. doi:10.1136/medethics-2012-101314

Grácio, J., Gonçalves-Pereira, M. & Leff, J., (2016). What do We Know about Family Interventions for Psychosis at the Process Level? A Systematic Review. *Family Process* 55, 79–90. doi:10.1111/famp.12155

Hieronymus, F., Lisinski, A., Nilsson, S. & Eriksson, E., (2017). Efficacy of selective serotonin reuptake inhibitors in the absence of side effects: a mega-analysis of citalopram and paroxetine in adult depression. *Molecular Psychiatry*. doi:10.1038/mp.2017.147

Hollon, S.D., Stewart, M.O. & Strunk, D., (2006). Enduring effects for cognitive behavior therapy in the treatment of depression and anxiety. *Annual Review of Psychology* 57, 285–315.

Hróbjartsson, A. & Gøtzsche, P.C., (2010). Placebo interventions for all clinical conditions, in: The Cochrane Collaboration (Ed.), *Cochrane Database of Systematic Reviews*. John Wiley & Sons, Ltd, Chichester, UK. doi:10.1002/14651858.CD003974.pub3

Jarrett, R.B., Schaffer, M., McIntire, D., Witt-Browder, A., Kraft, D. & Risser, R.C., (1999). Treatment of Atypical Depression with Cognitive Therapy or Phenelzine: A Double-blind, Placebo-Controlled Trial. *Archives of General Psychiatry* 56, 431. doi:10.1001/archpsyc.56.5.431

Johnson, C.F., Williams, B., MacGillivray, S.A., Dougall, N.J. & Maxwell, M., (2017). Doing the right thing?: factors influencing GP prescribing of antidepressants and prescribed doses. *BMC Family Practice* 18. doi:10.1186/s12875-017-0643-z

Kaptchuk, T.J., Friedlander, E., Kelley, J.M., Sanchez, M.N., Kokkotou, E., Singer, J.P., Kowalczykowski, M., Miller, F.G., Kirsch, I. & Lembo, A.J., (2010). Placebos without Deception: A Randomized Controlled Trial in Irritable Bowel Syndrome. PLoS ONE 5, e15591. doi:10.1371/journal.pone.0015591

Kelley, J.M., Kaptchuk, T.J., Cusin, C., Lipkin, S. & Fava, M., (2012). Open-Label Placebo for Major Depressive Disorder: A Pilot Randomized Controlled Trial. *Psychotherapy and Psychosomatics* 81, 312–314. doi:10.1159/000337053

Kirsch, I., Deacon, B.J., Huedo-Medina, T.B., Scoboria, A., Moore, T.J. & Johnson, B.T., (2008). Initial Severity and Antidepressant Benefits: A Meta-Analysis of Data Submitted to the Food and Drug Administration. *PLoS Medicine* 5, e45. doi:10.1371/journal.pmed.0050045

Kirsch, I., Sapirstein, G., (1998). Listening to Prozac but hearing placebo: A meta-analysis of antidepressant medication. *Prevention & Treatment 1.* doi:10.1037/1522-3736.1.1.12a

Lambert, M.J. & Barley, D.E., (2001). Research summary on the therapeutic relationship and psychotherapy outcome. *Psychotherapy: Theory, Research, Practice, Training* 38, 357–361. doi:10.1037/0033-3204.38.4.357

Lepping, P., Schönfeldt-Lecuona, C., Sambhi, R.S., Lanka, S.V.N., Lane, S., Whittington, R., Leucht, S. & Poole, R., (2014). A systematic review of the clinical relevance of repetitive transcranial magnetic stimulation. *Acta Psychiatrica Scandinavica* 130, 326–341. doi:10.1111/acps.12276

Liu, B., Zhang, Y., Zhang, L. & Li, L., (2014). Repetitive transcranial magnetic stimulation as an augmentative strategy for treatment-resistant depression, a meta-analysis of randomized, double-blind and sham-controlled study. *BMC Psychiatry* 14. doi:10.1186/s12888-014-0342-4

Mayberg, H.S., Silva, J.A., Brannan, S.K., Tekell, J.L., Mahurin, R.K., McGinnis, S. & Jerabek, P.A., (2002). The Functional Neuroanatomy of the Placebo Effect. *American Journal of Psychiatry* 159, 728–737. doi:10.1176/appi.ajp.159.5.728

Mead, N., Lester, H., Chew-Graham, C., Gask, L. & Bower, P., (2010). Effects of befriending on depressive symptoms and distress: systematic review and meta-analysis. *British Journal of Psychiatry* 196, 96–101. doi:10.1192/bjp.bp.109.064089

National Institute for Health and Clinical Excellence, (2015). *Repetitive transcranial magnetic stimulation for depression: Interventional procedures guidance [*IPG542].

National Institute for Health and Clinical Excellence, (2009). *Depression in adults: recognition and management. Clinical guideline (*CG90).

National Institute for Health and Clinical Excellence, (2003). *Guidance on the use of electroconvulsive therapy: Technology appraisal guidance.*

Nonpharmacological Versus Pharmacological Treatment for Patients with Major Depressive Disorder: Current State of the Evidence. (2016). *Medical Benefits* 33, 12–12.

Norcross, J.C., (2002). Empirically supported therapy relationships, in: *Psychotherapy Relationships That Work: Therapist Contributions and Responsiveness to Patients.* pp. 3–16.

Norcross, J.C. & Lambert, M.J., (2011). Psychotherapy relationships that work II. *Psychotherapy 48,* 4–8. doi:10.1037/a0022180

Oliveira-Maia, A.J., Press, D. & Pascual-Leone, A., (2017). Modulation of motor cortex excitability predicts antidepressant response to prefrontal cortex repetitive transcranial magnetic stimulation. *Brain Stimulation.* doi:10.1016/j.brs.2017.03.013

Pascual-Leone, A., Rubio, B., Pallardo, F. & Catala, M.D., (1996). *Rapid-rate transcranial magnetic stimulation of left dorsolateral prefrontal cortex in drug-resistant depression.* Lancet (London, England) 348, 233–237.

Patel, V., Simon, G., Chowdhary, N., Kaaya, S. & Araya, R., (2009). Packages of Care for Depression in Low- and Middle-Income Countries. *PLoS Medicine* 6, e1000159. doi:10.1371/journal.pmed.1000159

Patterson, C.H., (1985). What is the placebo in psychotherapy? *Psychotherapy: Theory, Research, Practice, Training* 22, 163–169. doi:10.1037/h0085489

Rauschenberg, V.M. & Crawford, J., (2003). ECT controversy. *American Journal of Nursing* 103, 63; author reply 64.

Reisner, A.D., (2003). The Electroconvulsive Therapy Controversy: Evidence and Ethics. *Neuropsychological Review* 13, 199–219.

Sackeim, H.A., (2014). Autobiographical memory and electroconvulsive therapy: do not throw out the baby. T*he Journal of ECT* 30, 177–186. doi:10.1097/YCT.0000000000000117

Saxena, S., Thornicroft, G., Knapp, M. & Whiteford, H., (2007). Resources for mental health: scarcity, inequity, and inefficiency. *The Lancet* 370, 878–889. doi:10.1016/S0140-6736(07)61239-2

Semkovska, M. & McLoughlin, D.M., (2010). Objective Cognitive Performance Associated with Electroconvulsive Therapy for Depression: A Systematic Review and Meta-Analysis. *Biological Psychiatry* 68, 568–577. doi:10.1016/j.biopsych.2010.06.009

Sensky, T., Turkington, D., Kingdon, D., Scott, J.L., Scott, J., Siddle, R., O'Carroll, M. & Barnes, T.R.E., (2000). A Randomized Controlled Trial of Cognitive-Behavioral Therapy for Persistent Symptoms in Schizophrenia Resistant to Medication. *Archives of General Psychiatry* 57, 165. doi:10.1001/archpsyc.57.2.165

Shapiro, A.K. & Shapiro, E., (1997). *The powerful placebo: from ancient priest to modern physician.* Johns Hopkins University Press, Baltimore.

Siette, J., Cassidy, M. & Priebe, S., (2017). Effectiveness of befriending interventions: a systematic review and meta-analysis. *BMJ Open* 7, e014304. doi:10.1136/bmjopen-2016-014304

Squire, L.R., (1977). ECT and memory loss. *American Journal of Psychiatry* 134, 997–1001. doi:10.1176/ajp.134.9.997

Sterling, P., (2000). ECT damage is easy to find if you look for it. *Nature 403,* 242. doi:10.1038/35002188

The UK ECT Review Group, (2003). Efficacy and safety of electroconvulsive therapy in depressive disorders: a systematic review and meta-analysis. *The Lancet* 361, 799–808. doi:10.1016/S0140-6736(03)12705-5

Vasco, A.B., (2018). Integração em Psicoterapia, in: Leal, I. (Ed.), *Psicoterapias.* Pactor, Lisboa.

Walsh, B.T., Seidman, S.N., Sysko, R., Gould, M., (2002). Placebo Response in Studies of Major Depression: Variable, Substantial, and Growing. *JAMA* 287, 1840. doi:10.1001/jama.287.14.1840

Whiteford, H.A., Ferrari, A.J., Degenhardt, L., Feigin, V. & Vos, T., (2015). The Global Burden of Mental, Neurological and Substance Use Disorders: An Analysis from the Global Burden of Disease Study 2010. *PLOS ONE* 10, e0116820. doi:10.1371/journal.pone.0116820

Wittchen, H.U., Jacobi, F., Rehm, J., Gustavsson, A., Svensson, M., Jönsson, B., Olesen, J., Allgulander, C., Alonso, J., Faravelli, C., Fratiglioni, L., Jennum, P., Lieb, R., Maercker, A., van Os, J., Preisig, M., Salvador-Carulla, L., Simon, R. & Steinhausen, H.-C., (2011). The size and burden of mental disorders and other disorders of the brain in Europe 2010. *European Neuropsychopharmacology* 21, 655–679. doi:10.1016/j.euroneuro.2011.07.018

World Health Organization, (2016). World Health Statistics 2016. World Health Organization.

World Health Organization, (2009). *Policies and practices for mental health in Europe: meeting the challenges.* WHO Regional Office of Europe, Copenhagen.

Zarbo, C., Tasca, G.A., Cattafi, F., Compare, A., (2016). Integrative Psychotherapy Works. *Frontiers in Psychology* 6. doi:10.3389/fpsyg.2015.02021

Zimmerman, M., Martinez, J.H., Young, D., Chelminski, I. & Dalrymple, K., (2013). Severity classification on the Hamilton depression rating scale. *Journal of Affective Disorders* 150, 384–388. doi:10.1016/j.jad.2013.04.028

Chapter 5

Psychiatric rehabilitation: the efforts for conceptualization and meaning construction – a review

Alexandre Gomes
Faro District Hospital, Portugal

The several evolutions in the mental health paradigm lead to the notion that the recovery of psychiatric patients is a reality, and that the psychiatric rehabilitation is a core concept and practice in this process. Our goal is to provide a historical understanding of the concept "rehabilitation", from the first studies to the consensus statements, focusing on the contexts where these practices have been developed, the meaning of "rehabilitation", and the central values involved.

Firstly, the method used is a non-systematic review based on the World Health Organization (WHO) and United States national reports and expert's primers on psychosocial rehabilitation. We then searched pubmed database for articles that presented "psychiatric" or "psychosocial" and "rehabilitation" in the title. Three authors independently read the abstracts, selecting 22 original papers where a historical or critical review was offered. We also consulted articles quoted by the aforementioned original papers when further clarification was needed. The results underlined that the origins of psychiatric rehabilitation are founded on several historical events, where the deinstitutionalization process stands out, as well as sociocultural features. It's essentially a field devoted to the recovery of psychiatric patients, with a broad and multidisciplinary nature, based on two main strategic axes: the person and the environment. In this dialectic, the psychiatrist gets an expanded role, beyond diagnosis and symptomatic control. Nevertheless, plenty of challenges remain, from the difficulties in funding, motivation and services coordination, adding to the prejudice, stigma and the lack of knowledge and professionals.

5.1 Introduction

The conceptualization of health as a "state of complete physical, mental and social well-being and not merely the absence of disease or infirmity" was settled by the World Health Organization since its very foundation in 1948. This conceptual positioning provides the foundations for the current holistic paradigm in health – which, by definition, highlights the necessity of a multidisciplinary, patient-centred approach, in addition to an implicit understanding that all systems exist in mutual connection and influence.

Mental Health is, as such, deemed crucial for the general wellbeing of individuals and communities. The profound impact of mental disorders is a Public Health concern, a fact that is highlighted by the World Health Organization report: 12 to 15% of the total disability adjusted life years in the world is due to mental disorders, a value that exceeds that of cardiovascular and neoplastic diseases. In addition, it represents about 30% of all years lived with disability.

Mental Health, as a subject, has undergone several paradigm shifts, incited by the progressive accumulation of evidence regarding different models of patient care and, equally important, in line with the evolution of sociocultural principles. The first paradigmatic period, usually called The Asylum Period, declined progressively from the mid-fifties on when in parallel the movement of deinstitutionalization emerged, the second period. The third and last one supports the view of balanced care involving the Hospital and the Community – and emphasizes the rehabilitation of people with psychiatric disorders.

Psychiatric rehabilitation thus started to come to light in the sequence of the process of deinstitutionalization and received increasing attention from the international scientific community. This was essentially due to three factors:

- The pressing need to solve some of the consequences of the deinstitutionalization – as in many places there had not been enough support and preparation both of the community and the patients.
- The generalization of the human rights of people with mental disabilities.
- The progress of psychiatric empirical knowledge – particularly where studies challenged the rooted idea that certain disorders, notably schizophrenia, were necessarily associated with poor prognosis. This was added by the recognition that the chronification of disabilities is not necessarily intrinsic to the

disorder itself, but otherwise strongly dependent on social variables – where intervention is possible, but with different strategies from those of the biomedical model.

The practical implementation of the rehabilitation principles relies on the development and propagation of evidence-based models that are economically viable. It also strongly relies on the motivation of the involved professionals and the facilitation of the communication and alliance between different sectors of activity. As such, a precise understanding of each process of intervention and of how to implement, register and measure them is essential in order to provide clear plans for rehabilitation.

The field of Psychiatric Rehabilitation is recent and broad and is an ongoing area both at the practical and theoretical levels. In this paper, we will be focusing on the path traced since the first studies were published, particularly regarding how the term rehabilitation has been conceived.

5.2 The origins of psychiatric rehabilitation and the first studies

The way people with mental or behavioural disturbances have been dealt with has ever since closely paralleled the prevailing social values, with drastic changes throughout the history – from respected messengers of the Gods to persecuted and publicly killed, to imprisonment in huge asylums (WHO, 2001).

The history of Mental Health Services has been divided into 3 periods: the asylum period, the decline of the asylum period, and the community-based care period, in which psychiatric rehabilitation started to gain its status.

Through the XIX century, in Europe, two diverging approaches were observed – on the one hand, mental disorders were starting to be recognized as a legitimate subject of scientific discussion and consequently people with a variety of disturbances treated as patients. On the other hand, the way to deal with these same people was to isolate them from society within the custody of big facilities, the asylums. This model then spread to Africa, America and Asia.

The Asylum Period is characterized by the segregation of patients into big custody facilities, usually remote from the communities. Despite the poor outcomes of this model, and despite representing a clear contradiction to the holistic and humanistic modern model of Mental Health, in many countries nowadays the mental health care is still provided exclusively in this way (Thornicroft, 2002).

The decline of the asylum started, in most developed countries, in the mid-fifties. This paradigm shift was essentially due to three independent factors:

- The progress in psychopharmacology – with the development of antidepressants and neuroleptics – as well as psychosocial interventions.
- The expansion of the Human Rights, which were soon to become a global event.
- The pressure made by the World Health Organization regarding the recognition of the importance of social and mental wellbeing in a complete definition of health.

These events, a mix of technical progress and socio-political positioning, majorly contributed to the adoption of a more open and flexible approach to psychiatry (WHO, 2001). Overall, in the field, the deinstitutionalization movement consisted of three major principles: prevention of unnecessary hospital admissions through the creation of community centres, discharge of patients with long inward periods, and the creation of community support systems for deinstitutionalized patients.

The third and last period emerged partially as an answer to the problems rose by the deinstitutionalization. The main goal at this stage was to develop a range of balanced care within local settings, as much as possible. The idea would be to ensure that the positive aspects of the asylum period are provided, and the negative ones not perpetuated. As such, the objectives were:

- To provide balanced care, with both community and hospital settings, but always in the local context, in services that are close to the residences, alongside with modern inward units integrated into the hospital with other medical specialties, adding to residential facilities inserted in the community.
- To intervene not only in the symptoms but in the disabilities as well.
- To direct the treatment and patient-care not just to the diagnosis, but essentially to the person global needs.
- To progressively substitute static services by movable units, including with the capacity to provide support in the patients' residencies (Thornicroft, 2002).

The origins of psychiatric rehabilitation were also founded on several historical events. Moral Treatment, an approach that emerged in the context of the age of Enlightenment, although severely criticized, first pointed to the importance of including elements such as work, leisure and social bonding as part of the treatment plan. This was followed by recognizing the therapeutic value of integrating patients in structured activities. Then the first attempts to include people with psychiatric disorders in programs of vocational orientation and the development of techniques for training competencies started (Anthony & Liberman, 1986).

The essential elements of psychiatric rehabilitation have been subliminally present in the various periods of the mental health history. Though first considered to be a form of tertiary prevention, these sparse practices evolved together to form a model, encompassing a broad body of knowledge including methodological and organizational aspects regarding prevention and reduction of the disability associated with severe mental disorders (WHO, 1996).

5.3 Conceptualizing rehabilitation

The references to "psychiatric" and "psychosocial rehabilitation" in scientific publications are profuse. The first publication including the terms "Psychiatric Rehabilitation" in its title dates from 1945 (Bartemeier, 1945). In this article, the rehabilitation is implicitly defined as some form of returning the war veterans (who were discharged from the war field due to mental problems) to a regular, active life. The first publication that mentions psychiatric rehabilitation as a specific practice was published in 1951 (Bailey, 1951). In this article, it is described as the "practice of dynamic therapeutic programs made in order to allow the chronic patient to achieve the highest level of physical, psychological and socioeconomic adjustment as possible." This definition, though sufficiently clear in enumerating the generic objectives of the field, does not constitute a real definition, as it does not make clear what is the method being advanced, nor where to draw the borders with the other subfields in mental health care. As such, a debate followed, for, on the one hand, to achieve the goals defined by a rehabilitation-oriented paradigm it would be necessary to invest in more research and training, but on the other hand, the loose conceptualization impeded the growth of specialized programs. Some of the ideas debated in these years – such as the need of a bridge between hospital and community, the importance of recognizing social motivation as a limiting factor for rehabilitation, and of assuming that the treatment is something that exists in a continuum – are still present in current debates.

The difficulties in clearly conceptualizing "psychiatric rehabilitation", however, lasted for decades. The trouble to draw borders between this new, ill-defined area and the several disciplines that, partially and disconnectedly, served a similar role, is seen as an essential reason for the delay in the development of stronger rehabilitation policies (Munich & Lang, 1993).

Finally, in 1996, the World Health Organization published a declaration of consensus regarding rehabilitation. The experts involved acknowledged that being a developing field, the concepts and methods have not yet achieved stability, and hence it was urgent to publish a consensus. This document, it is assumed, is not intended to be the last word on the subject. What the authors seek is to provide a common ground, to serve as a foundation for further conceptual refinements (WHO, 1996). The authors define psychosocial rehabilitation as a process that facilitates the opportunities for individuals - who are impaired, disabled or handicapped by a mental disorder – to reach their optimal level of independent functioning in the community. It implies both are improving individuals' competencies and introducing environmental changes in order to create a life of the best quality possible for people who have experienced a mental disorder, or who have an impairment of their mental capacity which produces a certain level of disability. Psychosocial rehabilitation thus aims to provide the optimal level of functioning of individuals and societies, and the minimization of disabilities and handicaps, stressing individuals' choices on how to live successfully in the community (WHO, 1996).

Starting from the same acknowledgement that a standard definition for psychiatric rehabilitation was urgent, in 1997, following the World of Health Organization Consensus Statement, the Board of United States Psychiatric Rehabilitation Association published a similar official document. In that report, the authors defined rehabilitation as an enabler of the recovery of people with psychiatric disabilities, of their integration in the community, aiming to maximize their quality of life, focusing particularly in those individuals with severe and disabling disorders. Rehabilitation should help the acquisition of competencies and access to the resources available, providing a meaningful and successful life in the living, work, learning and social spheres of preference of the individuals. As a paradigm, psychiatric rehabilitation expanded the acknowledgement of the negative consequences of mental disorders beyond symptoms, highlighting the handicap and disability as an essential part of the pathological spectrum and as an essential goal in the therapeutic journey (Anthony & Farkas, 2009).

Adding to the aforementioned consensus statements, other authors sought for the best way to conceptualize rehabilitation. Most mentions these years included:

- Enabler of the community integration of individuals adapting to a disability consequent of a mental disorder
- An emergent professional area, with goals, values and principles of its own
- Although overlapping with other disciplines involved in Mental Healthcare, such as psychology, social working, counselling, it focuses on a narrower group of patients, those with severe mental disorders
- Contrasting with the traditional approach or biomedical model, which focuses on symptom reduction, rehabilitation devotes itself to assist the individuals to identify their values and preferences, so that they can achieve satisfactory social roles in their own communities (Barret, 2005; Corrigan, 2008; Gill, 2009).

Other authors, still, underscore that psychiatric rehabilitation as a discipline is part of a more global process, the recovery, which focuses on the resources and competencies for the construction of a meaningful life, combined with the other facets of intervention (symptom control, crisis intervention, community support). As a paradigm, it stresses that the needs of individuals with severe mental disorders are beyond the reduction of symptoms and aim to the complete integration of the patient in his environment.

The retrospective reading of this struggle for a clear cut concept makes it clear that it has been a hard task. The difficulties include:

- The fact that the principles of rehabilitation have been present in the history of mental health throughout the various stages, more or less explicitly.
- The fact that the first so-called rehabilitation initiatives (such as industrial jobs, protected jobs (Denber, 1960), and the Fountain Houses (Norman, 2006) emerged out a conceptual framework, at the time non-existent.
- The word rehabilitation itself is prone to semantic confusion; the services working in the reform of Mental Health policies, for instance, were called rehabilitators (Anastácio, 2012). In general

> medicine, the notion of rehabilitation implies a deceiving simplicity – it means to return the patient to the previous state of functioning, prior to the "illness". Such simplicity is not suitable for the psychiatric homonymous.

The importance of framing rehabilitation as a field on its own rises from the recognition of the complexity of the processes involved, its central role as promoter of the recovery of psychiatric patients, the need for specialized training, research and investment of its own. Adding to this, the inclusion of individuals with severe mental disorders in society strongly contributes to the fight against prejudice and stigma (McReynolds, 2002).

Another aspect in discussion is to what patients should rehabilitation apply? Some authors argue that this is an intuitive definition and that every individual with a severe, disabling condition should benefit from the practices of rehabilitation. In that sense, it should include those patients with persistent psychopathology, marked instability, frequent exacerbations and difficulties in social interaction. With these characteristics, the diagnosis of schizophrenia surely stands out, but it can be extended to other psychotic and non-psychotic disorders (Hirdes & Kantorski, 2004; Rossler, 2006). Other authors present slightly distinct definitions of chronic mental disorders, usually sharing three aspects: an established diagnosis, a prolonged duration, and social maladaptation. More than 50% of the patients with these characteristics present a dual diagnosis, with ongoing or previous substance abuse, and that population is even harder to approach, with a worse prognosis as well. Still, other authors highlight the tremendous heterogeneity of the psychiatric patients, mentioning that even patients with the diagnosis of schizophrenia do not necessarily need specific rehabilitation, arguing that it is unlikely, and even counterproductive, to try to settle strictly what patients benefit and should receive rehabilitation oriented practices.

5.4 Principles, Goals and Strategies

The general philosophy of rehabilitation relies on two strategic intervention axes. The first is centred in the individual, with the overall goal of capacitating him with the necessary competencies to successfully interact with the environment. The second is ecological, therefore directed to the creation and potentiation of community resources and the minimization of stressors. Most patients will need an intervention in both (Rossler, 2006).

To achieve the goals of psychiatric rehabilitation, it must approach all spheres – social, emotional and intellectual. Living, learning and working inside the community with the least necessary professional support is one major goal.

Another important consideration is to answer such aims, it must focus on the individual as he is – that is, no clear generalized plans fit every individual. And the same goes for the environment. So, it is important to recruit professionals who know the particularities and resources available in the specific context of the patient.

The first step must be an assessment of motivation and an identification of personal objectives. Some interviews, as the motivational interview, facilitate a more precise view of the objectives, motivations and resistances of the patient, and are not just a mere list of objectives and requirements (Leona & Bachrach, 2000). Furthermore, it allows evaluating and approaching the readiness for change. The resulting plan must focus on the abilities of the individual, so to reinforce the hope, self-esteem, and empowerment, all essential aspects for achieving the defined goals (Corrigan et al., 2001; Rogers et al., 2001).

Another key aspect is that, by definition, the rehabilitation process cannot be imposed. The patient is the centre and the final decision maker of every aspect regarding treatment and rehabilitation. These same principles are also incorporated in the concept of recovery (Farkas et al., 2005; Rossler. 2006; William & Farkas, 2012).

As a general rule, individuals with severe and chronic mental disorders have the very same basic aspirations as the population of the cultural group they belong to. These are, generally speaking, to have a proper home, education and a professional career, social, intimate and significant relationships, and to participate in the community life with the same rights as anyone else (Barber, 2012). They prefer to be consumers of a mental health service, with an active interest in learning about their disorder and deciding every step of the therapeutic and rehabilitative process (Liberman, Kopelowicz & Young, 1994).

The autonomous participation in competitive jobs represents another major goal. However, historically, this constitutes one of the hardest aspects to achieve. The unemployment rates in chronic psychiatric patients rounded 88% in the nineties (Department for Education and Employment,1988; Perkins, 2002). Two independent meta-analysis revealed that the interventions of psychiatric rehabilitation do have an impact regarding employment – with around 58% of the patients achieving such a goal (Crowther, 2001; Twamley et al., 2003). Even so, the duration of the employment period is most frequently small (Becker et al., 1998). Pre-vocational training oriented models assume that a preparation period is necessary, including participation in institutional workshops, alongside with social skills training. Then it is also beneficial for most patients to have a transition job (most often this is offered by organizations with some financial relationship with the mental health services) and that has been proved to augment the stability in the following job (Gournellis et al., 2010).

Several studies pointed that the best result is obtained through strategies of protected employment – which presupposes that there is ongoing local support and the possibility of quick professional intervention in case of crisis - when comparing to other strategies that usually apply a long and staged approach (Bond, Drake & Becker, 2008; Crowther, 2001; Twamley, Jeste & Lehman, 2003).

Other central aspects, which, again, are simultaneously central challenges, are: first, the need for ongoing research and evidence; second, the design of models based on the evidence obtained; and third, the implementation of such models. This is a particularly sensitive matter, both because classic outcome measures are hard to apply in the context of rehabilitation (which is added by the lack of human resources) and the implementation of new models necessarily implicates the intervention of health policy decision-makers.

As there are a number of prejudices deeply ingrained in the population, as well as in medical professionals and decision-makers, the Psychiatric Rehabilitation Center of the University of Boston decided to publish the 10 evidence-based messages that are intended to be the cornerstones of rehabilitation. They also voice hope (Anthony & Farkas, 1982; Farkas et al., 2007).

1. Psychiatric Rehabilitation has been prejudiced by the poor outcomes, devaluated by irrelevant diagnostic systems, and precluded by inefficient therapeutical approaches adding to disorganized community support systems.

2. To orient people to find their life goals can positively affect the outcome of their rehabilitation.

3. Psychiatric diagnostic systems are neither predictors nor prescribers or descriptive of the adequate practices and outcomes of rehabilitation.

4. Essentially, there is no important connection between the patients' symptoms and their capacities.

5. Psychiatric Rehabilitation can be taught.
6. Rehabilitation services have a positive impact on mental health patients.
7. Persons with a psychiatric disease can and want to engage the education system.
8. Persons with a psychiatric disease can and want to work.
9. Persons with a psychiatric disease can and do rehabilitate.
10. Every support system can help improve the patients functioning in the life, learning and work of choice of the individual.

The recovery programs must encompass different categories of services, including treatment, crisis management, case management, rehabilitation, self-help and prevention. For each category, one should define specific contents (tasks, competencies, modifications of the environment), and the objective defined by the individual. These outcomes are, most frequently, distant goals. The recovery, therefore, consists in keeping in track, keeping in a way with goals. For that reason, and in order to mitigate frustration, the individual can be advised to define sub-goals, so that feedback is more easily obtained, and so that the individuals can realize the changes and progresses made (Anthony et al., 1983).

Most models present the stages of the psychiatric rehabilitation process as being divided into: Diagnostic. Planning; Intervention.

The diagnostic, differently from that in the traditional psychiatric model, consists in the self-determination and readiness for change evaluation, and in helping the patient to identify their competencies, strengths, supports and deficits, and from here work on establishing the goals for the treatment. This information serves as the basis on which the general plan is then construed, where the processes necessary to develop certain competencies and the resources need are specified. The intervention phase occurs through the direct acquisition and use of these competencies, and in the coordination and modification of resources. Different settings offer different opportunities for rehabilitation, from community health centres, to clubhouses, day care facilities, and other existing structures can be made part of the resources. As it is patient centred, different interventions can have different roles for different patients. For instance, a workshop on healthy eating can be an intervention of competencies acquisition for one individual, aiming towards an independent life, it can constitute a functional evaluation for another, and it can serve as a stimulus for program engaging for yet another patient. It is thus necessary to clearly establish with the patients what are the goals and meanings of these interventions (Farkas, Cohen & Nemec, 1988).

5.5 On the outcomes and motivation

The psychiatric rehabilitation paradigm offered a broader view on the reality of individuals with psychiatric disorders and implied a shift in the focus of actuation: from the symptoms to individual characteristics and functionality. As said, the outcomes are beyond symptom control and include social relationships, labour, leisure, quality of life, and family dynamics (Rossler, 2006).

Two major challenges that are frequently mentioned in the literature are: outcomes measurement and motivation. The patient must obviously be motivated before engaging a psychiatric rehabilitation process, as patent in the “readiness for change” scales. Not less important is the motivation of the professionals. Indeed, the journey towards recovery is far from straightforward, which adding to resources constriction, frequently strikes the motivation of professionals.

A close monitoring of the patient's progress is an essential part of the process; an outcome measuring is as well essential - particularly to provide quantitative data for cost-benefit deliberation - actually one critique frequently made to psychiatric rehabilitation is that most of these practices are not evidence-based. When attending to what those who work in the field have to say regarding this objection, two reasons for that arise: on the one hand, most services lack resources, particularly human resources - and therefore time. This leads to the second, and perhaps more interesting argument presented, which is "the difference that they are making for the patients are so obvious that it feels pointless to spend resources proving it." (Anthony & Farkas, 2009)

This is connected to motivation. Of course, the patient must be motivated to engage in such a program, and there are several tools to measure it as well. But the motivation of professionals is key when it comes to multidisciplinary approaches. The motivation and enthusiasm seem to be the better way to keep going with, most of the time, the lengthy path of psychiatric rehabilitation. (Anthony & Farkas, 2009).

5.6 Affective resources and empowerment

The therapeutic relationship is, of course, fundamental to the whole process – that of comprehension and mutual trust, which by itself may provide the solid basis for individual growth. Corrigan (2005) concluded that the most important factor in rehabilitation is peer support. As such, psychiatric rehabilitation is equally an exercise of reconstructing a social support network (Rossler, 2006; Anthony & Farkas, 2009; Borg et al., 2009).

According to Cancro (2000), 'a psychiatrist must be able to prescribe psychosocial intervention such as social skills training, as well as he prescribes psychopharmacological agents'. That surely does not mean that the psychiatrist must be able to perform all the tasks involved in the process, but at least he should be able to identify the specific needs of his patients, to know what services are available in his community, and help to direct the patient towards them (Rossler, 2006).

Monitoring the psychopharmacological intervention is of course part of the process. But in the light of the rehabilitation paradigm, as the focus is functionality, different attitudes can be performed – that is, the symptoms do not necessarily need to cease to exist if that is only achieved at the expense of functionality. The point of equilibrium must be found for each patient.

The rehabilitation must also, overall, be a place that provides affective resources. As such, all institutions, more or less formal, more or less structured, can contribute to the rehabilitation process. The professionals involved must also actively support the initiatives of the patients, as to provide autonomous decision making. That is, the professionals must adopt the position of a facilitator and not of a conductor, hence empowering the individuals.

As one of the major challenges relates to the perspective that the general population has regarding individuals with severe mental disorders, all agents involved in this field must assume an active role in transmitting the idea that recovery of individuals with severe, chronic and disabling psychiatric disorder is, indeed, possible. (Jorge-Monteiro & Matias, 2007; Saraceno, 2011).

5.7 Challenges and future contributions

Ongoing research has clearly demonstrated that the correct treatment of mental health disorders results in a socioeconomic improvement of the communities (Beddington et al., 2008). Nevertheless, there is a huge gap to be filled regarding the treatment of mental health patients, particularly but not limited to developing countries (Bass et al. 2012). Numerous challenges persist, as the inadequate allocation of resources, the influence of the cultural and subcultural contexts in the manifestation of the disorders, stigma and discrimination, adding to nosologic fragilities, etiological uncertainties, and pharmacological interventions with modest efficacy and non-negligible side effects. The centralization of mental health resources is another problem, as it secludes a huge number of patients (WHO, 2001; WHO, 2011; William & Farkas, 2012).

Indeed, as seen, paradigm changes happen in the context of technical and socio-political advances. The progressive advances in neurosciences may provide explanatory models of psychiatric disorders, and new pharmacologic treatments and prevention possibilities. In parallel, programs such as the "Great Push for Mental Health", a colligation between the World Federation of Mental Health and the World Health Organization which represents the concerted effort of local, national and global forces, also play a role. (WHO, 2011)

5.8 Conclusions

The exposed review was intended to globally present the history of psychiatric rehabilitation concept, as a model and practice that are core in the paradigm of recovery of individuals with psychiatric disorders.

Overall, from our review, we found no clear, universally accepted definition. Nevertheless, the consensus statement settled a basic ground that is generally concordant regarding the principles, objectives and strategies to be implemented, presenting rehabilitation as an essential part of the recovery of individuals with severe disabilities related to mental disorders, and viewing the rehabilitation as a proper, broad, multidisciplinary and dialectic field.

It seems clear that there's a growing interest in the scientific community regarding the Psychiatric Rehabilitation – which becomes evident by the growing body of research devoted to it, and in the several WHO Reports. These ideas transport a more optimistic view of severe psychiatric disorders, and in some points reveal a deep rupture to the prior biomedical model.

Contrasting with the optimism that rehabilitation advocates, there is the reality of most countries. It is clear that the challenges are enormous, from economic support to human resources. More outcome studies of well-characterized rehabilitation models are also in need, to provide evidence-based practices. The need for a consensus is still necessary, as the scientific investment in this area is prejudiced by the ill-definition of rehabilitation practices.

Perhaps the biggest challenge is still the stigma, both in the community and the professionals. Every effort to bring hope can thus make a difference.

Acknowledgements

I'd like to thank Dr. Margarida Passos and Prof. José Leão Queirós of Hospital de Magalhães Lemos for the valuable orientation in my Master Thesis, on which this paper is based.

5.9 References

Anastácio, J.P. (2012) Reabilitação Psicossocial e Recovery: conceitos e influências nos serviços oferecidos pelo sistema de saúde mental. *Cadernos Brasileiros de Saúde Mental.* 4(9), 72-83.

Anthony, W.A., Cohen, M. R. & Cohen, B. F. (1983) Philosophy, treatment process, and principles of the psychiatric rehabilitation approach. *New Directions for Mental Health Services.* 17(17), 67-79.

Anthony, W. & Farkas, M. (1982) A Client Outcome Planning Model for Assessing Psychiatric Rehabilitation Interventions. *Schizophrenia Bulletin.* 8(1), 13-38.

Anthony, W. & Liberman, R. (1986) The Practice of Psychiatric Rehabilitation: Historical, Conceptual, and Research Base. *Schizophrenia Bulletin.* 12(4), 542-559.

Anthony, W. A. & Farkas, M. D. (2009) *Primer on the Psychiatric Rehabilitation Process.* Boston University: Center for Psychiatric Rehabilitation.

Anthony, W., Cohen, M. & Vitalo, R. (1978) The Measurement of Rehabilitation Outcome. *Schizophrenia Bulletin.* 4(3), 365-383.

Bailey, P. (1951) Neurologic rehabilitation. *Neurology.* 1(6), 476-482.

Barber, M. (2012) Recovery as the New Medical Model for Psychiatry. *Psychiatric Services.* 63(3), 277-279.

Barrett, N., MacDonald-Wilson, K. & Nemec, P. (2005) Establishing a Consortium of Psychiatric Rehabilitation Educators. *American Journal of Psychiatric Rehabilitation.* 8(2), 121-133.

Bartemeier, L.H. (1945) Recent Civilian Experiences with Psychiatric Rehabilitation. *Proceedings of the Royal Society of Medicine.* 38(12), 680-681.

Bass, J., Bornemann, T., Burkey, M., Chehil, S., Chen, L., Copeland, J., Eaton, W., Ganju, V., Hayward, E., Hock, R., Kidwai, R., Kolappa, K., Lee, P., Minas, H., Or, F., Raviola, G., Saraceno, B. & Patel, V. (2012) A United Nations General Assembly Special Session for Mental, Neurological, and Substance Use Disorders: The Time Has Come. *PLoS Medicine.* 9(1), 100-159.

Becker, D.R., Drake, R.E., Bond, G.R., Xie, H., Dain, B.J., Harrison, K. (1998) Job terminations among persons with severe mental illness participating in supported employment. *Community Mental Health Journal.* 34(1), 71-82.

Beddington, J.C., Field, J., Goswami, U. & Huppert, F.A. (2008) The mental health nations. *Nature.* 455, 1057-1060.

Bond, G., Drake, R. & Becker, D. (2008) An update on randomized controlled trials of evidence-based supported employment. *Psychiatric Rehabilitation Journal.* 31(4), 280-290.

Borg, M., Karlsson, B., Tondora, J. & Davidson, L. (2009) Implementing person-centered care in psychiatric rehabilitation: what does this involve? *The Israel Journal of Psychiatry and Related Sciences.* 46(2), 84-93.

Cancro, R. (2000) The Introduction of Neuroleptics: A Psychiatric Revolution. *Psychiatric Services.* 51(3), 333-335.

Corrigan, P., Muesler, K., Bond, G., Drake, R. & Solomon, P. (2008) *Principles and Practice of Psychiatric Rehabilitation: An Empirical Approach.* New York, Guilford.

Corrigan, P., McCracken, S.G. & Holmes, E.P. (2001) Motivational interviews as goal assessment for persons with psychiatric disability. *Community Mental Health Journal.* 37(2), 113-122.

Corrigan, P., Slopen, N., Gracia, G., Phelan, S., Keogh, C. & Keck, L. (2005) Some Recovery Processes in Mutual-Help Groups for Persons with Mental Illness; II: Qualitative Analysis of Participant Interviews. *Community Mental Health Journal.* 41(6), 721-735.

Crowther, R. (2001) Helping people with severe mental illness to obtain work: systematic review. *British Medical Journal.* 322(7280), 204-208.

Denber, H. (1960) Industrial Workshop for Psychiatric Patients. *Psychiatric Services.* 11(6), 16-18.

Gregor, C. (2006). The UN Convention on the Rights of Persons with Disabilities. *Development.* 49(4), 158-160.

Department for Education and Employment. (1998) Labour Force Survey. Available from: http://webarchive.nationalarchives.gov.uk/20130323012706/https://www.education.gov.uk/publications/eOrderingDownload/AR99.pdf

Farkas, M. & Anthony, W. (2007) Bridging science to service: Using Rehabilitation Research and Training Center program to ensure that research-based knowledge makes a difference. *The Journal of Rehabilitation Research and Development.* 44(6), 879.

Farkas, M., Gagne, C., Anthony, W. & Chamberlin, J. (2005) Implementing Recovery Oriented Evidence Based Programs: Identifying the Critical Dimensions. *Community Mental Health Journal.* 41(2), 141-158.

Farkas, M., Cohen, M. & Nemec, P. (1988) Psychiatric rehabilitation programs: Putting concepts into practice? *Community Mental Health Journal.* 24(1), 7-21.

Gill, K.J. & Barrett, N.M. (2009) Psychiatric rehabilitation: an emerging academic discipline. *The Israel Journal of Psychiatry and Related Sciences.* 46(2), 94-102.

Gournellis, R., Triantafillou, E., Madianos, M., Tsinia, N., Ploumpidis, D. & Tomaras, V. (2010) Diachronic trends of employment outcome of prevocational training in psychiatric rehabilitation. *Annals of General Psychiatry.* 9(1), 2.

Hirdes, A.K. & Kantorski, L.P. (2004) Reabilitação Psicossocial: Objetivos, Princípios e Valores. *Revista de Enfermagem da Universidade do Estado do Rio de Janeiro.* 12, 217-221.

Jorge-Monteiro, Fátima & Matias, J. (2007) Atitudes face ao recovery na doença mental em utilizadores e profissionais de uma organização comunitária: Uma ajuda na planificação de intervenções efectivas? *Análise Psicológica.* 1(XXV), 111-125.

Leona, M. & Bachrach, L. (2000) Psychosocial rehabilitation and psychiatry in the treatment of schizophrenia - what are the boundaries? *Acta Psychiatrica Scandinavica.* 102(s407), 6-10.

Liberman, R., Kopelowicz, A. & Young, A. (1994) Biobehavioral treatment and rehabilitation of schizophrenia. *Behavior Therapy.* 25(1), 89-107.

McReynolds, C.J. (2002) Psychiatric Rehabilitation: The need for a specialized approach. *International Journal of Psychosocial Rehabilitation.* 7, 61-69.

Munich, R. & Lang, E. (1993) The Boundaries of Psychiatric Rehabilitation. *Psychiatric Services.* 44(7), 661-665.

Norman, C. (2006) The Fountain House movement, an alternative rehabilitation model for people with mental health problems, members' descriptions of what works. Scandinavian *Journal of Caring Sciences.* 20(2), 184-192.

Perkins, R. (2002) Unemployment rates among patients with long term mental health problems; a decade of rising unemployment. *Psychological Bulletin.* 26, 295-298.

Rogers, E.S., Martin, R., Anthony, W., Massaro, J., Danley, K., Crean, T. & Penk, W. (2001) Assessing readiness for change among persons with severe mental illness. *Community Mental Health Journal.* 37(2), 97-112.

Rossler, W. (2006) Psychiatric rehabilitation today: an overview. *World Psychiatry.* 5(3), 151-157.

Saraceno, B. (2011) Interview with Benedetto Saraceno by Domingos Savio Alves. *Ciência e Saúde Coletiva.* 16(12), 4695-4700.

Thornicroft, G. & Tansella, M. (2002) Balancing community-basedand hospital-based mental health care. *World Psychiatry.* 1(2), 84-90.

Twamley, E.W., Jeste, D.V. & Lehman, A. F. (2003) Vocational rehabilitation in schizophrenia and other psychotic disorders: a literature review and meta-analysis of randomized controlled trials. *Journal of Nervous and Mental Diseases.* 191(8), 515-523.

William, A., Farkas, M. (2012) Guia Orientador do Processo de Reabilitação Psiquiátrica. Porto, Livpsic.

World Health Organization. (1996) Psychosocial Rehabilitation - a Consensus Statement. Geneva. Available from: http://apps.who.int/iris/bitstream/10665/60630/1/WHO_MNH_MND_96.2.pd

World Health Organization. (2001) Atlas - Mental Health Resources in the World. Available from: http://www.who.int/mental_health/publications/mh_atlas_2001/en/

World Health Organization. (2001) Mental Health: New Understanding, New Hope. Available from: http://www.who.int/whr/2001/en/whr01_en.pdf?ua=1

World Health Organization. (2004) Global Burden of Disease. Available from: www.who.int/healthinfo/global_burden_disease/GBD_report_2004update_full.pdf

World Health Organization. (2011) Atlas - Mental Health Resources in the World. Available from: http://apps.who.int/iris/bitstream/10665/44697/1/9799241564359_eng.pdf

Second Part: The Effectiveness of Reflexive Mental Health Services: Innovation and Research

Chapter 6

Therapeutic communities for the future: surviving modernisation and staying at the radical edge

Rex Haigh
Nottingham University's School of Sociology and Social Policy, UK

Therapeutic Communities are intensive group-based therapeutic interventions that have been existence in different forms since the days of 'moral treatment' in the eighteen century, and possibly before. In the British wartime response to battle-shocked soldiers, the treatment model was refined and operationalised. In the 1960s they became associated with the 'antipsychiatry' movement and survived largely unchanged through to the 1980s. However, since then, major changes in the organisation and delivery of healthcare have made the 'old model' largely untenable in the face of a changing management culture throughout the world, increasing regulatory pressure, and requirements to comply with a standardised evidence base.

Therefore, changes were needed in order to survive – and to do so without losing the radical 'relational' focus, the emancipatory aim of authentic empowerment, and the group context of the work. This paper illustrates four areas of relevant activities in the field: regulation and quality assurance; learning through experiential processes; researching effectiveness using experimental methodology; and modifications of the basic values and principles to make the approach relevant in a wider range of settings. In conclusion, the paper argues that this 'humanisation' of controlling, bureaucratic and statutory processes is possible, as well as necessary, in order to prevent 'dehumanisation' through the machinery of modern management in mental health care.

6.1 Introduction

Therapeutic Communities (TCs) developed from models of treatment that broke from traditional practice, and posed profound challenges to dominant thinking at various times: at the end of the eighteenth century (Tuke, 1813), during the Second World War (Bion, 1961; Harrison, 1999), and in the 1960s (Foucault, 1965; Goffman, 1968; Laing, 1967; Rapoport, 1960). More recent descriptions encourage medication reduction and use of other radical ideas (Haigh, 2007). Their clinical work also takes account of the interface between the individual and their social and cultural setting (Jones, 1956). Therapeutic communities involve user empowerment, and the development of a sense of agency; democratisation, through user involvement in the day to day running of the community, shared decision-making, and peer therapy (Haigh, 2013). Therapeutic communities, both democratic and substance abuse, operate all over the world, in diverse countries and cultures, and have had to adapt to local demands (Haigh & Lees, 2008).

Many clinicians, academics and service users who understand therapeutic communities believe that they are guided by an underlying radical challenge to authority and oppression (Baron, 1987; Jones, 1979; Spandler, 2006, 2009) that was articulated in the 1960s (Cooper, 1968), and if that underlying therapeutic philosophy is lost, they will have moved from being an emancipatory movement for the dispossessed to becoming a mere therapeutic technology in a crowded field.

This intention of this paper is to illustrate the adaptations that democratic therapeutic communities have made in the last two or three decades in the United Kingdom. To turn from ideas of revolution and emancipation to the subject of survival, these changes have been undertaken in to cope with threats in the real world. Although those threats can be seen as ontological, ideological or existential, they usually impinge in a less abstract way: with the economically-driven closure of services, loss of clinicians' jobs and the sale of property and land. The response to the threats has been underway for more than twenty years and is presented in four different areas. To start with a 'what TCs need to do' list:

1. TCs need to... *Know what they are doing and doing it well*: quality improvement, such as the Royal College of Psychiatrists' *Community of Communities* project, started in 2002 in the UK. Also developed in Australian TCs since 2007 and in Sicily since 2012.

2. TCs need to... *Understand what they are doing and showing that it works:* this is producing the research evidence in the right way for people who run and pay for mental health services. Nick Manning, the eminent sociologist, calls it the 'politics of data': to what extent can we work with 'the system', and to what extent do we need to resist it?
3. TCs need to... *Offer training which can convey the unique experience* of having the sort of relationships people have in therapeutic communities, which is difficult to describe without experiencing it.
4. TCs need to... *Be willing to operate therapeutic communities differently,* to spread the ideas beyond therapeutic communities – and join forces with others who have a similar therapeutic philosophy.

These four themes are quality, research, training and innovation.

6.2 QUALITY: Knowing what TCs are doing and doing it well

The 'Community of Communities' started in London in 2002: members of the steering group of the Association of Therapeutic Communities (ATC) approached the Royal College of Psychiatrists to set up a project, initially funded by the National Lottery, to bring therapeutic communities together. The methodology reflected the beginning of the ATC in 1972 – with communities visiting each other and sharing their ideas and practice - but also incorporating the rigour of contemporary quality assurance. The aims were to reduce isolation, increase professional confidence in the TC ways of working, and provide support – as well as measuring and improving quality in a way that was accepted by the formal organisations and institutions which the field needed to relate to. It also had to be done in a way which was based on TCs' democratic and relational principles.

The idea of written standards was incorporated – to ensure sufficient objectivity and have the authority of the Royal College behind us. But the standards were democratically developed, using therapeutic community principles. All participating TCs were involved in the discussion, and in agreeing and voting on the first version of the standards. This ensured that the member TCs felt that they were meaningful and owned by 'the Community of Communities' (Haigh and Tucker, 2004). By contrast, national standards overseen by regulatory organisations are often seen as persecutory inspection undertaken in a mechanical way by people who did not understand the processes being inspected.

This project rapidly became a recognised quality improvement process, as well as a way for therapeutic communities not to feel isolated, or not knowing if 'they were doing it right'. Member communities became quite comfortable using language like 'sharing best practice', 'benchmarking' and 'action planning'. We had been 'modernised', done the right thing, called 'quality improvement' - but in a therapeutic community way, that we could live with. And several other British quality networks in mental health have used some of these ideas – such as how to co-create a process with service users being completely involved and develop a friendly informality within which the formal work is done.

By believing in the central importance of relationships, in process as well as content, we had also produced something that was more meaningful than a treatment manual. Perhaps this has introduced something of a Trojan horse, to spread a message of gentle subversion and quiet revolution, into the dry and technical world of the Royal College of Psychiatrists, and all their quality processes. There is only one quality network there that is older than us, and about twenty newer ones – covering everything from ECT (electroconvulsive therapy) suites to mother and baby units, and acute wards to psychological services. We introduced them all to working in a way that we call 'an antidote to inspections'.

The standards and the process are all available on the RCPsych website (Royal College of Psychiatrists, 2015) – and versions have now been adapted and used in Australia, Sicily and Northern Italy.

6.3 RESEARCH: Understanding what TCs are doing, and showing that it works

As a product of globalisation, medical research has become very standardised across the world – and the only 'language' that everybody understands is *numbers*. Statistics and graphs can be understood without much translation. However, those who work in complex interpersonal and relational systems do much that cannot be measured in numbers. Also, many 'scientific' or 'biomedical' psychiatrists say that the only evidence that is 'good enough' in the 'hierarchy of evidence' is 'Type 1 evidence'. That means a meta-analysis of high quality randomised controlled trials (RCTs). This is why somebody who is now a very senior government psychiatrist once called therapeutic communities an 'evidence-free zone'.

RCTs are difficult in TCs, because there are so many ethical problems and methodological difficulties in doing a meaningful experimental study on therapeutic communities – as well as the bluntness and coarseness of the polar 'yes or no' results that they give. But without these experimental studies, influential individuals and organisations can promote the idea that we are an 'evidence-free zone'. This specific requirement for experimental studies is at the heart of the government's National Institute for Clinical Excellence's (NICE) work, and it is a very powerful process in the British health system. Although its stated intentions are to provide 'guidance', this is often interpreted to mean 'instructions'. Its methods are also being adopted internationally.

In my work on the committee for drawing up guidelines for Borderline Personality Disorder, I soon saw that no equivocal evidence, or doubt, or shades of grey are tolerated in what the committee see as satisfactory evidence. The process adheres very strictly to the hierarchy of evidence. To me, this represents a 'zero tolerance' of uncertainty. In this specific guideline I was involved with, there was little 'high-quality evidence' for treatment of Borderline Personality Disorder, so a useful and 'common sense' guideline emerged which was not too prescriptive. That was because it was based on a good consensus among the experts by experience, the clinicians and the researchers on the panel. But this is called 'clinical consensus', which is very low-rating in the 'hierarchy of evidence'. Perhaps this shows that 'no high-quality evidence' can be quite a good thing for common sense and thoughtful clinical practice! (NICE, 2009)

However, the volume of 'acceptable' research is growing – particularly for psychotherapy treatments for borderline PD, and therapeutic communities would have soon become invisible unless we had started to produce our own acceptable evidence. We used to say that '*no evidence of effectiveness* does not mean *evidence of no effectiveness*', the only fact being that the research has not been done. Unfortunately, nobody who manages or commissions services takes much heed of this.

My own response to this has been to think that the whole 'evidence-based practice' principle is based on absolutely sound principles, but the spirit of the whole evidence initiative has been lost to an unrealistic quest for absolute certainty. As a result, the whole process is reduced to an arcane and annoying game, with rules too complex for everybody except those whose careers depend on it. Such meticulousness and precision might be understood for cancer chemotherapy, but it seems out of place in disciplines like psychiatry – where diagnosis and psychopathology are based on no equivalent basic science. However, we must play the game, with all the complex rules, in order to communicate the 'scientific fact' that therapeutic communities are an effective treatment, to the important officers and organisations they depend upon for their continued existence.

Many colleagues with long clinical careers in TCs are convinced that people benefit from time in them. This makes equipoise difficult when randomising subjects to treatment as usual. Also, most clinicians believe that the small proportion of people who might be harmed by them can be helped to get out, with due clinical sensitivity, before they are done any damage. However, many clinicians and service users remain unconvinced that RCTs and standardised questionnaires and protocols will pick 'outcome' up in a way that is meaningful or sufficient. Perhaps, in a digital world where the only possible answers are 'yes' or 'no', this is asking for more than experimental methods can achieve. At the very least we need to undertake good qualitative research alongside the data from RCTs. In this way, we might start to understand what the numbers do not tell us.

In the last year, the first ever RCT of non-residential democratic therapeutic communities has been completed and published - to the standards required of the medical research establishment (Pearce et al., 2017; Haigh, 2017a). The results are generally positive, and longer-term follow up results are expected in the next few years. However, others in TCs now need to become involved in serious research and to follow this up by undertaking whatever data collection or research they can (Lees et al., 2004). There are at least two reasons for this.

The first is that there will need to be more trials in the next few years to back up the first one. Research is never finished – there is always more to know, as well as needing to test out the first results, and gather enough data to do a meta-analysis of several studies. Therapeutic communities need to get together to plan this so that *the data* can make a real impact by doing large and multi-centre clinical trials.

And the second reason applies to all therapeutic community: unless you can justify the existence of your therapeutic community, with data in a hard-line objective and numerical terms, you may not be able to survive. It means collecting, analysing and communicating outcome data. It could start as service evaluation (such as an annual report to demonstrate what you are doing and how it helps) but it does need to develop into and join with, substantive and formal research programmes. Therapeutic communities as we know them will not survive in the long-term unless they show that they are doing something that works, in the format and language that the international health field demands.

In a very different frame of reference, Nick Manning, a prominent medical sociologist, asks difficult questions about what he calls 'the Politics of Data': where the real power is, how funding decisions are made, and what are the unstated assumptions and biases in the process (Manning, 2014). If TCs are to keep to their principles of challenging authority and holding it to account, they will need to become involved in these matters.

6.4 TRAINING: the only way is experiential: the "Living-Learning experience"

There is something about the 'atmosphere' or 'culture' of a therapeutic community that is very difficult to put into words, in any language. It is about the quality of relationships – the way people are with each other – that is experienced in a preverbal or non-verbal way. Some describe these qualities as 'ineffable' or 'numinous'. Any number of books, academic papers or lectures of can try to explain what it is – but can never recreate the experience itself. Personal therapy might give some idea, perhaps more so if it is in a group format, but this will still not give a realistic experience of what it is like in a therapeutic community. People who are trained in group analysis, psychoanalysis or existential psychotherapy will probably have some idea of it – but not the continuous 'edginess' of it, or the 'hurly burly' that it often feels like. Manualised and directive therapies have reasonably high levels of predictability: this is not so in therapeutic communities, where tolerance of not knowing what might happen in the next hour, day or week is an important therapeutic factor. The spontaneous and creative solutions that the group needs to find very often mark a turning point in how people relate to each other, and to themselves.

For staff, it is very useful training to experience this, but away from the heat of day-to-day clinical work. Hence the development of the 'Living-Learning Experience' workshops over the past thirty-odd years (Lees et al., 2016). These are three-day residential transient therapeutic communities for staff to attend as residents, which are conducted by senior TC staff and 'experts by experience' (people who have been through therapeutic communities themselves, as client members, and had relevant subsequent experience and training). The format is for 15-25 strangers (most of whom are staff from different therapeutic communities) and 3-5 workshop staff to conduct it. In England, it is run it several times per year in a large manor house in Kent, which is part of an organic farm study centre. Exactly the same workshops also take place regularly in Italy (Sicily, Rome and occasionally elsewhere), India (near Bengaluru), and irregularly in other countries (which include Australia, Northern Ireland and Portugal).

It is managed by a not-for-profit organisation called 'Living and Learning'. The organisation draws on fifteen suitably qualified and experienced staff members of different nationalities. The international group is called 'International Network of Living-Learning Experiences'

(Living and Learning, 2017). Basic information about the events and brief staff biographies are kept on the website.

By attending these events, and maybe only by doing so, TC staff can really come to understand what their patients and clients and service users feel like when they start in a therapeutic community and have to deal with the process of forming numerous new relationships in a short period of time. Participants need to make decisions about what to do together, to have small stranger groups (90 minutes each, five times – regular group analytic format) with other members with whom they live in the same house and make meals for each other. Usually, it is also quite playful, and people generally have fun together as well as experiencing intense and sometimes difficult emotions. Music is often enjoyed: one of the staff team regularly takes a guitar, and another musician on the staff team took a 'cello once – and he sat in the kitchen playing Vivaldi while the rest of us made lunch. Various other domestic and rural activities take place, in the building, grounds and surrounding landscape. All decisions about the activities are made in the morning and evening community meetings. On the third day, everybody tidies up after lunch, says goodbye and leaves - and might never see each other again.

Experience of the group over many years indicates an extraordinary intensity, scope and level of emotional and interpersonal work that many participants do: how far the individuals and the group come in such a small period of time.

Courses are always evaluated with questionnaires at the beginning and the end, and then for follow-up a few weeks later. With this, we hope to be able to understand how people feel that they benefit or not, and we can adjust the programme accordingly. As well as receiving predominantly positive comments, we have done some basic quantitative analysis. This has shown high and increasing levels of satisfaction. More recently, formal qualitative research has been undertaken using ethnographic methods. This has already shown that the courses meet the 'Community of Communities' standards (Rawlings, 2005), and the results are now starting to show interesting differences from other types of group relations events. Data is also being collected to demonstrate how LLE courses need to adapt to local cultures – such as the difference between (superficially) identical courses run in urban Sicily and rural India, for example. We hope this work will develop and grow into something that can be published in the scientific literature to help understand the theoretical basis of therapeutic communities and the acceptability of the method across different cultures.

6.5 INNOVATION: Using therapeutic community ideas in new ways

As a result of external pressure, as much ideological as economic, much has changed in the type of democratic therapeutic communities in mental health services. We have lost all the seven day 'total immersion' residential therapeutic communities – no more now exist in the UK. The Cassel Hospital continues to offer medium to long-term inpatient intensive psychoanalytic psychotherapy, but it no longer defines itself as a therapeutic community *per se*. The Henderson closed in 2010, and Francis Dixon Lodge has become a 3-day per week unit since then.

There are a few, including new types, growing up as charities and cooperatives. The state sector – the NHS in the UK – is very highly regulated and risk-managed and is becoming an increasingly difficult context for therapeutic communities to flourish, or even survive. But these charitable or social enterprises or sometimes commercial, therapeutic communities now sell their places to the state system when the local services have no suitable facilities to help. A more flexible range of facilities, with varying degrees of 'TC-ness', is a possible future development: different places for those who need a therapeutic environment with specified levels of physical containment, with types of therapy programmes suited to their particular circumstances. Not all like traditional therapeutic communities, but with some or many of the features and qualities – and the underlying values.

As well as closing the beds, non-residential therapeutic communities have also 'reduced the dose' over the last ten years – and TCs have evolved from 5 days per week to three, to one, or even less. A 'micro-TC' started in 2013 gives even less time together – it only meets together for two and a half hours per week but is connected to numerous other therapeutic and social activities that are available in the town (Haigh, 2017b). Some are run by health services, others by social services and some are in the charitable and voluntary sector. Members of the 'micro-TC' make their own 'pick-and-mix' therapeutic programme. In this way, various therapeutic elements (psychotherapy, sociotherapy as well as the traditional psychiatric treatment) are distributed across the whole town, but with the continuity established between them - through people feeling that the whole system is joined together. This gives 'continuity of care' rather than 'fragments of services' which is otherwise so often the case. There is an interesting debate about what to call it – is it a 'micro therapeutic community' because it only meets for 2½ hours a week, or 'macro therapeutic community' as it is like it is a therapeutic community without walls, that involves elements from across the whole town?

Perhaps this is a 'post-post-modern' therapeutic community, and it shares some features with a faith community on the south coast. This is a church-based therapeutic community without buildings or even staff: members meet in various places in the town, and sometimes live in each other's houses. What seems to matter most, and is the common factor with long-established TCs, is 'the therapeutic community in the head'. That means experiencing the *nature and quality of relationships* around you that make you feel safe enough to be able to do the work you need, even if they are only physically present for part of the time and are with different people.

An important consideration for the future of TCs is perhaps to better recognise this 'therapeutic community in the head' (Haigh, 2013). In this way, the experience of being with each other is the essential element, and not the timetable or buildings. The physical space and regular timing become less prominent. This could also lead to the development of virtual, or partially virtual, therapeutic communities. Most crucially, members feel they have a *sense of belonging* in the community for 24 hours a day and seven days a week.

A different innovation has arisen from the work of the Community of Communities (see 'quality', above). In 2007, a set of 'core standards' was defined – the specific factors that exist in all therapeutic communities but are not generally present in other types of service. From these, a set of 'core values' was established. These were the values that underlay the core standards – you did not need to be a formal 'therapeutic community' to hold those values, but all therapeutic communities did– as did many other places that looked and felt like therapeutic environments (Haigh et al., 2012).

Here are ten the values which are thus defined for 'Enabling Environments': recognizing the importance of RELATIONSHIPS is always a central requirement; BOUNDARIES are what keeps a space safe; all behaviour can be seen as COMMUNICATION; DEVELOPMENT needs spontaneity for creativity and growth; INVOLVEMENT means everybody is responsible for the place, and each other as well as themselves; SAFETY is created through support; STRUCTURE gives a purpose to people; EMPOWERMENT comes through authority being open to question and discussion; the LEADERSHIP is responsible for making sure the environment is enabling, and external relationships are sought and valued in a culture of OPENNESS.

Standards were then written to embody these values, with several flexible criteria for each. The criteria show how to demonstrate that a unit or service is meeting the value-based standards. To work towards the award itself, there is a process of compiling a portfolio, which is often done with great pride and energy, and submitting it. As we might expect in a therapeutic type of activity, the process of doing this preparation work is more important and enjoyable, and meaningful, than simply getting the award at the end (Royal College of Psychiatrists, 2016).

The Enabling Environments project now has nearly two hundred units as part of its network. It has been particularly successful in the criminal justice world – where prison staff are finding a new sense of purpose in their jobs, there is much less violence on prison wings, reoffending is reduced, and morale of prisoners as well as staff is demonstrably higher. Because of its success, the head of the UK Prison Service has publicly supported the project, setting it as an aim for all probation hostels, and as an aspiration of values across the prison estate. This is at a time when British prisons are under great pressure and strain, where this project and its outcomes are seen as an area of hope in a grim landscape. Wider scope for enabling environments, particularly in the health and social care systems, is now being planned.

These examples illustrate some examples of how TCs can be adapted to fit well within modern highly-specified services - by modifications to the traditional TC programme and using the longstanding therapeutic community ideas as the fundamental principles behind all services where relationships are important. It is likely that there are many other people, projects and organisations doing similar things, and we all need to draw support from each other.

6.6 Discussion

Global managerial processes - particularly through 'New Public Management' (Gruening, 2001) have led to health and social care systems being more standardised, risk-averse and regulated. While this is undoubtedly of benefit in terms of efficiency and effectiveness in areas such as cancer care or ischaemic heart disease, it produces a mechanical and technological framework for mental health services which allows little recognition of the importance of creativity, spontaneity or playfulness. It can also support care that is devoid of meaningful human interaction, compassion or kindness (Ballatt and Campling, 2011). Well-functioning therapeutic communities thrive on such elements and cannot function well without them. Most significantly, they prize relational practice – where every individual is treated as an individual with their own personality, talents and problems recognised and respected - above procedural practice, with performance-managed adherence to closely-defined policies, targets and goals (Hoggett, 2004).

It is often a natural reaction of any team, in times of staff reduction, service cuts and efficiency savings, to 'batten down the hatches' and protect what they have, rather than try to change it. However, in therapeutic communities, this has led to a climate of fear which has brought about rigidity, inflexibility and an inability to grasp the size and scale of changes needed – simply to survive. Therefore, many TCs have not survived: there are now less than twenty residential therapeutic community beds in the National Health Service, and nearly all the non-residential day-TCs have reduced from five-day programmes to three, two or one days per week.

But this could be seen as not so much a decline, as an accommodation – with its own distinctive clinical advantages (Haigh, 2007). Although, while we in a state of apprehension and uncertainty about survival, it is natural to fight to continue what we are doing, that is unlikely to increase the probability of that survival. The TC 'movement' needs to be more willing to adapt and change – while keeping to the therapeutic philosophy described above – in order to keep the underlying emancipatory ideal alive. It is not helpful to fight for the detailed structures, programmes, buildings and guidelines that worked fifty – or even five – years ago. The individual elements that are worth preserving include experiences that are difficult to isolate or enumerate: such as emotional safety, interpersonal conduct, continuity, feeling empowered, allowed to act authentically – and, dare I say – about having fun. Therapeutic communities value the playful aspects of the work, although there seems to be little corporate recognition of its value in any public services.

Modernisation may also be a reaction to some fundamental fear of 'other' in a very diverse and globalised world: with lack of trust in relationships is growing with the global tensions and responses we see around us, we retreat into technical solutions, certainty and individualism. In contrast, Foulkes, the founder of group analytic psychotherapy, noted that *"each individual is an abstraction, determined by the world, the community and the group we live in"* (Foulkes, 1946). This idea of fundamental human experience being more about our relationships than about us as individuals can be used to draw an analogy with the well-known phenomenon of wave-particle duality in physics. Although the two seem incompatible, both are valid, and electrons can only be understood if we understand their nature as both a probability wave function and a particle with uncertain position or momentum. Particles could represent the individual view of human existence and wave the relational aspect. We need to accept both. Reflecting on the recent discovery of the Higg's

Boson: if Higg's Bosons are the things that give matter its mass, then relationships could be seen as the things that give humans their meaning.

Perhaps TCs can leapfrog over 'modernisation' – and get to a more holistic, wholesome and complex way of organising our services (Haigh, 2005a). This is beyond modernism as we know it: not minimalist, sterile and automated, but human, complex, uncertain and authentic. We can incorporate some of the 'efficiency benefits' of modernisation (such as flexible use of new technology), but the new focus will be relational, not performative, transactional or procedural. In that way, it could be seen as post-modern. This has a close parallel with Bracken and Thomas's coinage of 'post-psychiatry' (Bracken and Thomas, 2001). Because the therapeutic community field has been working progressively for many years, in ways such as transparency, democracy and service user partnership, it is familiar territory. Using these principles, along with the underlying emancipatory philosophy, could be even more exciting and worthwhile than being part of traditional therapeutic communities, and trying to maintain what they have always done. The placing of relationships at the centre perhaps moves even further – beyond the post-modern dogma of relativism, 'anything goes' and 'no grand narratives'. The relational framework itself could be described as a new 'grand narrative'.

However, leadership to promote such ideas is not common in state-commissioned services, and an accommodation must be found to 'keep the flame alive' while getting on with the business of doing effective clinical work - in an increasingly regulated and managed system. Manning (1989) clearly articulated this in 'Charisma and Routinisation': having a powerful charismatic leader is not a stable or sustainable position, while becoming routinized and bureaucratic is no better. It is a fine line to tread, and I have previously argued that the ideas behind therapeutic communities carry plenty of their own charisma – and that the key is to place relationships (at every level) at the centre of the work and the development of the field, rather than have celebrity-type individuals as its powerful leaders (Haigh, 2005a), or to turn to bureaucratic and administrative solutions (Haigh, 2005b).

We who work in mental health know that we can only ever work in uncertainty and probability, but the world in which we now exist increasingly demands certainty. A current example in the UK is the 'zero tolerance of suicide'. In the digital tyranny that computers demand of us all, only zeroes and ones are allowed, and no fuzzy indeterminacy is permitted. This 'untrusting performativity' characterises the 'New Public Management' and now permeates everything we do in health services and universities – and other public sector organisations. It is also becoming dominant in the voluntary sector, for charities and social enterprises. Unless one shows corporate understanding of the layers of governance, risk management and compliance checks, funding is unlikely to be granted.

This process fundamentally does not trust relationships: this is not acceptable as the core foundation for human services, particularly mental health. Mental Health needs to be seen again as a social movement as it was in the 1790s (with moral treatment) and again in the 1950s (with antipsychiatry) – with the challenge to authority and oppression which that needs. The work and thinking of therapeutic communities needs to be a part of it. With much of the current culture based on mistrust and fear, therapeutic communities, and other relational approaches, need to work hard to demonstrate that there is a different way to live our lives, and a better way to be with each other.

6.7 References

Ballatt, J. & Campling, P. (2011) *Intelligent kindness: reforming the culture of healthcare.* London, RCPsych publications.

Baron, C. (1987) *Asylum to anarchy.* Free Association Books.

Bion, W.R. (1961) *Experiences in groups: And other papers.* London, Tavistock Publishers.

Bracken, P. & Thomas, P. (2001) Postpsychiatry: a new direction for mental health. *British medical journal.* 322 (7288), 724.

Cooper, D. (1968) *The Dialectics of Liberation.* Harmondsworth, Penguin.

Foucault, M. (1965) *Madness and Civilization: A History of Insanity in the Age of Reason.* New York, Pantheon.

Foulkes, S.H. (1946) On group analysis. *The International journal of psycho-analysis.* 27 (1–2), 46–51.

Goffman, E. (1968) *Asylums: Essays on the social situation of mental patients and other inmates.* AldineTransaction.

Gruening, G. (2001) Origin and theoretical basis of New Public Management. *International public management journal.* 4 (1), 1–25.

Haigh, R. (2017a) A New Biopsychosocial Programme for Emotional Instability: the Slough Model. *The Psychotherapist.* 66 (Journal Article), 27–29.

Haigh, R. (2005a) Charismatic Ideas--Coming or Going? *Therapeutic Communities.* (Journal Article).

Haigh, R. (2007) The New Day TCs: Five Radical Features. *Therapeutic Communities.* 28 (2), 111–126.

Haigh, R. (2013) The quintessence of a therapeutic environment. *Therapeutic Communities: The International Journal of Therapeutic Communities.* 34 (1), 6–15.

Haigh, R. (2005b) The trouble with modernisation: we need better relationships, not policies and procedures. *Mental Health Review Journal.* 10 (3), 3–7.

Haigh, R. (2017b) Therapeutic communities enter the world of evidence-based practice. *British Journal of Psychiatry.* 210, 313–314.

Haigh, R., Harrison, T., Johnson, R., Paget, S., et al. (2012) Psychologically informed environments and the "Enabling Environments" initiative. *Housing, Care and Support.* 15 (1), 34–42.

Haigh, R. & Lees, J. (2008) Fusion TCs: Divergent histories, converging challenges. *Therapeutic Communities.* 29 (4), 347–374.

Haigh, R. & Tucker, S. (2004) Democratic development of standards: the community of communities—a quality network of therapeutic communities. *Psychiatric Quarterly.* 75 (3), 263–277.

Harrison, T. (1999) A momentous experiment. *1999]"Therapeutic Communities: Past, present and future", London, Jessica Kingsley.* (Journal Article).

Hoggett, P. (2004) Shame and performativity: Thoughts on the psychology of neoliberalism. *Psychoanalysis, Culture & Society.* 1–19.

Jones, M. (1956) The concept of a therapeutic community. *American Journal of Psychiatry.* 112 (8), 647–650.

Jones, M. (1979) The therapeutic community, social learning and social change. In: R. D. Hinshelwood & N. Manning (eds.). *Therapeutic Community – Reflections and Progress.* London, Routledge & Kegan Paul. p.

Laing, R.D. (1967) *The Politics of Experience.* Harmondsworth, Penguin UK.

Lees, J., Haigh, R., Lombardo, A., Rawlings, B., et al. (2016) Transient therapeutic communities: the 'Living-Learning Experience' trainings. *Therapeutic Communities: The International Journal of Therapeutic Communities.* 37 (2).

Lees, J., Manning, N. & Rawlings, B. (2004) A culture of enquiry: Research evidence and the therapeutic community. *Psychiatric Quarterly.* 75 (3), 279–294.

Living and Learning (2017) *The International Network of Living Learning Experiences.* [Online] Available from: www.inlle.net.

Manning, N. (1989) *The therapeutic community movement: Charisma and routinisation.* Routledge.

Manning, Nick (2014) *Politics of data (personal correspondence).*

NICE (2009) *Borderline Personality Disorder: Treatment and Management.* [Online]. National Institute for Health and Clinical Excellence: Guidance. Available from: http://europepmc.org/books/NBK55403.

Pearce, S., Scott, L., Attwood, G., Saunders, K., et al. (2017) Democratic therapeutic community treatment for personality disorder: randomised controlled trial. *The British journal of Psychiatry.* 210 (2), 149–156.

Rapoport, R. (1960) *Community as Doctor: New Perspectives on a Therapeutic Community.* London, Tavistock Publications.

Rawlings, B. (2005) The Temporary Therapeutic Community: a qualitative evaluation of an ATC Training Weekend. *Therapeutic Communities.* 26 (1), 6–18.

Royal College of Psychiatrists (2016) *What is the Enabling Environments award?* [Online] Available from: http://www.enablingenvironments.com.

Royal College of Psychiatrists, E.P., S. Thorne, J., Das, A. (2015) *Service Standards for Therapeutic Communities, 9th edition.* [Online] Available from: http://www.rcpsych.ac.uk/pdf/Service%20Standards%20for%20Therapeutic%20Communities%209th%20Ed%20FINAL%20-%20For%20%20Website%20-%20Copy.pdf.

Spandler, H. (2006) *Asylum to action: Paddington day hospital, therapeutic communities and beyond.* Jessica Kingsley Publishers.

Spandler, H. (2009) Spaces of psychiatric contention: a case study of a therapeutic community. *Health & place.* 15 (3), 672–678.

Tuke, S. (1813) *Description of The Retreat, an institution near York, for Insane Persons of the Society of Friends.* York, The Retreat.

Chapter 7

Open dialogues principles and dialogical meetings for psychosis

Jaakko Seikkula
Department of Psychology, University of Jyväskylä, Finland

Birgitta Alakare
Health District of Länsi-Pohja, Finland

The name open dialogue was first used in 1995 to describe the entire family and social network-centred treatment approach that was first initiated in the Finnish Western Lapland. The approach includes two elements: first, open meetings in which all relevant members in the actual situation – including both the social network of the client and the professionals working as a team - participate from the outset to generate new understanding by dialogue; second the guiding principles for the entire system of psychiatric practice in one geographic catchment area. The main forum for dialogues is the open treatment meeting where all management plans and decisions are made with everyone present. Referring to the genuine idea of combining the treatment planning and therapeutic meaning of the dialogues Alanen (1997) concluded that the treatment meeting has three functions: 1) to gather information about the problem; 2) to build a treatment plan and make all decisions necessary based on the problem that was described in the conversation; and 3) to generate a psychotherapeutic dialogue. In what follows, we first describe the comprehensive system of care with the core elements and in the end focus on the description of the dialogical sessions. We also present results of a large outcome study with the first episode psychotic patients. In five years, more than 80% had returned to full employment and 81% did not have any remaining psychotic symptoms, which illustrate significantly better outcomes compared to treatment as usual. After 19 years the differences in favor of Open Dialogue treatment were still visible.

7.1. Open dialogues in organizing psychiatric practice

At the beginning of 1980's when we began to develop the acute psychiatric inpatient system at Keropudas Hospital in Tornio, we had two primary interests. First, we were interested in individual psychotherapy for patients diagnosed with schizophrenia. At that time Keropudas Hospital was occupied by dozens of long-term patients who had been considered 'incurable'. What was important in the Need-Adapted tradition was the shift to a more optimistic treatment model and learning how to work with the psychological resources of patients with psychotic problems; finding ways to make use of patients' own psychological resources in our treatments has proved to be crucial.

In Finland, psychotherapeutic practice has long been a part of public health care and especially important has been the development and research in the Turku psychiatric clinic since the 1960s by Professor Yrjö Alanen and his team (Alanen, 1997). Starting with individual psychodynamic psychotherapy, the Turku team integrated systemic family therapy into their treatments in the late 1970s and called the approach Need-Adapted Treatment. This emphasised that every treatment process is unique and should be adapted to the varying needs of each patient. To make it possible to find the unique needs of every patient and the family they generated the genuine innovation of the open treatment meeting that we adopted to our clinical work at 1984.

When starting to work with the open treatment meetings, we first met successful and unexpected outcomes. At the same time, however, we were faced with confusing experiences, which we could not understand and realized that we needed both more training of the staff and finding ways of analysing the new practice concerning both the content of the processes and the outcomes. Several effectiveness and treatment process evaluations of the open dialogue approach have been completed employing an action research methodology (Aaltonen et al., 1997; Haarakangas, 1997; Keränen, 1992; Seikkula, 1991, 1994; Seikkula et al., 2003, 2006). In one of the studies (Aaltonen et al., 1997; Aaltonen, Seikkula & Lehtinen, 2011) we summarized the optimal elements of the approach, in which the main part of the psychiatric crises was taken care by organizing mobile crisis intervention teams having the first meeting at patients home instead of hospitalization. The main principles of an optimal treatment process were: 1 immediate response; 2 a social networks perspective; 3 flexibility and mobility; 4 responsibility; 5 psychological continuity; 6 tolerance of uncertainty; and 7 dialogism. These principles of the open dialogue approach are enlarged upon below. It is worth noting that these principles came out of the research and were not principles planned before and then followed. Later on, more general ideas about good treatment were added. Although most of the studies have focused on the treatment of psychotic problems, they are not diagnosis specific but describe an entire network-based treatment that is especially practical for initiating in crisis situations.

Immediate response. In a crisis it is vital to act immediately, and not, for instance, to wait for the patient with psychosis to become more coherent before a family meeting. It is preferable that the first response be initiated within 24 hours. The meeting is organized regardless of who first contacted the response unit. In addition, a 24-hour crisis service ought to be set up. One aim of the immediate response is to prevent hospitalization in as many cases as possible. Everyone, including the patient, participate in the very first meetings during the most intense psychotic period. The patients usually seem to be experiencing something that has been unappreciated or unacknowledged by the rest of the family. Although the patient's comments may sound incomprehensible in the first meetings after a while it becomes apparent that the patient was actually speaking of real incidents in their lives. Often these incidents include some terrifying issues or a threat that they have not been able to articulate before the crisis. Psychotic experiences most often include real incidents and the patient is bringing forth themes that have not previously been verbalized. This is also the case in other forms of difficult behaviour. In extreme anger, or depression, or anxiety, the patient is speaking of previously unspoken themes. In this way, the main person in the crisis, the patient, reaches for something that had not been touched by others in their surroundings. The aim of the treatment becomes the expression of these experiences that had not previously have words especially ones that had not developed a shared language. During the first couple of days of a crisis, it seems possible to speak of things that are difficult to discuss later.

During the first days, hallucinations may be handled and reflected upon, but the opportunity to deal with them may not reappear until after several months of individual therapy. It is as if the window for the extreme experiences only stays open for the first days (Falloon et al., 2002). If the team manages to create a safe enough atmosphere by responding rapidly and listening carefully to all the themes the clients bring up, then important themes may find a space where they can be handled, and the prognosis improves.

Including the social network. The patients, their families, and other key members of their social network are always invited to the first meetings to mobilize support for the patient and the family. The other key members may be representatives of other authorities, such as state employment and insurance agencies, vocational rehabilitation services, fellow workers or the supervisor at the patient's workplace, neighbours or friends. Social networks are relevant in helping to define the problem(s). A problem is one that has been defined as such in the language of either those closest to the patient or by the patients themselves. In the most severe crises, the first notion of a problem often emerges in the definition of those closest to the patients after they note that some forms of behaviour no longer conform to their expectations: for example, if a young member of the family is suspected of using drugs. The young person will seldom see using drugs as a problem, but their parents can be terrified by the first signs of possible drug misuse. From a network perspective, all these individuals should be included in the process. It is helpful to adopt a simple way of deciding who should be invited to meetings. It can be done, for instance, by asking the person who made the contact in the crisis: 1) Who is concerned about the situation or who has been involved? 2) Who could be of help and is able to participate in the first meeting? 3) Who would be the best person to invite them, the one who contacted the services or the treatment team? By doing it this way, the participation of those closest to the patient is suggested as part of an everyday conversation, which decreases any possible suspicion about the invitation. Also, the one who has made contact with the services can decide who they do not want to participate in the meetings. If the proposal for a joint meeting is done in an official tone, by asking, for instance, `Will you allow us to contact your family and invite them to a meeting?' problems may arise in motivating both the patients and those close to them. Another factor in deciding about the relevant participants is to find out whether the clients have contacted any other professionals either in the current situation or previously. If the other professionals cannot attend the first meetings, a joint meeting can be arranged later. The people in the client's social network can be included in many ways. They can be present, or if some of them cannot manage to attend meetings, then the clients can be asked if they want to invite others who know of their situation and who could possibly help. Some member of the network can be given the task of contacting them after the meeting and relaying the absent person's comments to the next joint meeting. Those present can be asked, for instance, `What would Uncle Mark have said if he was present in this conversation? What would your answer be? And what would he say to that?'

Flexibility. Flexibility is guaranteed by adapting the treatment so that it is a response to the specific and changing needs of each patient and his or her family using the therapeutic methods best suited to each family, their specific language and their way of living. The approach and the length of treatment should fit the actual problem instead of applying a generic programme without variation from case to case. During the first ten to twelve days of a crisis, the need is quite different compared with the need three weeks later. For instance, during the acutest phase, it is advisable to possibly have a meeting every day, which will no longer be necessary once the situation has stabilized. In that later period, families generally know how frequently they should be meeting. The best place for the meeting, if the family approves, might be the patient's home. However, meeting in an emergency department or a psychiatric outpatient clinic is an option if the family sees this as more suitable. Home meetings seem to prevent unnecessary hospitalizations, since the family's own resources are more accessible in a home setting (Keränen,1992; Seikkula, 1991). Families can easily refuse to participate in treatment, if the invitation is made formally or if the professionals follow too rigid ways of working with families (Friis et al., 2003). However, the need-adapted approach with its emphasis on taking into account the uniqueness of each treatment process has been more successful in engaging with families. It seems to suit the Nordic system, in which every psychiatric unit has total responsibility for all clients in its catchment area.

Responsibility. Organizing a crisis service in a catchment area is difficult if all the professionals involved are not committed to providing an immediate response. A good rule of thumb is to follow the principle that whoever is contacted takes responsibility for organizing the first meeting and inviting the team. The one contacting the professional could be, for example, the patient, a family member, a referring practitioner or other authorities, such as a family doctor or a school nurse. Organizing an acute team is one possibility with all staff members knowing who to contact to organize an immediate first meeting. This means that it is no longer possible to respond to a request for help by saying `this has nothing to do with us, please contact the other clinic'. It is important to reassure the family member contacting the service that they have come to the right place and they will take care of organizing a meeting. One can say, for instance, `It sounds to me that alcohol misuse may be involved in your son's problem. Would you allow me to invite someone from the alcohol misuse clinic to join us in the meeting tomorrow?' In the meetings, decisions are made as to who will best form the team that will be responsible for the treatment. In multi-problem situations, the best team is formed with professionals from different units, for instance, one from social care, one from a psychiatric outpatient clinic and one from the hospital ward. The team mobilized for the first meeting should take all the responsibility needed for analysing the current problem and planning the treatment. Everything needed for an adequate response is available in the room, there is no other authority elsewhere that will know better what to do. This means that all team members should take care of gathering the information they need for the best possible decisions to be made. If the doctor was not able to attend the meetings, this individual should be consulted by phone, and if there is a difference of opinion about certain decisions, a joint meeting is advisable to discuss the choices in the presence of the family. This empowers family members to participate more in the decision-making process.

Guaranteeing psychological continuity. The team takes responsibility for the treatment for as long as needed in both outpatient and inpatient settings. This is the best way to guarantee psychological continuity. Forming a multidisciplinary team early increases the possibility of crossing the boundaries of different treatment facilities and preventing people dropping out. In the first meeting it is impossible to know how long the treatment will continue. In some instances, one or two meetings are enough, but in others, intensive treatment for two years may be needed. Problems may occur if the crisis intervention team meets three or five times and then refers the patient to other authorities. In these circumstances, even in the first meetings, there is a risk of too much focus on actions to be taken and not enough on the process of dialogue itself. Representatives of the patient's social network participate in the treatment meetings for the entire treatment sequence, including when other therapeutic methods are applied. One part of psychological continuity is to integrate different therapeutic methods into a cohesive treatment process where different methods complement each other. For instance, if individual psychotherapy is recommended for the patient, psychological continuity is easily guaranteed by having one of the team members act as the individual psychotherapist. If this is not possible or advisable, the psychotherapist could be invited to one or two joint meetings, in which ideas are generated that can serve as the basis for an individual therapy process. The therapist should be invited every now and then to meetings with the team and the family. Problems may occur if the individual psychotherapist does not want to participate in the joint meetings. This can intensify the family's suspicion towards the therapy, sometimes affecting the entire treatment process. This is particularly important to consider in the case of children and adolescents.

Tolerating uncertainty. The first task for professionals in a crisis is to increase the safety of the situation, when no one yet knows the reasons for the problem nor what the solutions will be. The aim is to mobilize the psychological resources of the patient and those nearest to him or her so as to increase the agency in their own life, by generating new stories about their most extreme experiences. This is supported by building up a sense of trust in the joint process. For instance, in psychotic crises, an adequate sense of security can be generated by meeting every day at least for the first ten to twelve days. After this, meetings can be organized on a regular basis according to the wishes of the family. Usually no detailed therapeutic contract is made in the crisis phase, but instead, at every meeting it is decided if and when the next meeting will take place. In this way, premature conclusions and treatment decisions are avoided. For instance, neuroleptic drugs are not commenced during the first few weeks. This allows for more time to understand the problem and the whole situation. There is also time for spontaneous recovery and, in some cases, the problem can dissolve by itself. A recommendation for neuroleptic drugs should be discussed in at least three meetings before implementation to clarify whether those present think the drugs are necessary. This approach is in contrast with illness-oriented approaches that during the early phase of treatment focus on trying to remove symptoms with drugs. For psychotic patients, these are typically neuroleptics. Psychiatric drugs can affect the intensity of hallucinations, of course, but the risk is that they decrease psychological resources at the same time. Neuroleptic drugs often have a sedative effect that calms psychological activity and thus may be a hindrance to psychological work. The challenge is to create a process that increases safety and encourages personal work. In our study only 33% of acutely psychotic patients used neuroleptics at all during the five-year follow-up period.

Besides the practical aspects of seeing that the family is not left alone with its problems, increasing safety means generating a quality in the therapeutic conversation such that everyone can be heard. Working as a team is one prerequisite in guaranteeing safety in a crisis with loaded emotions. One team member may start to listen more carefully to what the son says when he is saying that he does not have any problems, it is his parents who need the treatment. The other team member may become more interested in the family's burden of not being successful at stopping his drug misuse. Already in the very first meeting, it is good to reserve some time for reflective discussion among the team in the presence of the family to guarantee the different and often contradictory perspectives to be heard and consequently, every family member having a feeling being recognized and accepted. If the team members can listen to each other, it may increase the possibility for the family members to listen to each other as well. A situation in which professionals are in a hurry to get to the next meeting and therefore propose a rapid decision is not the best use of the family members' psychological resources. It would be better to note that important issues have been discussed, but no firm conclusions can be made, and thus the situation is defined as open. One way to put it into words might be: `We have now discussed this for about an hour, but we have not reached any firm conclusion of what this is all about or the best option to address it. However, we have discussed very important issues. Why not leave this open and continue tomorrow?' After that, concrete steps should be agreed on before the next meeting to guarantee that family members know what they should do if they need help.

Dialogicity. In meetings, the focus is primarily on promoting dialogue and only secondarily on promoting change in the patient or in the family. Dialogue is the forum through which families and patients can acquire more agency in their own lives by discussing the problems (Haarakangas, 1997). A new understanding is generated in dialogue (Andersen, 1995; Bakhtin, 1984, Voloshinov, 1996). For a professional, this means eliciting new aspects of being an expert in whom clients can trust. Professionals must become skillful in promoting dialogues through which their specific expert knowledge becomes rooted in the context.

Overall, the starting point for the dialogue in the open treatment meeting is the language of the family, how each family has, in their own language, described the patient's problem. All persons present speak in their own voices. The stance of the therapists is different compared with the traditional one in which it is the therapist who makes the interventions. While many family therapy schools are especially interested in creating specific forms of interviewing, in focusing on generating dialogue, listening and responsively responding becomes more important than the manner of interviewing. Team members can comment on what they hear with each other as a reflective discussion while the family listens (Andersen, 1995).

The meeting takes place in an open forum. All participants sit in a circle in the same room. The team members who have taken the initiative for calling the meeting take charge of opening the dialogue. On some occasions, there is no prior planning regarding who takes charge of the questioning and thus all staff members can participate in interviewing. On other occasions, the team can decide in advance who will conduct the interview if the team members are used to more structured action in the therapeutic sessions. The first questions are as open-ended as possible, to guarantee that family members and the rest of the social network can begin to talk about the issues that are most relevant at the moment for them. The team does not plan the themes of the meeting in advance. From the very beginning, the task of the clinicians is to adapt their answers to whatever the clients say. Most often, the team's answer takes the form of a further question that is based on what the client and family members have said. Often this means repeating word for word some part of the utterance and encouraging saying more about the issue.

In a psychotic crisis, it may happen that the patient does not want to participate in the meeting or runs out of the meeting room suddenly. The team needs to discuss carefully with the family members whether or not to continue the meeting. If the family wants to continue a staff member informs the patient that she or he can return if she or he wants. During this discussion, we do not make any other decisions concerning the patient.

Everyone present has the right to comment whenever she or he is willing to do so but comments should not interrupt an ongoing dialogue. Every new speaker should adapt his or her utterance to what was previously said. For the professionals, this means they can comment either by inquiring further about the theme under discussion or by commenting reflectively to the other professionals about their thoughts in response to what is being said. Most often, in those comments, professionals may repeat word by word a part of the utterance thus verifying the client's most difficult experiences. In every meeting, there is also a need to handle issues that are relevant to the unit, and it is advisable to focus on these issues toward the end of the meeting after family members have spoken about what are the most compelling issues for them. After deciding that the important issues for the meeting have been addressed, the team member in charge suggests that the meeting be adjourned. It is important, however, to close the meeting by referring to the client's own words and by asking, for instance, `I wonder if we could begin to close the meeting. Before doing so, however, is there anything else we should discuss?' At the end of the meeting, it is helpful to briefly summarize the themes of the meeting, especially whether or not decisions have been made, and if so, what they were. The length of meetings can vary, but usually ninety minutes is adequate.

7.2 The long term effectiveness of the open dialogues approach

In Western Lapland, the effectiveness of open dialogue has been assessed in follow-up studies for first-episode psychotic patients. In one quasi experimental study, the schizophrenia patients were selected to compare with another psychiatric unit in another place in Finland (Seikkula et al., 2003). This study was part of the Finnish API (Integrated treatment of acute psychosis study (Lehtinen et al., 2000). Patients diagnosed with schizophrenia were hospitalized significantly less compared with usual treatment, an average of fourteen days/person compared with one hundred and seventeen over a two-year period in a treatment as usual setting (Seikkula et al., 2003). Only 33% used neuroleptics during some phase of treatment compared with 100% in the comparison group. Families were actively involved in all the cases, having an average of twenty-six meetings per patient over two years. When comparing outcomes, open dialogue patients diagnosed with schizophrenia seem to recover better from their crises. At least one relapse occurred in 71% of the comparison group of patients compared with 24% in the open dialogue group. Only 17% of the open dialogue patients had at least occasional mild symptoms, compared with 50% in the comparison group. As many as 81% had returned to full employment compared with only 43% in the comparison group.

The results with open dialogue patients remained positive at five-year follow-up, in which all the non-affective psychotic patients were included (Seikkula et al., 2006). Only 29% of open dialogue patients experienced one or more relapses, and 82% of open dialogue patients had no residual psychotic symptoms. Employment status was better than in any other outcome study, with 81% of the open dialogue patients returning to their studies, work or to active job-seeking. When replicating the same design of the study ten years after the preliminary studies, it was found that the outcomes in Western Lapland had stayed about the same (Seikkula, Alakare & Aaltonen, 2011). Out of the first episode patients, 84% had returned to full employment or active job seeking in two years follow-up period. At the same time, some interesting changes also had occurred. The duration of the untreated period of psychosis (DUP) before the start of the treatment had declined to three weeks in Western Lapland. Perhaps in relation to this, the incidence of schizophrenia has declined in Western Lapland from 33 new schizophrenia patients per 100 000 inhabitants at 1985 to 2 per 100 000 inhabitants at 2005 (Aaltonen, Seikkula & Lehtinen, 2011; Seikkula, Alakare & Aaltonen, 2011).

In the most recent research project with the same patients included between 1992 to 2005 the 20-year outcome of the first episode psychotic patients are analyzed (Bergström et al., 2017). The first analysis showed that about one third are currently using psychosis medication. When comparing to the treatment as usual of psychotic patients in Finland, significant differences emerge in relation to the use of psychosis medication, in the mortality and in the retirement rate. In Western Lapland, the number of patients living on a disability allowance is the lowest in Finland (35%) when the rate in the rest of Finland is about twice as high (Kiviniemi, 2014).

7.3 Conclusions and reflections

The results show a remarkable change in the prognosis for severe mental health problems. This suggests that our approach to psychiatric crisis should change. We are used to thinking of psychosis as a sign of schizophrenia and schizophrenia as a relatively stable state that affects the patient throughout the entire life. For instance, a third of the patients with schizophrenia are said to need ongoing treatment, a third will need intermittent treatment, and a third will fully recover and actively work. In the few long-term follow-up studies of first-time psychotic patients after five years treated by traditional methods more than a half, often about 60% are said to be living on a disability pension (Lenior et al., 2001; Svedberg, Mesterton & Cullberg, 2001). The positive outcomes using the open dialogue approach may indicate that psychosis no longer needs to be seen as a sign of illness but can be viewed as one way of dealing with a crisis, and after this crisis, many or most people are capable of returning to their active social life. And given so few actually need psychosis drugs, we can ask whether our understanding of the underlying problem needs to change. New ways of thinking about psychoses have emerged through the introduction of open dialogues. Does this mean that we should re-think the way psychiatric services are organized? Instead of primarily focusing on having control over the symptoms and removing the symptoms as rapidly as possible, the attention could be on organizing meetings for those involved, including family members and other relevant individuals from the personal social network and from professional networks. It seems to mean that in these meetings we should be more interested in generating dialogues by following what family members are saying rather than in planning interventions aimed at change in the patient or in the family. If so, the training of professionals should be restructured to include these new aspects: not only to read books about medical interventions but also to reflect upon the philosophy of our human views and how to generate dialogue and listen to people instead of dominating the therapeutic process.

7.4 References

Aaltonen, J., Seikkula, J., Alakare, B., Haarakangas, K., Keränen, J. and Sutela, M. (1997). Western Lapland Project: A Comprehensive Family- and Network Centered Community Psychiatric Project. *ISPS Abstracts and Lectures 12 -16 October 1997, p. 124.* London: ISPS.

Aaltonen, J., Seikkula, J., & Lehtinen, K. (2011). Comprehensive open-dialogue approach I: Developing a comprehensive culture of need-adapted approach in a psychiatric public health catchment area the Western Lapland Project. *Psychosis*, 3, 179-191

Alanen, Y. (1997). *Schizophrenia. Its Origins and Need-Adapted-Treatment.* London: Karnac Books.

Alanen, Y., Lehtinen, K., Räkköläinen, V. and Aaltonen, J. (1991). Need-adapted treatment of new schizophrenic patients: Experiences and results of the Turku Project. *Acta Psychiatrica Scandinavica,* 83, 363- 372.

Andersen, T. (1995). Reflecting processes. Acts of informing and forming. In S. Friedman (Ed.) *The Reflective Team in Action* (pp. 11- 37). New York: Guilford.

Bakhtin, M. (1984). *Problems of Dostojevskij's Poetics*: Theory and History of Literature (vol. 8). Manchester: Manchester University Press.

Bergström, B., Alakare, B., Aaltonen, J., Mäki, P., Köngäs Saviaro, P., Taskila, J. & Seikkula, J. (2017) The long-term use of psychiatric services within the Open Dialogue treatment system after first-episode psychosis. *Psychosis,* 9, 310-32

Falloon I., Roncone, R., Giosiu, P., Mela, M., Held, T., Coverdale, J. (2002). The Integration of Innovative Biomedical and Psychosocial Strategies into Clinical Practice. In: Kashima H.,

Falloon I.R.H., Mizuno M., Asai M. (eds) *Comprehensive Treatment of Schizophrenia.* Keio University International Symposia for Life Sciences and Medicine, vol 8. Springer, Tokyo.

Friis, S., Larsen, T. K. and Melle, I. (2003). Terapi ved psykoses. Tidsskr Nor Laegeforen, 123, 1393.

Gergen, K. (1994). *Realities and Relationships. Soundings in Social Construction.* Cambridge, MA: Harvard University Press.

Gergen, K. (1999). *An Invitation to Social Construction.* London: Sage.

Haarakangas, K. (1997), Hoitokokouksen äänet. [The voices in treatment meeting. A dialogical analysis of the treatment meeting conversations in family-centred psychiatric treatment process in regard to the team activity. English Summary.] *Jyväskylä Studies in Education, Psychology and Social Research,* 130, 119 - 126.

Keränen, J. (1992). The choice between outpatient and inpatient treatment in a family centred psychiatric treatment system. English summary. *Jyväskylä Studies in Education, Psychology and Social Research,* 93, 124 - /129.

Kiviniemi, M (2014) *Mortality, disability, psychiatric treatment and medication in first-onset schizophrenia in Finland: the register linkage study.* Acta Universitatis Ouluensis. D, Medica.

Lenior, M., Dingemans, P., Linszen, D., de Haan, L. and Schene, A. (2001). Social functioning and the course of early-onset schizophrenia: Five-year follow-up of a psychosocial intervention. *British Journal of Psychiatry,* 179, 53 - 58.

Lehtinen, V., Aaltonen, J., Koffert, T., Räkköläinen, V. & Syvälahti, E. (2000). Two-year outcome in first-episode psychosis treated according to an integrated model. Is immediate neuroleptisation always needed? *European Psychiatry,*15, 312–320.

Seikkula, J. (1991). Family-hospital boundary system in the social network. English summary. *Jyväskylä Studies in Education, Psychology and Social Research,* 80, 227 - 232.

Seikkula, J., Aaltonen, J., Alakare, B., Haarakangas, K., Keränen, J., & Sutela, M. (1995). Treating psychosis by means of Open Dialogue. In S. Friedman (Ed.), *The reflective team in action: Collaborative practice in family therapy* (pp. 62–80). New York: Guilford Press.

Seikkula, J., Alakare, B., Aaltonen, J., Holma, J., Rasinkangas, A., & Lehtinen, V. (2003). Open Dialogue approach: Treatment principles and preliminary results of a two-year follow-up on first episode schizophrenia. *Ethical Human Sciences and Services, 5*(3), 163–182.

Seikkula, J., Alakare, B., Aaltonen, J., Haarakangas, K., Keränen, J., & Lehtinen, K. (2006). 5 years experiences of first-episode non-affective psychosis in Open Dialogue approach: Treatment principles, follow-up outcomes and two case analyses. *Psychotherapy Research, 16*, 214–228.

Seikkula, J., Alakare, B., & Aaltonen, J. (2011). The comprehensive open-dialogue approach (II). Long-term stability of acute psychosis outcomes in advanced community care: The Western Lapland Project. *Psychosis, 3*, 192-204.

Svedberg, B., Mesterton, A., & Cullberg, J. (2001). First-episode non-affective psychosis in a total urban population: A 5-year follow-up. *Social Psychiatry*, 36, 332/337.

Voloshinov, V. (1996). Marxism and the philosophy of language. Cambridge, MA: Harvard University Press.

Chapter 8

Mentalizing through the early stages of the Psychosis Continuum

Martin Debbané
Developmental Clinical Psychology Research Unit, Faculty of Psychology and Educational Sciences, University of Geneva, Switzerland; University College London, UK

Elodie Toffel
University of Geneva, Switzerland

Recent case reports depicting fruitful attempts to provide mentalization-based psychotherapy to individuals suffering from psychosis have fuelled interest in further specifying the rationale for clinical work focusing of the psychotic patient's ability to mentalize (Brent, 2009, Brent, 2015, Debbané et al., 2016a), within and beyond the clinical realm of psychosis. In this chapter, we provide an overview of a mentalization-informed perspective on the wider continuum of psychotic phenomena. We first proceed by exposing two key assumptions: 1) Psychotic phenomena can be understood from a continuum perspective, heuristically divided into four putative stages; 2) Linking this continuum perspective to a developmental framework affords the opportunity to examine the factors promoting or protecting against the evolution of the disorder from premorbid risk to clinical psychosis. From this basis, we further elaborate on the evidence that trauma and attachment insecurity constitute two key characteristics that hamper the developmental course of mentalizing. We frame mentalization as a key developmental acquisition that protects against the exacerbation of risk for psychosis. Finally, we conclude by formulating a possible clinical account of the relationship between trauma, attachment and mentalizing.

8.1 Key assumptions: the psychosis continuum from a developmental perspective

Two key assumptions are necessary to keep in mind. First, from a phenomenological point of view, we, along with others (Linscott and van Os, 2010, Debbané and Barrantes-Vidal, 2015), suggest that psychotic phenomena are expressed along a continuum ranging from non-clinical to more serious clinically psychotic manifestations. For a taxometric view of the psychosis continuum, please refer to Beauchaine et al. (2008). The continuum ranges from benign to moderately dissociative or pre-clinical symptoms, evolving towards more severe experiences of psychotic nature, which can eventually lead to significant withdrawal from reality. Assuming a continuum perspective offers the opportunity to gain knowledge on the nature of psychosis not only from deeply disturbed individuals, but also from individuals reporting subclinical symptoms of psychosis. This opportunity is key to disentangling effects that originate from psychosis itself and its psychosocial consequences in the contemporary society (hospitalization, medication, declining social role, etc.), from effects that play a role in the emergence of the disorder.

The second key assumption, based on the observation that schizophrenia spectrum disorders usually emerge during late adolescence and early adulthood (Kessler et al., 2007), assumes that a developmental perspective is critical to understanding the borders between non-clinical and clinical phenomena, or in other words, how psychotic-like phenomena turns into full blown psychosis. This leads us to a focus on the developmental transitions between four putative stages of the continuum, where pathophysiology gains ground from premorbid to clinical high-risk states, and then from these states to full-blown episode(s) of psychosis, with a possibility thereafter of trajectories of repetitive relapses and/or chronic schizophrenia.

Figure 8.1. Stages along the psychosis continuum

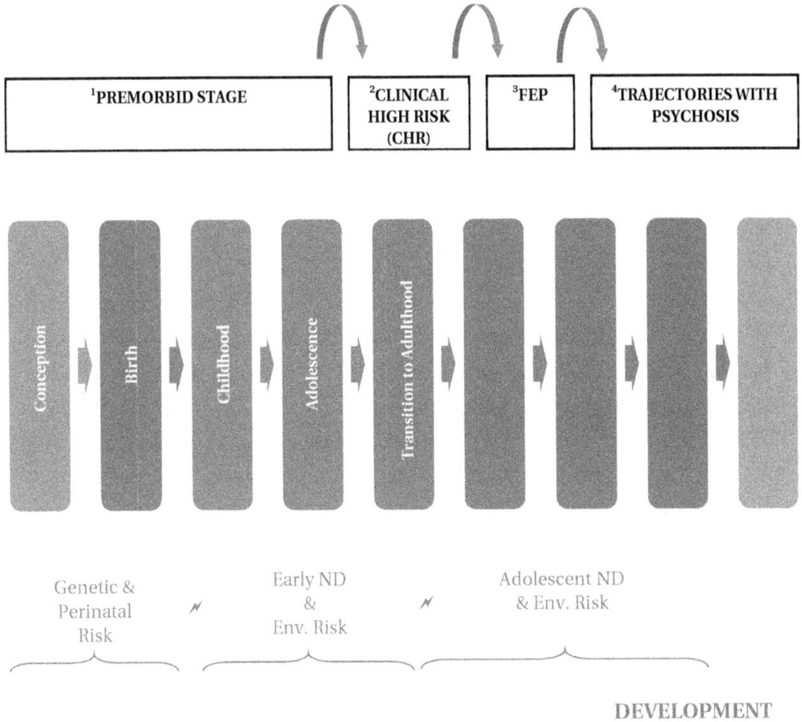

From such a developmental perspective, let us consider the four different stages along the psychosis continuum (Figure 8.1). In the premorbid stage, several risk factors such as genetic and perinatal risks around conception and birth, childhood stress factors (e.g. environmental risk, early neurodevelopment) and adolescence stress factors including bullying, use of substances, but also normal stress factors associated to developmental tasks (e.g. socialization, intimate relationships, etc.), can combine to augment risk to psychopathological transition. Following the premorbid stage, the second stage is best defined by the Clinical High Risk (CHR) concept (Fusar-Poli et al., 2013). Many studies have shown that CHR symptoms significantly predict transition to the First Episode of Psychosis (FEP; stage 3) (Schultze-Lutter et al., 2015). In figure 8.1., the red arrows represent the epochs where people effectively transition from one stage to the next stage. For each arrow, only a small proportion in individuals actually transition; for example, 9 out of 10 individuals at genetic risk for psychosis (with one first-degree relative with schizophrenia) will not develop any psychotic disorder. Therefore, factors sustaining transition, but also protecting against transition, are critical to identify. Indeed, such a developmental perspective leads us to both learn about factors of transition to psychosis as well as resilience factors (e.g. why do the people not develop the disorder). Before we consider such resilience factors, namely mentalization, allow us now to take a closer look at the critical CHR stage, where indicated preventive treatment may already be warranted.

8.2 Clinical High Risk (CHR)

Research in the past three decades has put forward two approaches to CHR states, namely the Ultra High Risk approach (UHR) and the basic symptoms approach (Schultze-Lutter, 2009). Following the UHR concept, McGlashan developed the Structured Interview for Prodromal Syndromes (SIPS) (McGlashan, 2001). This interview is designed to evaluate unusual thought content, delusional ideas (P1), suspicion, persecutory ideas (P2), grandiosity (P3), perceptual abnormalities / hallucinations (P4) and disorganized communication (P5) as subclinical symptoms of ultra high risk to develop psychosis. Their subclinical nature refers to the conserved ability of individuals to criticize their symptoms, yet still report emotional and functional disturbances resulting from these symptoms. As can be appreciated here, the UHR approach predominantly assesses the positive dimensions of psychotic symptomatology, and as a result may be more sensitive to imminent risk for psychosis. Between 20 and 35% of individuals who are diagnosed with a UHR risk state go on to develop a psychotic disorder, the majority of cases within 24 months of the initial prodromal diagnosis (Nelson et al., 2013).

The complementary approach to assess CHR states relates to basic symptoms (BS), which consist in subjectively reported sense of disturbance in affective, cognitive and or perceptual processes (Schultze-Lutter et al., 2016). The diagnosis of symptoms, therefore, relies entirely on the subjective report of the patient (not the observation of the clinical assessor), the frequency of the symptom, and the novel, ego-dystonic nature of the experience it engenders for the subject (Schultze-Lutter, 2009). The basic symptoms assessed are cognitive and perceptual in nature, including but not restricted to thought interference, unstable ideas of reference, derealisation, visual or auditory echoes. It is required that these symptoms occur at least weekly during the past three months to be considered as true basic symptoms. Indeed, while anyone reading this text might have experienced such symptoms occasionally, it is only when the individual reports an emergent, weekly occurrence of the symptom that it may signify a CHR state for psychosis.

It is possible, and clinically useful, to combine both assessments (UHR + BS) when evaluating CHR for psychosis. A study carried out by Schultze-Lutter, Klosterkötter & Ruhrmann (2014) examined the value of combining UHR and BS criteria to predict cumulative conversion rates of participants followed up forty-eight months after initial risk CHR diagnosis (Schultze-Lutter et al., 2014) (Figure 8.2.). For individuals who exclusively presented the UHR criteria or the BS COGDIS criteria, the authors reported a convergence proportion around 30%. However, those diagnosed with both subclinical attenuated positive symptoms and cognitive basic symptoms, 70% of them converged to psychotic disorders. It thus appears that these clinical tools, in combination, may decisively help the clinician assess risk for psychosis.

Figure 8.2. Risk for psychosis

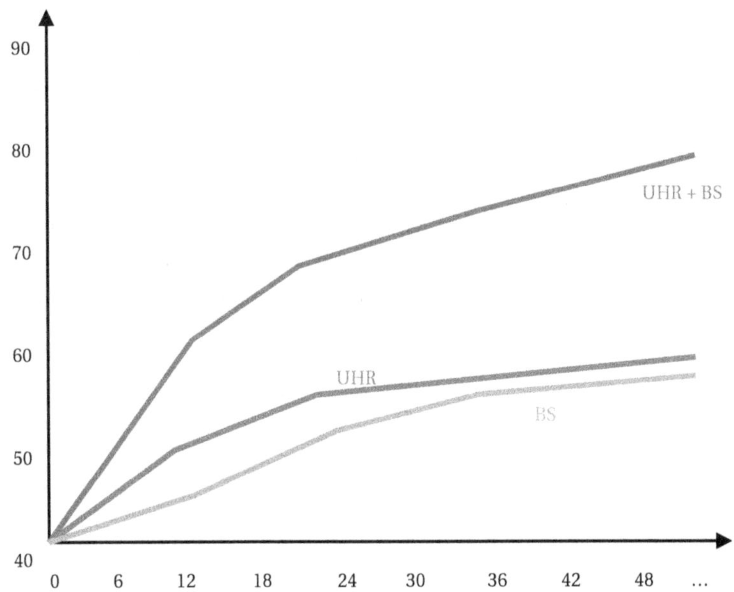

Returning to our developmental perspective, it becomes clearer how the idea of catching the pathology as it unfolds, during its pre-clinical course, is becoming more of a reality for practitioners and researchers alike. It further fuels the hope that intervening in the early stages may reduce the conversion rate to psychotic disorders; if psychosis was nevertheless to set in, early treatment may have carried an effect of reducing the functional disability associated with the clinical course of psychosis. Indeed, the greatest clinical challenge today is not so much harnessing psychotic symptoms, but rather enabling individuals to regain a functional role in the interpersonal, professional and social domains, a problem we now discuss as the symptom-disability gap.

8. 3. The symptom-disability gap: hinting towards a crucial role for social cognition in CHR states

The expression "symptom-disability gap" captures our current difficulty in treating individuals suffering from a psychotic disorder: while significant therapeutic gains may be achieved through symptomatic remission, our therapeutic effects (both psychotherapeutic and psychopharmacological) most often fail to impact recovery in the domains of interpersonal and social functioning (Birchwood et al., 2013). As depicted in Figure 8.3., disability in the functional domain increases (blue line) during the early premorbid and CHR states, where the first pre-clinical signs emerge. Disability then follows a stable trajectory while symptoms peak in severity around the onset of psychosis. Unfortunately, as therapeutic efforts start effecting positive signs by decreasing symptomatology (green line), in a majority of patients, the functional disability tends to persist, showing its most tenacious and pernicious effects in the interpersonal and social functioning domains (Carbon and Correll, 2014). The gap between the symptomatic and functional courses in psychotic disorders represents a clinical puzzle: why would functional disability persist in the absence of psychotic symptoms?

Figure 8.3

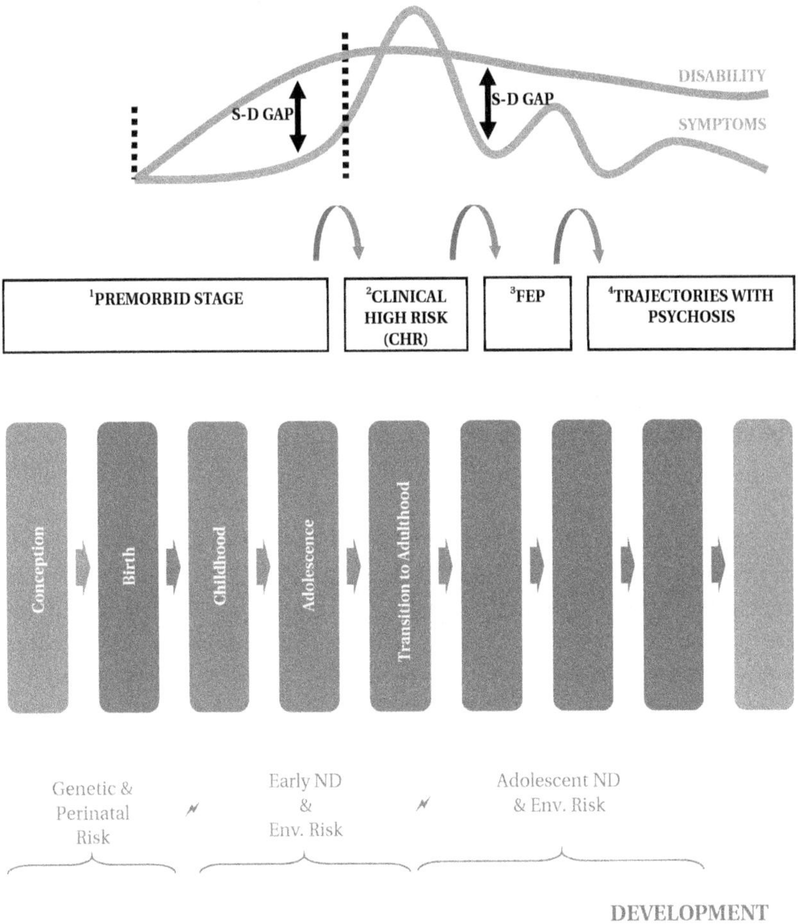

Obviously, a plethora of hypotheses may be offered to solve the clinical riddle of the symptom-disability gap in psychosis. Following our developmental approach, together with the evidence that subtle social cognitive impairments are already present during adolescence and early adulthood (Olsen and Rosenbaum, 2006), we hypothesize that the symptom-disability gap may at least, in part, be linked to this critical developmental window of social cognition. This period is now recognized as critical for cerebral specialization of higher order cognitive processes (Badoud et al., 2016), namely executive functions underlying self-regulation, inhibition and planification (Steinberg, 2008), as well as social cognitive processes, such as mentalizing, governing the interpretation of behaviours as a function of different contexts. Here we note that the earliest manifestations of psychotic-like experiences are associated with impairments in theory of mind i.e. mentalizing others' mental states. This period is also key for developing socialization skills, on which depend both the integration into social groups as well as the development for autonomous navigation of the social world. We further note that social withdrawal (Mason et al., 2004, Addington et al., 2008) is a key symptom in high risk for psychosis profile and may indeed constitute an obstacle to socialization. Last but not least, attachment plays an important role during adolescence, especially in the interpretation of social feedback (Vrticka et al., 2014) and the formation of key-dependence relationships, which will form the basis for adult emotional growth. This is relevant to our hypothesis, especially since research shows that social relationships deplete in the face of emerging psychosis (Gayer-Anderson and Morgan, 2013). Within this context of adolescent and early adulthood developmental tasks, driven by both genetic and environmental influences (Debbané and Barrantes-Vidal, 2015), the premorbid and CHR pathophysiological path represent potent obstacles to socio-emotional growth (Lincoln et al., 2017).

Vulnerability to psychosis, even in its most premorbid form, is associated with poorer executive functioning (Reichenberg et al., 2010) and difficulty in interpreting self and others' behaviours (i.e. *mentalizing*) as driven by intentional mental states (for a review, see Debbané et al. 2016b). The pervasiveness of impairment in higher order cognitive processes (HOC) along the psychosis continuum make them likely contributors to the symptom-disability (Debbané et al., 2016a). For the purpose of this chapter, we will focus on the HOC processes subsumed under the concept of mentalizing, that is the suite of social cognitive processes contributing to interpreting self and others' behaviours as driven by beliefs, intentions, desires, emotions, motives, and the like.

8.4 What is mentalizing?

Mentalizing is a higher order cognitive process that allows us to see ourselves from the outside and others from the inside, and to have in mind the *mental quality of experience* that is concealed in actions and behaviours (Fonagy et al., 2002). This process unfolds as a developmental acquisition. It critically involves imagination, which is employed to make sense of human behaviour (self-other) as based on mental states such as beliefs, desires, goals, feelings, etc. The learning of mentalizing starts with the initial attachment bond (Fonagy et al., 1996). Attachment theory and research have revealed how the early caregiving relationships will, to a degree that can be variable but almost always significant, determine the way we relate to significant others during our development (Fonagy et al., 2002). Through the establishment of internal working models of self and others, we may more or less trust what people have to offer, both in emotional as well as in cognitive terms (Corriveau et al., 2009). In other words, attachment relationships lead to trust, and perhaps more profoundly to have faith that what others conceal, their thoughts and emotions, may be trustworthy and beneficial to our own development (Fonagy and Allison, 2014). This awareness comes to modulate the engrained vigilance (Sperber et al., 2010) towards others in favour of approach and cooperation (Fonagy et al., 2015). Trust and interest in others' mental states, that is the curiosity to find out about mental states, is easily disrupted by significant arousal, and crucially disrupts interpersonal and social functioning as demonstrated in personality disorders (PD) (Fonagy and Allison, 2014). In this context, because the developmental history of individuals along the continuum is fraught with trauma and attachment insecurity, it is likely that the curiosity to think about mental states is impaired in psychotic disorders, and most probably along the different stages of the psychotic continuum (Debbané et al., 2016b). Before we consider this developmental history and its links to mentalizing along the psychosis continuum, let us briefly clarify some conceptual issues that may have arisen in the reader's mind.

8.4.1 Mentalizing and its conceptual cousins

The mentalizing process arises out of binding information, and continuously considering information coming from a number of different sources (Fonagy and Luyten, 2009). Mentalization as a clinical concept has been criticised as being too vague; indeed, it stands at the crossroad of other, more clearly defined concepts, such as theory of mind, mindfulness, psychological mindedness, empathy and affect consciousness (Choi-Kain and Gunderson, 2008). However, mentalization as an integrative concept captures more of the overall process involved when, for example, one uses both empathy, theory of mind, and mindfulness to understand an interpersonal event. In this way, mentalizing is preferable to mentalization when we wish to address the process. We argue that mentalizing as a clinical construct is also more useful clinically because it is less restrictive. For example, in contrast to mindfulness that typically focuses on slow, controlled, self-directed processes, with little theory devoted to interpersonal, automatic, other-oriented processing, mentalizing embraces how mindfulness connects with other HOC such as theory of mind, or compassion, at the risk of being too vague as a construct and with the difficulty of quantifying mentalizing. Indeed, whether measuring mentalization skills, or mentalizing processes, measurement still requires much refinement. As we will see next, much of what we know about mentalizing in psychosis comes from studies using measurements designed for conceptual cousins.

8.4.2 Mentalization and Psychosis

In this section, we consider the evidence that mentalizing impairments can be observed from the first stage along the psychosis continuum to its final, most pathological state. Given the paucity of measures in mentalizing, most of the evidence originates from research on theory of mind, reality and self-monitoring, insight or perspective taking, in samples that vary from the premorbid, to CHR, to FEP and full blown schizophrenia (for a review, see Debbané et al. 2016b).

Mentalizing impairments are observable in youths with presenting "premorbid" risk for psychosis. The earliest associations between disturbances in the process of mentalizing and preclinical psychotic manifestations can be found in the studies performed by Clemmensen and colleagues (Clemmensen et al., 2015, Clemmensen et al., 2014), with samples of children aged 9-12 years old reporting auditory verbal hallucinations (AVHs). Using a mental state attribution task, the authors observe that children reporting AVHs generated overly mentalized explanations for behaviours, which the authors characterized as hyper-theory-of-mind. This concept is close to that of hyper-mentalizing (Sharp et al., 2013), which can also be observed in adolescent teenagers with borderline features. What the excessive use of mentalizing translates is perhaps an early, subtle deflection in the developmental process of mentalizing, which exposes the youth to experience social interactions as very real, complex, and potentially overwhelming. As such, hyper-mentalizing may consist in a trans-diagnostic marker of social cognitive vulnerability that undermines resilience in the face of adolescent social stress.

Adolescents can also experience transitory auditory verbal hallucinations. Sandra Escher and Marius Romme (2002) from the Netherlands followed such adolescents to demonstrate that most of AVHs disappear without any real therapeutic treatment. It may be that the development of social cognitive processes such as mentalizing during adolescence, may favour an innocuous outcome for AVH. Along these lines, Bartels-Velthuis et al. (2011) observe that adolescents with good theory of mind skills are much less likely to develop delusional thinking following their early auditory verbal hallucinations. These results are in line with the "protective" role mentalizing can play to preserve against progression of risk to clinical states.

Later on during adulthood, phenomena such as AVH can exist in the absence of a psychotic disorder. In fact, communities of people experiencing such phenomena have gathered around Europe and elsewhere as "the Voice Hearers" movement (https://www.hearing-voices.org/about-us/). A recent study involving voice hearing subjects by Peters et al. (2016) showed that such individuals have acquired superior mindfulness skills during their development, in comparison to both non-voice-hearing controls as well as controls with a psychotic disorder. This study illustrates how we may conceive of mentalizing as a protective factor, which helps regulate the impact of psychotic phenomena within one's subjective experience. Mentalizing helps put psychotic symptoms into perspective, thereby decreasing the psychological bearing symptoms may hold on one's interpretation of reality.

Taken together, the results with children, adolescents and adults in the premorbid stage of psychosis, experiencing non-pathological symptoms of psychosis, may indicate that different components of mentalizing represent protective factors that prevent or attenuate the pathological valence of such psychotic experiences. Impairments in mentalizing do not represent "risk factors" per se, as they would be rather unspecific to psychotic psychopathology. Rather, good mentalizing represents a generic protective factor for the onset of psychopathology (Fonagy et al., 1994). In the case of genetic and environmental vulnerability to psychosis mentalization may, to an extent that remains to be demonstrated, safeguard against transitioning to a clinical state of psychosis.

8.4.3 Trauma in psychosis

The prospective studies on adolescents experiencing AVH, mentioned in the previous section, further point out an important aspect of the participants' developmental history: 86.3% reported experienced traumatic and stressful life events (Escher et al., 2002). Furthermore, prospective evidence links traumatic and stress-inducing events in these youths with the persistence and severity of auditory hallucinations, the emergence of new hallucinatory experiences in early adolescence, and the secondary development of delusional ideation (Bartels-Velthuis et al., 2012). Furthermore, the exposure to these traumatic experiences early in life appears to be highly predictive of the development of psychotic experiences over time, with specific associations reported in relation to severity of trauma (Kelleher et al., 2013) and to childhood abuse characterized by harmful intentions (Arseneault et al., 2010). As these studies attest, trauma can become intimately tied to the development of risk for psychosis, and this brings us to consider the link between attachment, trauma and mentalizing in the developmental framework of psychotic disorders. Indeed, trauma is consistently reported at every stage of the psychosis continuum. In samples of individuals with clinical high risk, approximately 90% of participants with CHR report experiences of maltreatment and neglect in early childhood (Bechdolf et al., 2010). In studies involving individuals diagnosed with schizophrenia, rates as high as 82% (Duhig et al., 2015) and 85% (Larsson et al., 2013) signal the relevance of trauma in developmental history. Fisher & al (2011) examined the reliability of retrospective assessments of childhood trauma by adults suffering from schizophrenia and found that their reporting seems to be quite concurrently valid with parental bonding measures as well as clinical case notes when they consulted as children (Fisher et al., 2011). Overall, it has become impossible to ignore the relevance of trauma in the developmental history of psychotic disorders.

8.4.4 Attachment along the stages of psychotic unfolding

Concurrently to research on trauma and psychosis, attachment research with individuals suffering from schizophrenia has brought clinically relevant insights to understanding the grounds of perturbations in interpersonal relationships so characteristic of the disorder. High rates of insecure attachment have been reported among patients experiencing a first psychotic episode (MacBeth et al., 2011) (Figure 8.1.: FEP), as well as those suffering from chronic psychosis (Mickelson et al., 1997). Studies also found associations between adult attachment avoidance with both positive and negative symptomatology, with more modest evidence for an association between positive symptoms and attachment anxiety/preoccupation (Berry et al., 2007b, Gumley et al., 2014). Avoidant attachment has most consistently been linked with reduced help-seeking behaviours and poor use of treatment, diminished therapeutic alliance as well as increased service disengagement (Debbané et al., 2016b). By contrast attachment security acts as a protective factor, because it favours help-seeking and treatment engagement, as well as stronger therapeutic alliance among individuals with established schizophrenia (Berry et al., 2007b).

More recently, research during the premorbid and CHR phases has let us appreciate how attachment may influence the developmental trajectories of individuals with vulnerability to develop psychosis. In individuals meeting the CHR criteria, as many as 80% of CHR young adults showed evidence of insecure attachment (Gajwani et al., 2013). In a sample of 31 participants, Quijada et al. (2012) found that most participants were fearfully attached (71%), followed by preoccupied (16.1%) and dismissing attachment (6.4%) (Quijada et al., 2012). In a follow-up study, the same authors report that patients with lower levels of fearful and dismissing/avoidant styles at baseline experienced better clinical outcomes (Quijada et al., 2015). For individuals within the premorbid phase, measures of schizotypy are commonly employed to reflect vulnerability to develop schizophrenia spectrum disorders (Debbané et al., 2016b). Consistently with evidence from more clinically advanced samples, insecure avoidant attachment is associated with schizotypy (Sheinbaum et al., 2013, Sheinbaum et al., 2014). Cognitive perceptual symptoms, or positive schizotypy, has been associated with both attachment avoidance and attachment anxiety, while interpersonal symptoms, or negative schizotypy, such as social withdrawal, flat affect and social anxiety, appears to more directly relate to attachment avoidance (Berry et al., 2007a).

To recapitulate, the literature linking insecure attachment to psychotic phenomena appears to underline two main findings. The first developmental observation suggests that already during childhood and adolescent development, attachment insecurity is associated with emergent psychotic phenomena. Second, the clinical course of psychosis, from risk states to diagnosable syndromes, may also be partly determined by attachment, to the degree that security and avoidance play a role in engaging with carers and mental health professionals. From an attachment point of view, it may be that when positive internal working models of self and/or others are still operating in individuals with psychosis, these models open the possibility for trusting that carers may make a difference in one's life thereby sustaining help-seeking and treatment adherence in those patients.

8.4.5 From insecure attachment to psychotic states of mind

Following the previous sections on trauma and attachment along the psychosis continuum, we attempt a clinical formulation linking to the developing process of mentalizing. To begin, we know that trauma during childhood represents a non-specific risk factor for psychopathology, and affects in important ways the development of the brain, especially the stress regulation system. The pathogenic effects of trauma, as well as insecure attachment, have been documented both at the psychological and neurodevelopmental level (McCrory et al., 2017). At the interpersonal level, trauma implicates the early caregiving environment in at least two possible forms. In the majority of circumstances, evidence shows that harmful intentions and actions of abused children sadly originate from caregivers (Sedlak et al., 2010). In other instances, where abuse is not perpetrated within family, caregivers cannot but feel profound guilt for not having kept their child from harm's way, which can affect how they relate to their child (Liotti, 2017). In this sense, trauma most often inserts itself within the attachment system, similarly to a slow-acting poison with potentially devastating effects. In clinical terms, abused children face and potentially fall into an affective double bind, which we consider here.

If we remind ourselves of the typical workings of a child's attachment system (Fonagy et al., 2002), we know that threat or danger leads the child to automatically trigger attachment needs, which is adaptive to activate safety-seeking behaviours and ensure security. In other words, threat triggers seeking (approach behaviour) the child's caregiver. The abused child, however, faces incongruity when activated by a threat, as s/he is automatically compelled to approach the caregiver that can actually be a source of neglect and/or maltreatment. How does the child manage this superposition in source of threat and safety? Giovanni Liotti and others have observed that quite often, children dissociate in the face of abusive threats. Recent research demonstrates that dissociation mediates the relationship between trauma and later psychological distress (Ensink et al., 2017). In such circumstances, the attachment system of the child will often developmentally skew into profound insecurity, or even disorganization (Liotti, 2004). Hypervigilant and/or avoidant behaviours of the child towards the caregiver are common responses that translate the child's attempts to predict or control threatening states of mind within the caregiver. Most drastically, in a minority of children, a shutdown of attachment needs may ensue, which will carry profound and long-lasting consequences on the child's psychological development. From a mentalization-informed point of view, the abused child's subjective experience of trauma faces the paradoxically confusing mental states of the parent, who can hold both neglectful/malevolent intentions, inextricably entangled with the same parent's caring intentions for the child. In the face of this paradox, the very development of mentalizing during childhood and adolescence can be compromised.

8.5. Conclusion

From these considerations, we can better imagine how trauma and attachment insecurity influence a developmental trajectory that strays away from investing in mentalizing skills, and that promotes a dissociative coping style in individuals having experienced hostile and sometimes traumatic early environments. Adding to this genetically endowed vulnerability for psychosis, the premorbid path of the continuum is opened through a number of different neurobiological pathways (HPA axis dysfunction and dopamine dysregulation, but also oxytonergic reduction, oxidative stress damage and vulnerability to neuroinflammation (Debbané et al., 2016b), which will interact with adolescent stress maturation factors to potentially evolve into clinical expression (Brent and Fonagy, 2014). Along this trajectory, the compromised social cognitive development of mentalizing acts as a facilitator for the pathophysiology to evolve along the continuum of risk into states demanding clinical attention. If this account reflects at least part of the process of emerging psychosis, then our knowledge on the developing trajectories of mentalization may become extremely useful in designing indicated preventive interventions to safeguard those at risk. We also acknowledge the hypothetical stature of the approach presented along this chapter. There remains a lack of empirical instantiating these hypotheses, and we anticipate that results will come to modify some of our positions. We need to include other potential protective factors of psychosis, such as executive functions, as well as better delineate how the bottom-up, early-embodied processes may contribute to the later maturing higher cognitive processes such as mentalizing. The field is wide open and ready for research trying to concretely bridge the cerebral and psychological processes into a developmental model that helps us further understand the enigmatic nature of psychotic onset and its clinical course.

8.6 References

Addington, J., Penn, D., Woods, S. W., Addington, D. & Perkins, D. O. (2008). Social functioning in individuals at clinical high risk for psychosis. *Schizophrenia Research,* 99, 119-24.

Arseneault, L., Cannon, M., Fisher, H. L., Polanczyk, G., Moffitt, T. E. & Caspi, A. (2010). Childhood Trauma and Children's Emerging Psychotic Symptoms: A Genetically Sensitive Longitudinal Cohort Study. *American Journal of Psychiatry.*

Badoud, D., Speranza, M. & Debbané, M. (2016). Vers un modèle du développement de la mentalisation à l'adolescence - Towards a developmental model of the mentalization dimensions during adolescence. *Revue Québécoise de Psychologie,* 49-68.

Bartels-Velthuis, A. A., Blijd-Hoogewys, E. M. & van Os, J. (2011). Better theory-of-mind skills in children hearing voices mitigate the risk of secondary delusion formation. *Acta Psychiatrica Scandinavica,* 124, 193-7.

Bartels-Velthuis, A. A., van de Willige, G., Jenner, J. A., Wiersma, D. & van Os, J. (2012). Auditory hallucinations in childhood: associations with adversity and delusional ideation. *Psychological Medicine,* 42, 583-93.

Beauchaine, T., Lenzenweger, M. F. & Waller, N. G. (2008). Schizotypy, taxometrics, and disconfirming theories in soft science: Comment on Rawlings, Williams, Haslam, and Claridge. *Personality and Individual Differences,* 44, 1652-1662.

Bechdolf, A., Thompson, A., Nelson, B., Cotton, S., Simmons, M. B., Amminger, G. P., Leicester, S., Francey, S. M., McNab, C., Krstev, H., Sidis, A., McGorry, P. D. & Yung, A. R. (2010). Experience of trauma and conversion to psychosis in an ultra-high-risk (prodromal) group. *Acta Psychiatrica Scandinavica,* 121, 377-84.

Berry, K., Band, R., Corcoran, R., Barrowclough, C. & Wearden, A. (2007a). Attachment styles, earlier interpersonal relationships and schizotypy in a non-clinical sample. *Psychology and Psychotherapy,* 80, 563-76.

Berry, K., Barrowclough, C. & Wearden, A. (2007b). A review of the role of adult attachment style in psychosis: Unexplored issues and questions for further research. *Clinical Psychology Review,* 27, 458-475.

Birchwood, M., Connor, C., Lester, H., Patterson, P., Freemantle, N., Marshall, M., Fowler, D., Lewis, S., Jones, P., Amos, T., Everard, L. & Singh, S. P. (2013). Reducing duration of untreated psychosis: care pathways to early intervention in psychosis services. *British Journal of Psychiatry,* 203, 58-64.

Brent, B. (2009). Mentalization-based psychodynamic psychotherapy for psychosis. *Journal of Clinical Psychology,* 65, 803-14.

Brent, B. & Fonagy, P. (2014). A Mentalization-Based Treatment Approach to Disturbances of Social Understanding in Schizophrenia. In: Lysaker, P. H., Dimaggio, G., Brüne, M. (ed.) *Social Cognition and metacognition in Schizophrenia.* San Diego, Elsevier, pp. 245-247.

Brent, B. K. (2015). A mentalization-based approach to the development of the therapeutic alliance in the treatment of schizophrenia. *Journal of Clinical Psychology,* 71, 146-56.

Carbon, M. & Correll, C. U. (2014). Clinical predictors of therapeutic response to antipsychotics in schizophrenia. *Dialogues in Clinical Neuroscience,* 16, 505-24.

Choi-Kain, L. W. & Gunderson, J. G. (2008). Mentalization: ontogeny, assessment, and application in the treatment of borderline personality disorder. *American Journal of Psychiatry,* 165, 1127-35.

Clemmensen, L., van Os, J., Drukker, M., Munkholm, A., Rimvall, M. K., Vaever, M., Rask, C. U., Bartels-Velthuis, A. A., Skovgaard, A. M. & Jeppesen, P. (2015). A Longitudinal General Population Birth Cohort Study of Psychotic Experiences, Hyper-Theory of Mind, and Putative Antecedents and Correlates in Preadolescence. *Schizophrenia Bulletin,* 41, S136-S137.

Clemmensen, L., van Os, J., Skovgaard, A. M., Vaever, M., Blijd-Hoogewys, E. M. A., Bartels-Velthuis, A. A. & Jeppesen, P. (2014). Hyper-Theory-of-Mind in Children with Psychotic Experiences. *Plos One,* 9. https://doi.org/10.1371/journal.pone.0113082

Corriveau, K. H., Harris, P. L., Meins, E., Fernyhough, C., Arnott, B., Elliott, L., Liddle, B., Hearn, A., Vittorini, L. & de Rosnay, M. (2009). Young Children's Trust in Their Mother's Claims: Longitudinal Links with Attachment Security in Infancy. *Child Development,* 80, 750-761.

Debbané, M. & Barrantes-Vidal, N. (2015). Schizotypy from a developmental perspective. *Schizophrenia bulletin,* 41 Suppl 2, S386-95.

Debbané, M., Benmiloud, J., Salaminios, G., Solida-Tozzi, A., Armando, M., Fonagy, P. & Bateman, A. (2016a). Mentalization-Based Treatment in Clinical High-Risk for Psychosis: A Rationale and Clinical Illustration. *Journal of Contemporary Psychotherapy,* 46, 217-225.

Debbané, M., Salaminios, G., Luyten, P., Badoud, D., Armando, M., Solida Tozzi, A., Fonagy, P. & Brent, B. K. (2016b). Attachment, Neurobiology, and Mentalizing along the Psychosis Continuum. *Frontiers in Human Neuroscience,* 10, 406.

Duhig, M., Patterson, S., Connell, M., Foley, S., Capra, C., Dark, F., Gordon, A., Singh, S., Hides, L., McGrath, J. J. & Scott, J. (2015). The prevalence and correlates of childhood trauma in patients with early psychosis. *Australian & New Zealand Journal of Psychiatry,* 49, 651-9.

Ensink, K., Berthelot, N., Begin, M., Maheux, J. & Normandin, L. (2017). Dissociation mediates the relationship between sexual abuse and child psychological difficulties. *Child Abuse & Neglect,* 69, 116-124.

Escher, S., Romme, M., Buiks, A., Delespaul, P. & Van Os, J. (2002). Independent course of childhood auditory hallucinations: a sequential 3-year follow-up study. *British Journal of Psychiatry - Supplements,* 43, s10-8.

Fisher, H., Schreier, A., Zammit, S., Lewis, G. & Wolke, D. (2011). Prospective Investigation of Psychological and Affective Pathways from Childhood Trauma to Psychosis-Like Experiences in a UK Birth Cohort. *European Journal of Psychotraumatology,* 2.

Fonagy, P. & Allison, E. (2014). The role of mentalizing and epistemic trust in the therapeutic relationship. *Psychotherapy (Chic),* 51, 372-80.

Fonagy, P., Gergely, G., Jurist, E. L. & Target, M. (2002). *Affect Regulation, Mentalization, and the Development of the Self,* New York, Other Press.

Fonagy, P., Leigh, T., Steele, M., Steele, H., Kennedy, R., Mattoon, G., Target, M. & Gerber, A. (1996). The relation of attachment status, psychiatric classification, and response to psychotherapy. *Journal of Consulting and Clinical Psychology,* 64, 22-31.

Fonagy, P. & Luyten, P. (2009). A developmental, mentalization-based approach to the understanding and treatment of borderline personality disorder. *Development and Psychopathology,* 21, 1355-81.

Fonagy, P., Luyten, P. & Allison, E. (2015). Epistemic Petrification and the Restoration of Epistemic Trust: A New Conceptualization of Borderline Personality Disorder and Its Psychosocial Treatment. *Journal of Personality Disorders,* 29, 575-609.

Fonagy, P., Steele, M., Steele, H., Higgitt, A. & Target, M. (1994). The Emanuel Miller Memorial Lecture 1992. The theory and practice of resilience. *Journal of Child Psychology and Psychiatry,* 35, 231-57.

Fusar-Poli, P., Borgwardt, S., Bechdolf, A., Addington, J., Riecher-Rossler, A., Schultze-Lutter, F., Keshavan, M., Wood, S., Ruhrmann, S., Seidman, L. J., Valmaggia, L., Cannon, T., Velthorst, E., De Haan, L., Cornblatt, B., Bonoldi, I., Birchwood, M., McGlashan, T., Carpenter, W., McGorry, P., Klosterkotter, J., McGuire, P. & Yung, A. (2013). The psychosis high-risk state: a comprehensive state-of-the-art review. *JAMA Psychiatry,* 70, 107-20.

Gajwani, R., Patterson, P. & Birchwood, M. (2013). Attachment: developmental pathways to affective dysregulation in young people at ultra-high risk of developing psychosis. *British Journal of Clinical Psychology* 52, 424-37.

Gayer-Anderson, C. & Morgan, C. (2013). Social networks, support and early psychosis: a systematic review. *Epidemiology and Psychiatric Sciences,* 22, 131-46.

Gumley, A. I., Taylor, H. E., Schwannauer, M. & MacBeth, A. (2014). A systematic review of attachment and psychosis: measurement, construct validity and outcomes. *Acta Psychiatrica Scandinavica,* 129, 257-74.

Kelleher, I., Keeley, H., Corcoran, P., Ramsay, H., Wasserman, C., Carli, V., Sarchiapone, M., Hoven, C., Wasserman, D. & Cannon, M. (2013). Childhood trauma and psychosis in a prospective cohort study: cause, effect, and directionality. *American Journal of Psychiatry* 170, 734-41.

Kessler, R. C., Amminger, G. P., Aguilar-Gaxiola, S., Alonso, J., Lee, S. & Ustun, T. B. (2007). Age of onset of mental disorders: a review of recent literature. *Current Opinion in Psychiatry,* 20, 359-64.

Larsson, S., Andreassen, O. A., Aas, M., Rossberg, J. I., Mork, E., Steen, N. E., Barrett, E. A., Lagerberg, T. V., Peleikis, D., Agartz, I., Melle, I. & Lorentzen, S. (2013). High prevalence of childhood trauma in patients with schizophrenia spectrum and affective disorder. *Comprehensive Psychiatry,* 54, 123-7.

Lincoln, S. H., Norkett, E. M., Frost, K. H., Gonzalez-Heydrich, J. & D'Angelo, E. J. (2017). A Developmental Perspective on Social-Cognition Difficulties in Youth at Clinical High Risk for Psychosis. *Harvard Review of Psychiatry,* 25, 4-14.

Linscott, R. J. & van Os, J. (2010). Systematic reviews of categorical versus continuum models in psychosis: evidence for discontinuous subpopulations underlying a psychometric continuum. Implications for DSM-V, DSM-VI, and DSM-VII. *Annual Review of Clinical Psychology,* 6, 391-419.

Liotti, G. (2004). Trauma, dissociation, and disorganized attachment: Three strands of a single braid. *Psychotherapy,* 41, 472-486.

Liotti, G. (2017). Conflicts between motivational systems related to attachment trauma: Key to understanding the intra-family relationship between abused children and their abusers. *Journal of Trauma & Dissociation,* 18, 304-318.

MacBeth, A., Gumley, A., Schwannauer, M. & Fisher, R. (2011). Attachment states of mind, mentalization, and their correlates in a first-episode psychosis sample. *Psychology and Psychotherapy-Theory Research and Practice,* 84, 42-57.

Mason, O., Startup, M., Halpin, S., Schall, U., Conrad, A. & Carr, V. (2004). Risk factors for transition to first episode psychosis among individuals with 'at-risk mental states'. *Schizophrenia research,* 71, 227-37.

McCrory, E. J., Gerin, M. I. & Viding, E. (2017). Annual Research Review: Childhood maltreatment, latent vulnerability and the shift to preventative psychiatry - the contribution of functional brain imaging. *Journal of Child Psychology and Psychiatry,* 58, 338-357.

McGlashan, T. (2001). *Structured Interview for Prodromal Syndromes (SIPS),* New Haven.

Mickelson, K. D., Kessler, R. C. & Shaver, P. R. (1997). Adult attachment in a nationally representative sample. *Journal of Personality and Social Psychology,* 73, 1092-1106.

Nelson, B., Yuen, H. P., Wood, S. J., Lin, A., Spiliotacopoulos, D., Bruxner, A., Broussard, C., Simmons, M., Foley, D. L., Brewer, W. J., Francey, S. M., Amminger, G. P., Thompson, A., McGorry, P. D. & Yung, A. R. (2013). Long-term follow-up of a group at ultra high risk ("prodromal") for psychosis: the PACE 400 study. *JAMA Psychiatry,* 70, 793-802.

Olsen, K. A. & Rosenbaum, B. (2006). Prospective investigations of the prodromal state of schizophrenia: review of studies. *Acta Psychiatrica Scandinavica* 113, 247-72.

Peters, E., Ward, T., Jackson, M., Morgan, C., Charalambides, M., McGuire, P., Woodruff, P., Jacobsen, P., Chadwick, P. & Garety, P. A. (2016). Clinical, socio-demographic and psychological characteristics in individuals with persistent psychotic experiences with and without a "need for care". *World Psychiatry,* 15, 41-52.

Quijada, Y., Kwapil, T. R., Tizon, J., Sheinbaum, T. & Barrantes-Vidal, N. (2015). Impact of attachment style on the 1-year outcome of persons with an at-risk mental state for psychosis. *Psychiatric Research,* 228, 849-56.

Quijada, Y., Tizon, J. L., Artigue, J., Kwapil, T. R. & Barrantes-Vidal, N. (2012). Attachment style predicts 6-month improvement in psychoticism in persons with at-risk mental states for psychosis. *Early Intervention in Psychiatry,* 6, 442-9.

Reichenberg, A., Caspi, A., Harrington, H., Houts, R., Keefe, R. S., Murray, R. M., Poulton, R. & Moffitt, T. E. (2010). Static and dynamic cognitive deficits in childhood preceding adult schizophrenia: a 30-year study. *American Journal of Psychiatry,* 167, 160-9.

Schultze-Lutter, F. (2009). Subjective symptoms of schizophrenia in research and the clinic: the basic symptom concept. *Schizophrenia Bulletin,* 35, 5-8.

Schultze-Lutter, F., Debbane, M., Theodoridou, A., Wood, S. J., Raballo, A., Michel, C., Schmidt, S. J., Kindler, J., Ruhrmann, S. & Uhlhaas, P. J. (2016). Revisiting the Basic Symptom Concept: Toward Translating Risk Symptoms for Psychosis into Neurobiological Targets. *Frontiers in Psychiatry,* 7, 9.

Schultze-Lutter, F., Klosterkotter, J. & Ruhrmann, S. (2014). Improving the clinical prediction of psychosis by combining ultra-high risk criteria and cognitive basic symptoms. *Schizophrenia Research,* 154, 100-6.

Schultze-Lutter, F., Michel, C., Schmidt, S. J., Schimmelmann, B. G., Maric, N. P., Salokangas, R. K., Riecher-Rossler, A., van der Gaag, M., Nordentoft, M., Raballo, A., Meneghelli, A., Marshall, M., Morrison, A., Ruhrmann, S. & Klosterkotter, J. (2015). EPA guidance on the early detection of clinical high risk states of psychoses. *European Psychiatry,* 30, 405-16.

Sedlak, A. J., Mettenburg, J., Basena, M., Peta, I., McPherson, K. & Greene, A. (2010). Fourth National Incidence Study of Child Abuse and Neglect (NIS-4), *Washington, DC: US Department of Health and Human Services.*

Sharp, C., Ha, C., Carbone, C., Kim, S., Perry, K., Williams, L. & Fonagy, P. (2013). Hypermentalizing in adolescent inpatients: treatment effects and association with borderline traits. *Journal of Personality Disorders,* 27, 3-18.

Sheinbaum, T., Bedoya, E., Ros-Morente, A., Kwapil, T. R. & Barrantes-Vidal, N. (2013). Association between attachment prototypes and schizotypy dimensions in two independent non-clinical samples of Spanish and American young adults. *Psychiatry Research,* 210, 408-13.

Sheinbaum, T., Kwapil, T. R. & Barrantes-Vidal, N. (2014). Fearful attachment mediates the association of childhood trauma with schizotypy and psychotic-like experiences. *Psychiatry Research,* 220, 691-3.

Sperber, D., Clement, F., Heintz, C., Mascaro, O., Mercier, H., Origgi, G. & Wilson, D. (2010). Epistemic Vigilance. *Mind & Language,* 25, 359-393.

Steinberg, L. (2008). A Social Neuroscience Perspective on Adolescent Risk-Taking. *Developmental review,* 28, 78-106.

Vrticka, P., Sander, D., Anderson, B., Badoud, D., Eliez, S. & Debbane, M. (2014). Social feedback processing from early to late adolescence: influence of sex, age, and attachment style. *Brain and Behavior,* 4, 703-20.

Chapter 9

A Mentalization-Based and Neuroscience-Informed Model of Severe and Persistent Psychopathology

Tobias Nolte
Wellcome Trust Center for Neuroimaging, University College London, UK; Anna Freud National Centre for Children and Families, London, UK

Chloe Campbell
Research Department of Clinical, Educational and Health Psychology, University College London, UK

Peter Fonagy
Anna Freud National Center for Children and Families, London, UCL; Research Department of Clinical, Educational and Health Psychology, University College London, UK

This chapter describes a recent reconceptualization of psychopathology associated with personality disorders (PD) that speaks to the conceptual and diagnostic challenges currently encountered in the field (Fonagy, Luyten, Allison, & Campbell, 2017a, 2017b[1]). It is an approach that integrates our thinking about mentalizing (i.e. understanding ourselves and others in terms of intentional mental states) and epistemic trust (i.e. openness to the reception of social communication that is personally relevant and of generalizable significance) with recent findings on the structure of both adult and child psychopathology (Caspi et al., 2014) and resilience (Kalisch, Müller & Tuscher, 2015). This chapter seeks to bring an additional dimension to this thinking through the elaboration of the neuroscientific developments that underpin this theoretical framework. In particular, we will highlight how advances in computational

[1] This chapter is based upon previous work by Fonagy, Luyten, Allison & Campbell (2017) as well as by Debanné and Nolte (2018).

neurosciences can be harnessed to contribute a mechanistic understanding of the relationships between higher order social cognition, mentalizing and epistemic trust in generating psychopathology.

In the first part of this chapter we will discuss an emerging body of work on a general psychopathology (or "p") factor, which may provide a compelling explanation for the extensive comorbidity among disorders, as well as many of the other features associated with chronic or severe mental disorder. We will then suggest that a general factor for psychopathology might be most usefully described as an absence of resilience, rather than as reflecting specific vulnerability factors (although these may well play an important role and may be primarily responsible for the phenomenological heterogeneity observed among and within disorders). We outline the theory that factor "p" may reflect an absence of resilience resulting from difficulties in social communication emerging from disruptions in epistemic trust, and in problems imagining one's own or other people's thoughts in the face of interpersonal stressors, particularly when they arise in the context of attachment relationships.

9.1 A general factor in psychopathology

As has been observed by many in the field, the utility and meaning of the traditional categorical structure of psychopathology are significantly undermined by the overweening presence of comorbidity (concurrent and sequential over time), recurrence and the noisy overabundance of diagnostic disorders. The promise of potential clarity in the face of this clinical and conceptual confusion has been suggested by recent work on a general factor of psychopathology (p factor) – defined by Caspi and colleagues as 'one underlying dimension that summarized individuals' propensity to develop any and all forms of common psychopathologies' (Caspi et al., 2014, p. 131). Caspi and colleagues scrutinised the mental health trajectory of over a thousand individuals in the New Zealand city of Dunedin, across twenty years, from adolescence to mid-life, examining the persistence, co-occurrence and comorbidity of mental health disorder. They found that vulnerability to mental disorder was more accurately described by one general psychopathology factor than by three high-order (spectral) factors delineating internalizing, externalizing and thought disorder. A higher p factor score was associated with 'more life impairment, greater familiality, worse developmental histories, and more compromised early-life brain function' (Caspi et al., 2014, p. 131). Several studies have replicated this higher-order p factor (e.g. Laceulle, Vollebergh & Ormel, 2015; Lahey, Rathouz, Keenan, Stepp, Loeber, & Hipwell, 2015; Murray, Eisner, & Ribeaud, 2016; Del Giudice, 2016, Gibbon, Nolte,

Constantinou, Montague & Fonagy, in preparation). Importantly, the p factor concept may thus also explain why there has been such limited success for the field in locating precise causes, consequences or biomarkers and developing tailored treatments for specific psychiatric disorders (Caspi et al., 2014).

More specifically in relation to personality disorders (PD), Sharp and colleagues have considered the question of whether a general factor for psychopathology exists in the context of PD diagnosis (Sharp et al., 2015). In a series of exploratory factor analyses based on a large sample of 966 inpatients, only four of the six PDs (avoidant, schizotypal, narcissistic, and antisocial) examined formed factors with 75% of the criteria that mark their respective factors. Half the obsessive-compulsive PD criteria loaded with the narcissistic PD criteria, and the other half split across two other factors. However, Sharp et al. found that (a) a Borderline PD (BPD) factor included primary loadings from just over half (55.6%) of the BPD items, of which three had notable cross-loadings, each on a different factor; (b) nearly half (44.4%) of BPD items loaded most strongly on three non-BPD factors (although two had notable cross-loadings on the BPD factor); and (c) the BPD factor was also marked by a narcissistic PD item and had notable additional cross-loadings by other narcissistic as well as avoidant and schizotypal PD items. In the same study, Sharp et al. evaluated a bi-factor model of PD pathology in which a general factor and several specific factors of personality pathology account for the covariance among PD criteria. In the bi-factor model, it was found that all BPD criteria loaded only on to the general factor. Other PDs loaded either on to both the general and a specific factor or largely only on to a specific factor. The implication of this is that BPD criteria may capture the core of personality pathology or may be most representative of all PDs. To compound more widely the salient status of BPD traits, Caspi et al., in their work on the p factor, found that in terms of personality information, individuals who scored highly on the general psychopathology scale were characterized by 'three traits that compromise processes by which people maintain stability – low Agreeableness, low Conscientiousness and high Neuroticism; that is, high-p individuals experience difficulties in regulation/control when dealing with others, the environment and the self' (Caspi et al., 2014, p. 131).

The question that follows from this is what does the p factor represent, beyond a statistical construct? Currently, we can only speculate about the nature of this generic aetiological factor, but one association to be investigated may be childhood adversity. Studies indeed suggest that maltreatment, like p, increases the chance of most types of mental illness in adulthood (Scott, McLaughlin, Smith, & Ellis, 2012) and worsens the course of mental illness (Nanni, Uher & Danese, 2012). It has been recently suggested that childhood maltreatment may be an ecophenotype associated with an earlier age at onset of psychopathology, greater symptom severity, higher levels of comorbidity, greater risk for suicide and, importantly, a poorer response to treatment (Teicher & Samson, 2013). In our opinion, research findings on maltreatment, although still too narrow, point the way to understanding some of the mechanisms underlying the association between the p-factor and vulnerability to (severe) psychopathology. This emphasis on the role of adversity should not be associated with a narrowly environmental position on the relationship between adversity and BPD (to take a commonly used example in our work). Such a position would stand counter to the growing evidence for a substantial genetic determinant of BPD. Research showing the familial nature of BPD (White, Gunderson, Zanarini & Hudson, 2003; Zanarini, Barison, Frankenburg, Reich & Hudson, 2009), and classical twin studies that place heritability of BPD at around 40–50% (Bornovalova, Hicks, Iacono & McGue, 2009; Distel et al., 2008; Kendler et al., 2008; Torgersen et al., 2000) have been borne out further by more complex behaviour–genetic models that take into account siblings, spouses and twins (Torgersen et al., 2008). We would rather position the role of adversity as one possible – and not uncommon – way in which the psychological "immune system" of an individual becomes compromised. To extend the metaphor, difficulties in social communication arising from early adversity can limit the individual's capacity to mobilise defences (through recursive cooperative access to one's own and other people's thoughts) to stave off potentially destabilising incursions from the environment.

9.2 Resilience, mentalizing and reappraisal

In illuminating the nature of the relationship between mentalizing and the p factor, Kalisch and colleagues' (2015) conceptual framework for the neurobiology of resilience is valuable. Diverse accounts of resilience, often advanced at radically different levels of explanation – from socioeconomic through to genetic – can be unified within the positive appraisal style theory of resilience (PASTOR) conceptual framework presented by this group. According to the formulation, the mechanism underlying resilience is driven by top-down processes in the form of the representation of adverse experience with subsequent appraisal that is made of a stressful stimulus. The external and social factors that have been associated with resilience (such as social support or a secure attachment history) affect resilience either directly or indirectly in that they shape the individual's appraisal approach (including defence mechanisms) or minimize exposure to stressors. Kalisch et al. (2015) argue that psychological resilience is not an absence of disease processes, but a reflection of the work of active, biologically based mechanisms. The appraisal theory of resilience is based on a specific understanding of the nature of higher-order cognition (Folkman, Lazarus, Dunkel-Schetter, DeLongis, & Gruen, 1986).

Kalisch describes the three underpinning appraisal mechanisms that determine resilient behaviour and responses (Kalisch et al., 2015), as follows:

1. Positive situation classification. This refers to the manner of immediate appraisal of a situation at the moment of encountering it (e.g. 'What is the person approaching me carrying in their hand?'). In the case of an insignificant threat, a positive appraisal style enables the individual appropriately to view it as non-menacing. Clearly, in the context of an adverse event, a negative appraisal and stress response are called for. In such situations, resilience can be subsequently promoted through the second and third forms of appraisal. The mentalizing profile typically associated with individuals with BPD tends to lean towards the automatic, non-reflective, emotional poles on the mentalizing dimension (Fonagy & Luyten, 2009). They have also been found to be more likely to view characters/behaviours as negative or aggressive (Barnow et al., 2009); to have an impaired view of neutral faces in the context of anger or disgust (Daros, Zakzanis, & Ruocco, 2013); and to react with hostility to neutral social interactions (Domes, Czieschnek, Weidler, Berger, Fast, & Herpertz, 2008) – all suggestive of the negative appraisal style described by Kalisch and colleagues (2015). The mentalizing profile characteristic of an individual with BPD may therefore lead to an oversensitivity to possibly difficult/threatening social interactions (because distortions in mentalizing are more likely to result in mistaken interpretations of others' behaviour and motivation). Further, in the aftermath of a challenging or stressful interaction, it is difficult for the individual to make sense of, contextualize or put aside potentially upsetting memories of experiences, leaving them more vulnerable to emotional storms. A capacity for explicit, reflective mentalizing in particular serves a dual interpretive (appraisal-strengthening) and self-regulatory role. The absence of this capacity deprives the individual of a fundamental tool in reducing stress.

2. The retrospective reappraisal of threat. Whether a traumatic event results in post-traumatic stress disorder, for example, is dependent on how it is retrospectively reappraised (Gross, 1998; Lazarus & Folkman, 1984). This, as Kalisch et al. describe it, 'shifts the emphasis from the external situation (or changes in the situation) to the individual's ability to flexibly adjust current negative appraisal or to implement new, more positive appraisals and then to maintain those. Both processes have to occur in the face of interference from automatic and uncontrolled negative appraisals and the accompanying aversive emotional states' (Kalisch et al., 2015, p. 14). Patients with BPD or high on 'p' have a specific problem in relation to reappraisal proper because they find it challenging to generate second-order representations of mental states that might be modified to constitute more positive reappraisals of experiences or to modify and thus mitigate (adjust) negative appraisals. In essence, this lies at the core of Gunderson and Lyons-Ruth's interpersonal hypersensitivity theory of BPD (Gunderson & Lyons-Ruth, 2008). Hypersensitivity is the likely consequence of a failure of reappraisal following stressful social interactions. In the absence of being able to mentalize in a balanced way (Fonagy & Luyten, 2009), an event or a relationship can be endlessly discussed and dissected in an apparent attempt at reappraisal, but such attempts have an unreal quality. Complicated inferences about mental states are made, but they might have little connection with reality. We term this *pseudomentalizing*, or in extreme, *hypermentalizing*.

3. Inhibition of retraumatizing triggers. This mechanism enables the individual to inhibit the threat-associated sensations that might be experienced when remembering a traumatic event. An incapacity to inhibit these responses can serve to reinforce, perpetuate and generalize the sense of threat or distress. Individuals with the diagnosis of BPD may often find it difficult to limit the incursion of conflictive negative appraisals and find themselves overwhelmed by interfering emotional reactions to information processing. They cannot cognitively inhibit retraumatizing triggers, leaving them vulnerable to threat-associated sensations that might be experienced when remembering a traumatic event, which serve to reinforce the sense of threat or of being overwhelmed. It is not possible for these individuals to access mentalizing if the self is flooded by negative interference that impairs normal cognitive function.

> This is congruent with the view that emotional dysregulation is the fundamental problem in BPD (Siever & Davis, 1991; Linehan, 1993; Crowell, Beauchaine, & Linehan, 2009).

As described above, the appropriate functioning of higher-order cognition crucially depends on appropriate judgements about social contexts. In this sense, social cognition is part of the mechanism of higher-order cognition, although social cognition itself is made up of a set of processes that are monitored by the metacognitive evaluations that higher-order cognition performs: as in any feedback system, there is an inherent circularity in this conceptualization. This is inevitable given that we are describing the extent to which a system is capable of reorganizing its own functioning. Similarly, the modes of operationalizing the self and the identification of self-awareness are strongly shaped by developmental contributions from the social environment – parents, sibling, peers and significant others. In other words, the abstract algorithm that creates personal consciousness cannot be separated from social interactions and the neural updating processes that occur at the intersection between external reality, interoception and predictive capacities (see further below).

We consider this an important perspective but a narrow interpretation of what may be considered higher-order cognition. The outputs of neural processing intrinsically depend on the processing units that take input from the output of other units, perform specific functions, and generate output that in turn becomes the input of other processes. In most models of brain function, any psychological capacity is underpinned by a large number of such hypothetical processes (Rumelhart & McClelland, 1986; Marr, 1982). We are still faced with an important theoretical dilemma: where does this absence of positive reappraisal mechanisms originate from? How can we understand the inflexibility in social communicative processes, which seems to render these individuals with severe personality pathology so 'hard to reach'? How did this inflexibility develop over time? We believe that the answers to these questions lie in an evolutionarily informed account of developmental psychopathology.

9.3 Epistemic trust and social communication

Bringing together the threads of the argument we have built up so far, we propose that the common variance revealed by bi-factor studies of psychopathology indicates a shared variance in resisting socially expectable adversity. Moreover, persistent psychological distress associated with personality disorder has as some common element diagnostic criteria that we may particularly expect to see in BPD, thereby making BPD characteristics (namely compromised mentalising of the self in relation to others) the core features linked to persistence of psychiatric problems. So far, we have outlined a model that inverses this vulnerability from one focused on the common characteristics of the pathological condition to an alternative perspective that highlights the absence of resilience as the shared cause. Following Kalisch et al.'s persuasive model of resilience (2015), we argued that the persistence of psychopathology, as observed prototypically in BPD, results from a pervasive limitation on the appraisal of stressful social experience, which could be linked to limitations in the capacity to mentalize. It thus emerges that rather than searching for the clinical indicators of a transdiagnostic concept such as p, we may be wiser to conceptualize p as an indication of the absence of resilience and focusing on identifying mechanisms that 'normally' protect individuals from harsh conditions. Perhaps p may be more appropriately considered as pointing to protection (or rather the absence thereof).

An important question remains unanswered: What may explain this absence of a protective capacity to reappraise stressful social experiences? A core component of the model outlined here is that the type of functioning associated with many forms of psychopathology might best be understood as an evolutionarily driven form of entrenched adaptation to stimuli from the social environment – often in interaction with genetic propensity (this thinking is very much informed by the work of, among others, Belsky et al., 2012, Del Guidice, etc). It is the heft of these evolutionary imperatives that creates the quality of entrenchment that is typically regarded as characteristic of PD: the severe problems with social communication typically associated with such pathology, should be, according to this perspective, considered to be a socially generated outcome of a particular context or developmental niche.

This is where we suggest that evolutionary and developmental accounts of how the capacity for epistemic trust evolves in humans may provide some important answers. To summarise, the notion of epistemic trust posits that the human infant – most usually first within the context of early attachment relationships – is instinctively inclined to develop an openness to the reception of social communications from their primary caregivers. Epistemic trust is thus an adaptation allowing the infant to receive social knowledge from their better-informed elders (Gergely, Egyed & Kiraly, 2007), enabling them to benefit from the complex edifice of human knowledge that their immediate culture (the attachment dyad and the wider caregiving network) has available to them.

Trusting the communicator means that the learner does not have to go back to first principles each time they encounter novelty: a strange-looking tool without a self-evident purpose is accepted as being used as described by a trusted elder, because they have said so (Recanati, 1997). Being told in this way is enough and saves time and effort (and thereby processing capacity), and indeed possibly allows the infant to grow up and build upon or revolutionize the use of the tool in question. This capacity to teach and learn social knowledge largely underpins the evolution of human culture (Wilson, 1976): it has been proposed that this form of cultural evolution, based on the transmission of knowledge via epistemically trusted communication channels, emerged during the late Pleistocene era (Wilson & Wilson, 2007).

The internalization of knowledge about the social world constitutes a particular kind of learning: it involves encoding the piece of knowledge as *significant, relevant* to the recipient and *socially generalizable* – that is, as an accepted and reusable piece of cultural currency. This specific form of learning is stimulated by ostensive cues generated by the communicator (Sperber & Wilson, 1995; Russell, 1940). Such cues trigger a pedagogic stance in the recipient, priming them to regard forthcoming communications as significant and as worthwhile to be paid attention to. Human infants display species-specific sensitivity and deference to non-verbal ostensive cues, such as eye contact, turn-taking contingent reactivity, being called by their name, and the use of a special tone of voice ('motherese') by the communicator (Csibra & Gergely, 2009; Csibra & Gergely, 2011). These ostensive cues have in common the quality that the recipient is recognized as a subjective, agentive self – (s)he is being mentalized. What is required, in other words, is a second order capacity by the communicator: he or she must take into account which cues will indicate to the receiver the intention of communicating to this specific person. Once epistemic trust is stimulated in this way, the channel for the transmission of knowledge is opened. Mimicry may be protected by human evolution because it generates epistemic trust, inevitably signalling recognition in the child by the imitating adult. A social smile (recognition of the self by the other) probably increases the tendency for imitation because the smile generates epistemic trust and opens the communication channel to receive knowledge.

It has been argued that this mechanism for opening the epistemic channel exists because it cannot be left open by default: it is adaptive for humans to adopt a position of epistemic vigilance unless they are reassured that it is safe to do otherwise (Sperber et al., 2010; Recanati, 1997). The notion that children are promiscuously credulous to those around them has been disproved by ample evidence suggesting the ways in which dubious social signifiers and poor past performance may render a social communicator suspect and their assertions about the world regarded with scepticism (Brosseau-Liard, Cassels, & Birch, 2014; Koenig & Harris, 2005; Shneidman & Woodward, 2016). Epistemic vigilance is a necessary tool to protect against misinformation, whether as a consequence of malicious intent or incompetence on the part of the communicator (Sperber et al., 2010). Therefore, although the purpose of epistemic trust is the transmission of data, its application is a highly psychological and relational process, dependent on calculations about who is trustworthy, authoritative and knowledgeable –in other words, about whose information is worthy of being encoded as relevant and culturally significant to the self.

9.4 Epistemic mistrust and developmental psychopathology

In situations where a young learner experiences early caregivers as unreliable communicators, the opening of epistemic trust becomes problematic. In the face of an abusive and hostile caregiver, whose intentions towards the infant or child are not benign, epistemic mistrust becomes entrenched as an appropriate adaptation that has been prepared by natural selection. It may be more adaptive to remain persistently vigilant about, or even closed off to, the communication of social knowledge (under which we also include the "learning" about mental states when being contingently mirrored in a marked way.

The resulting impairments in epistemic trust are a further, and perhaps more damaging, long-term sequela of the experience of childhood maltreatment. Epistemic hypervigilance can manifest as the overinterpretation of motives, which can take the form of hypermentalizing (Sharp et al., 2011; Sharp et al., 2013), or pseudomentalizing (Asen & Fonagy, 2012). There is significant evidence that the quality of the relationship of a child to a given communicator determines the extent to which they acquire and generalize information from that communicator (Lane & Harris, 2015; Shafto, Eaves, Navarro, & Perfors, 2012; Mascaro & Sperber, 2009). When in a state of epistemic hypervigilance, the recipient of a communication assumes that the communicator's intentions are other than those declared, and the information is therefore not treated as being from a deferential source. Most typically, epistemic mistrust manifests as the misattribution of intention and the assumption of malevolent motives behind another person's actions and therefore treating them with epistemic hypervigilance (or conversely, in some instances, excessive, inappropriate epistemic trust).

The consequence is that the regular process of modifying one's stable beliefs about the world in response to social communication (and in the light of new data) is closed down or disrupted. This generates the quality of rigidity and being 'hard to reach' that therapists have often described in their work in the field of PD (Fonagy, Luyten & Allison, 2015). Change cannot happen in the therapeutic setting because, although the patient can hear and understand the communications transmitted to them by the therapist, the information cannot be accepted as relevant to them and generalizable to other social contexts. PD may, therefore, be best understood as a failure of communication arising from a breakdown in the capacity to forge trust-based learning relationships.

Furthermore, the role of attachment in our conception of personality has shifted as we have increasingly come to regard the conceptualization of linear causation in psychopathology as unhelpful; instead, we conceptualize the perpetuation of PD being driven by loosely coupled interacting systems working in a circular way. A linear approach would posit that the capacity for mentalizing is vulnerable because of the social-emotional quality of early attachment experiences; partial, erratic mentalizing turns into an interpersonal vulnerability whereby a person feels interpersonally brittle because they cannot reliably process the psychological meaning of social experience, and vulnerable because they cannot process their own emotional reactions to these experiences.

Evidence suggests that attachment stress derails mentalizing judgments (Nolte, Bolling, Hudac, Fonagy, Mayes, & Pelphrey, 2013); working in the other direction, attachment schemas predict mentalizing in adolescence (Dykas & Cassidy, 2011; Sharp, Fonagy & Allen, 2012). According to this model, mentalizing and affect regulation compete, and attachment insecurity or disorganization have a catalytic role in disrupting the development of optimal mentalizing capacity.

Wherever this cycle starts, mentalizing problems lead to interpersonal conflict and social difficulties, which generate intense (social) affect such as shame or rage, which is inadequately contextualized because of the failure of social cognition. This affect further undermines the capacity to mentalize, which in turn leads to social challenges, generating interpersonal conflict that will inevitably lead to higher emotional arousal. The upregulated arousal is poorly modulated and causes further disruptions of social cognition as part of a recursive process, the final outcome of which is an individual lacking the higher-order cognitive capacity (e.g. perspective-taking leading to reappraisal) necessary to withstand even everyday social challenge.

The failure of communication described occurs at a number of levels. First, the social disruption associated with interpersonal conflict will itself compromise the processes of social learning and, in particular, of salutogenesis (the capacity to learn and benefit from the social environment). This is a systemic failure of communication that may characterize a family, the members of a social group such as a gang, a social subculture, or indeed an entire culture. Second, the loss of balanced mentalizing triggered by interpersonal conflict generally lessens interest in the content of communication and social information exchange. There is a pervasive loss of interest in intentionality; observable outcomes are gradually prioritized indicators of attitudes and the general tenor of verbal communication is perceived as meaningless 'psychobabble' with few or no substantive implications for the life of the individual.

Third, social dysfunction, as well as the misinterpretation of social signals associated with the loss of mentalizing, leads to a probable failure to appropriately identify ostension – the sense that a communication is of personal relevance and imbued with the potential for meaning making.

These factors (and probably many others) contribute to the individual's failure to develop epistemic trust in personally relevant communications. Again, we are keen to point out that this is not inherently a maladaptive process. The failure to develop epistemic trust leaves the natural function of epistemic vigilance in place. It is, in fact, an efficient adaptation and an indication that the individual is exercising appropriate caution in relation to social influence, which we see as manifesting in the undesirable persistence of antisocial expectations or schemata and the individual's relative imperviousness to social influence.

However, the absence of epistemic trust sets a limit upon social learning. This can render the individual increasingly ill-equipped to function effectively within their social environment as the higher order cognition underpinning resilience, we believe, develops within and is iteratively sustained by the social context. Disruption of the social network within which the individual could (or perhaps should) function leaves them increasingly isolated and prone to further social/interpersonal dysfunction.

Based on the above considerations, we propose that there are three distinct processes of communication that cumulatively make psychotherapy effective:

Communication System 1: The teaching and learning of content

The different therapeutic schools belong to this system. They may be effective primarily because they involve the therapist conveying to the patient a model for understanding the mind that the patient can understand as involving a convincing recognition and identification of his/her own state. This may in itself lower the patient's epistemic vigilance and create a minimal interest in psychic reality.

Communication System 2: The re-emergence of robust mentalizing

When the patient is once again open to social communication in contexts that had previously been blighted by epistemic hypervigilance, he/she shows increased interest in the therapist's mind and the therapist's use and modulation of thoughts and feelings, which stimulates and strengthens the patient's capacity for mentalizing. Improvements in mentalizing or social cognition may thus be a common factor across different interventions.

Communication System 3: The re-emergence of social learning

The relaxation of the patient's hypervigilance via the first two systems of communication enables the patient to become open to social learning beyond the dyad. This allows the patient to apply his/her new mentalizing and communicative capabilities to wider social learning, outside the consulting room. This final part of the process depends upon the patient having a sufficiently benign social environment to allow him/her to gain the necessary experiences to validate and bolster the improved mentalizing and to continue to facilitate relaxation of epistemic mistrust, in the wider social world.

9.5 Basic Principles in Computational Neuroscience

As shown elsewhere (Debbané & Nolte, 2018), it is our view that experience-dependent reorganisation of the brain's structure-function relationships is critical for lifelong adaptation. Resilience, from that perspective, is governed by higher-order cognition (HOC) comprising a range of psychological processes such as perspective taking, metacognition, executive functions, attention, memory, general intelligence, and self-awareness. In the brain, HOC can be seen as an overarching re-routing of structural and functional connections between ensembles of neuronal populations or networks (e.g. Rudrauf 2014, Fonagy & Bateman, 2016; Fonagy, Luyten, Alison & Campbell, 2017) – also described as a particular form of neural plasticity. As a result, new routes for information processing between regions are created (for instance after structural damage due to a neurological event). The resulting optimised neural architecture seeks to compensate for the compromised structure by generating an alternative set of brain connections to sustain *functional* resilience in the face of partial organ failure.

Mentalizing in a broader sense, we postulate, is a prime candidate in order to dynamically preserve or recover adaptive functioning, i.e. it supports psychological resilience and robustness in the face of increased allostatic load impacting the brain. This can arise from either external stressors or changes in the environment, or through interpersonal and intrapsychic conflict. When both neural integrity, as well as that of the felt self, are under threat (for instance, during childhood adversity, illness, prolonged distress in maladaptive relationships or traumatic experiences), HOC can act as a compensatory cognitive reserve (Stern, 2009) to rearrange computations in the brain in order to ensure "business as usual" (Fonagy & Bateman, 2016; Fonagy, Luyten, Alison & Campbell, 2017). Heuristically speaking, following our notion of resilience, it is the human capacity for imaginative activity or perspective taking that constitutes the crucial mechanism enabling abstraction, separation of thoughts from feelings, and reappraisal which is key to flexible and resilient thinking.

A partial failure of resilience arises when the individual is unable to change processing systems in a sufficiently flexible manner so that an optimal outcome under changed circumstances can be maintained. When an individual cannot disengage a processing system that is no longer appropriate to the task – for example, a child whose perfectionistic attitudes serve them well during a period of knowledge acquisition and relatively simple tasks but cause great problems when task complexity has increased to a point where perfection is impossible – this inflexibility can generate vulnerability. Insensitivity does not create risk; the sensitivity of higher-order cognition is what provides protection through the appropriate appraisal of the functioning of neural structures in interaction with the environment.

Recently, computational psychiatry, which focuses on understanding the mechanistic aspects of psychopathology, has developed as a compelling possible approach to investigating these phenomena of the self in exchange with its social context in a rigorous empirical way. As an emerging discipline, computational psychiatry is located at the intersection between computational neuroscience, psychology and psychiatry. It seeks to provide insight into the key mechanisms underpinning mental functioning by examining the associations between brain function, interplay with the environment the brain is trying to model or act upon, and psychiatric symptoms that arise from adaptational failures via formulised computational processes (Montague et al., 2012, Friston 2015; Stephan & Mathys, 2015 and Adams, Huys & Roiser, 2016).

The notion of *active inference* is one of the main principles underpinning this novel approach. It posits that the brain performs so-called *Bayesian* inference processes. In the heuristics of this framework, the brain is viewed not just as a passive filter of its input, but rather as a 'statistical organ that generates hypotheses or fantasies that are tested against sensory evidence' (Friston, Stephan, Montague & Dolan 2014, p. 148). This is of particular importance as navigating the social world requires constant modelling of other people, as well as approximating to what intentional states might underpin their behaviours or our own. As a result of such *recursive modelling* coherent self-with-other experiences are generated.

Recognising discrepancies between one's assumptions (our prior beliefs about the causes of what we perceive) and reality generates 'surprise' or 'unbound energy'. Such a mismatch then leads to a neural alert signals, the so-called *prediction error,* which in turn incline the neural system to update the relevant working hypotheses as one accumulates more and new evidence or information' (Debbané & Nolte, 2018). Through pre-existing mental structures, our generative models of self and other – the brain's working hypotheses – are afforded greater or lesser precision via top-down neuromodulatory gain control mechanisms. These prescribe how prone the revisions of our belief are to rigidity or indeed flexibility. We may draw a bridge to the notion of *learning from experience* here, and the accompanying capacity to hold alternative viewpoints – a core capacity required for effective reappraisal as postulated in the PASTOR framework.

The key notion of this specific approach in computational neuroscience is that neural functioning shapes sensory input into experience according to Bayesian probabilistic calculus. As captured by Holmes (2017), in healthy minds, 'discrepancies between prediction and input are: (a) identified, via 'tolerating uncertainty', (b) explored, by action aimed at reducing uncertainty, (c) leading, finally, to cognitive restructuring and updating of 'prior' probabilities'. A mentalizing therapist supports patients in bearing the inevitable uncertainty or opaqueness that is intrinsic to any knowledge about mental states. He or she, furthermore, models affect regulation, can stimulate reflection, and thereby scaffolds the exploration of alternative perspectives so that the process of thinking and imagining can be rekindled, constantly reshaped and dynamically updated through new experiences.

Developmentally speaking, the generating of meaning in infancy within the biobehavioral (a-)synchronies with caregivers and the physical environment, the pressures to regulate allostatic challenges (both continuous life stressors as well as particular traumatic events) and the reappraisal, i.e. mentalising of caregiving experiences – all can be captured from the point of view of Bayesian principles. The accompanying basic generative models or beliefs, in the attachment relationship, undergo constant revision and refinement. In other words, 'a baby's brain is in the game of minimizing surprise, while acting epistemically to reduce expected surprise (i.e. uncertainty)' about the novel events it is constantly encountering (Friston, 2017). We can understand the baby's need to adapt to its specific developmental niche as 'conforming to one imperative; namely, to maximize the evidence for its own existence' (Friston, 2017). This is captured by the notion of a self-evidencing brain that strives to make sense of what is going on inside and around her/him.

In early development, it is the HOCs 'borrowed' from the parents' minds, through their caregiving disposition, that provide most of the infant's adjustment and the accompanying affect regulation. Shai & Belsky postulate that parental *embodied* mentalizing is initially non-verbal and implicit (Shai & Belsky, 2011). Critically, these embodied gestures present a model of the infant from the caregiver's perspective, which at both the neural and experiential level, assist the infant to integrate the primary sensory and multisensory signals (Fotopoukou & Tsakiris, 2017). Predictive models will spatially, affectively, and cognitively orient the infant to explore the environment and the mind of others. This is echoed in computational neuroscience: if the brain is designed to elaborate hierarchical generative models of the world, then the infant brain will fundamentally seek to forge and to differentiate between self versus non-self causes of sensations. 'In other words, the self-evidencing baby critically requires predictable input from the caregiver to integrate self-other boundaries. As a result of contingent and marked mirroring a continuity of selfhood is then established. When this input is missing or when it is conveyed in an affectively overwhelming or un- "marked" fashion, it carries the opposite effect: that of blurring the self-other boundary' (Debbané & Nolte, 2018) with the resulting risk for developmental disorders.

Following this line of thinking, psychopathology and compromised resilience, in a contemporary neuroscience framework, can be understood in terms of atypical and impaired developmental patterns underpinning *computational* aspects of such inference and prediction of caregiving behaviour.

The modelling of computations sustaining ongoing experience-dependent adaptation in life can generate empirical tools that will inform the mentalizing framework to better conceptualize processes that contribute to both health and to psychopathology. Significant conceptual contributions from this approach all bear relevance for a deepened understanding of *how* aspects of mentalizing are implemented neurally. In other words; they extend beyond identifying which network of brain areas is involved when a particular process is being called into action, but rather they seek to uncover the algorithms underlying how neural populations perform their computations. Recent advances in this area are, for instance, concerned with the inference on self and other representations (Moutoussis et al., 2014) including the development of caregiving-attachment contingencies (Cittern, Nolte, Friston & Edalat, in press), interoception and homeostasis of the self when engaged with the outside world (Gu et al., 2013; Seth, 2014), compulsivity (Hauser et al., 2016), and deep cognitive phenotyping (Xiang et al., 2012; Hula, Dayan & Montague, 2015). They can further be complemented by tangible applications focused on social learning (Diaconescu et al., 2017) and theory of mind in two-party interactions, for instance in the multi-round Trust Game (Hula, Vilares, Dayan & Montague, 2017;).

9.6 Conclusion: P Factor and the Brain

In line with the contemporary developmental account of severe personality impairment outlined above, computational neuroscience may shed light on both environmental and constitutional contributions to the general vulnerability or propensity of this persistent psychopathology. Furthermore, it can help identify key computational processes and brain-behaviour interactions of compromised functioning, for instance, sub-optimal recruitment of HOCs. With initial evidence emerging that casts "p" as strongly associated with high levels of childhood maltreatment and low mentalising capacities (Caspi et al., 2014; Gibbon, Nolte, Constantinou, Montague & Fonagy, in preparation) the neural underpinnings of mental state processing will illuminate transdiagnostic brain-behaviour relationships that may facilitate an understanding of why, for instance, certain symptoms cluster together or follow a sequential order.

With regards to psychopathology, such computational models can be informative for identifying where a single patient is located along a continuum of (interpersonal) problems (Stephan et al., 2017). They can also drive the identification of endophenotypes associated with higher-order mental processes that cut across traditional diagnoses. Most importantly, these new approaches may shed light on aspects of brain structure-function relationships and plasticity underpinning resilient adaptation (Kalisch, Müller & Tüscher, 2015) such as the epistemic trust-modulated broaden and build cycles – i.e. the individual's capacity to use social support and to learn from other mentalizing minds (Fonagy, Luyten & Allison, 2014). Thereby, computational neuroscience may further our still limited knowledge of the three therapeutic communication systems outlined above. Most importantly it may help to illuminate how insights gained in the consultation room are generalized in the patient's wider social environment, and how the patient's 'inferential organ' can be made more receptive to external influence and thus more amenable to updating on the basis of new interpersonal experience.

A better mechanistic understanding of how hierarchically organised microcircuits of the mentalising neural network in the brain orchestrate and update beliefs about relational experiences (including those schemata that are recruited without a person's awareness, i.e. unconsciously) will help to identify computational building blocks of self-other inference and HOC-related processes. This includes technological advances that will allow for new insights into how two or more mentalising brains influence one another e.g. (Bolis & Schilbach, 2017). 'As we know, entraining flexible thinking (balanced mentalising) and thereby an openness to psychological knowledge through epistemic trust results in learning from a social environment experienced as benign and offering support. Thus, a final extension to such between-brain processes and their neural machinery (with parent-infant or therapist-patient as paradigmatic dyads) has to be the investigation of how broader social networks and the wider social environment are represented and experienced as well as how the communicative processes outlined above are implemented neurally (Debbané & Nolte, 2018).

9.7 References

Adams, R.A., Huys, Q.J. and Roiser, J.P., (2015). Computational Psychiatry: towards a mathematically informed understanding of mental illness. *Journal of Neurolology, Neurosurgery and Psychiatry,* doi: 10.1136/jnnp-2015-310737.

Asen, E. & Fonagy, P. (2012). Mentalization-based therapeutic interventions for families. *Journal of Family Therapy.* 34 (4), 347-370.

Barnow, S., Stopsack, M., Grabe, H.J., Meinke, C., Spitzer, C., Kronmuller, K. & Sieswerda, S. (2009). Interpersonal evaluation bias in borderline personality disorder. *Behaviour Research and Therapy.* 47 (5), 359-365.

Belsky, D.W., Caspi, A., Arseneault, L., Bleidorn, W., Fonagy, P., Goodman, M., Houts, R. & Moffitt, T.E. (2012). Etiological features of borderline personality related characteristics in a birth cohort of 12-year-old children. *Development and Psychopathology.* 24 (1), 251-265.

Bolis, D. & Shilbach, L. (2017). Beyond one Bayesian brain: Modelling intra- and inter-personal processes during social interaction: Commentary on "Mentalizing Homeostasis: The social origins of interoceptive inference" by Fotopoulou, A. & Tsakiris, M. *Neuropsychoanalysis.* 19 (1), 35-38.

Bornovalova, M.A., Hicks, B.M., Iacono, W.G. & Mcgue, M. (2009). Stability, change, and heritability of borderline personality disorder traits from adolescence to adulthood: a longitudinal twin study. *Development and Psychopathology.* 21 (4), 1335-1353.

Brosseau-Liard, P., Cassels, T. & Birch, S. (2014). You seem certain but you were wrong before: developmental change in perschoolers' relative trust in accurate versus confident speakers. *PLoS One.* 9 (9): e108308.

Caspi, A., Houts, R.M., Belsky, D.W., Goldman-Mellor, S.J., Harrington, H., Israel, S., Meier, M.H., Ramrakha, S., Shalev, I., Poulton, R. & Moffitt, T.E. (2014). The p factor: One general psychopathology factor in the structure of psychiatric disorders? *Clinical Psychological Science.* 2 (2), 119-137.

Cittern, D., Nolte, T., Friston, K. & Edalat, A. (in press, *PloS one*). Intrinsic and extrinsic motivators of attachment under active inference.

Crowell, S.E., Beauchaine, T.P. & Linehan, M.M. (2009). A biosocial developmental model of borderline personality: elaborating and extending Linehan's theory. *Psychological Bulletin.* 135 (3), 495-510.

Csibra, G. & Gergely, G. (2011). Natural pedagogy as evolutionary adaptation. *Philosophical Transactions of the Royal Society B.* 366, 1149-1157. Available from: doi: 10.1098/rstb.2010.0319.

Daros, A.R., Zakzanis, K.K. & Ruocco, A.C. (2013). Facial emotion recognition in borderline personality disorder. *Psychological Medicine.* 43 (9), 1953-1963.

Debbané, M. & Nolte, T. (2018, in press). The neurobiology of mentalising. In: Fonagy, P & Bateman, A. (eds). *Handbook of Mentalizing in Mental Health Practice, 2nd edn.*

Del Giudice, M. (2016). The life history model of psychopathology explains the structure of psychiatric disorders and the emergence of the p factor: a simulation study. *Clinical Psychological Science.* 4 (2), 299-311.

Diaconescu A.O. Mathys, C., Weber, L.A.E., Kasper, L., Mauer, J. & Stephan, K.E. (2017). Hierarchical prediction errors in midbrain and septum during social learning. *Social Cognitive and Affective Neuroscience.* 12 (4), 618-634.

Distel, M.A., Trull, T.J., Derom, C.A., Thiery, E.W., Grimmer, M.A., Martin, N.G., Willemsen, G. & Boomsma, D.I. (2008). Heritability of borderline personality disorder features is similar across three countries. *Psychological Medicine.* 38 (9), 1219-1229.

Domes, G., Czieschnek, D., Weidler, F., Berger, C., Fast, K. & Herpertz, S.C. (2008). Recognition of facial affect in borderline personality disorder. *Journal of Personality Disorders.* 22 (2), 135-147.

Dykas, M.J. & Cassidy, J. (2011). Attachment and the processing of social information across the life span: theory and evidence. *Psychological Bulletin.* 137 (1), 19-46.

Folkman, S., Lazarus, R.S., Dunkel-Schetter, C., DeLongis, A. & Gruen, R.J. (1986). Dynamics of a stressful encounter: cognitive appraisal, coping, and encounter outcomes. *Journal of Personality and Social Psychology.* 50 (5), 992-1003.

Fonagy, P. & Luyten, P. (2009). A developmental, mentalization-based approach to the understanding and treatment of borderline personality disorder. *Development and Psychopathology.* 21 (4), 1355-1381.

Fonagy, P., Luyten, P. & Allison, E. (2015). Epistemic petrification and the restoration of epistemic trust: a new conceptualization of borderline personality disorder and its psychosocial treatment. *Journal of Personality Disorders.* 29 (5), 575-609.

Fonagy, P. & Bateman, A.W. (2016). Adversity, attachment, and mentalizing. *Comprehensive Psychiatry.* 64, 59-66.

Fonagy, P., Luyten, P., Allison, E. & Campbell, C. (2017a). What we have changed our minds about: Part 1. Borderline personality disorder as a limitation of resilience. *Borderline Personality Disorder and Emotion Dysregulation.* 4 (11). Available from: doi: 10.1186/s40479-017-0061-9.

Fonagy, P., Luyten, P., Allison, E. & Campbell, C. (2017b). What we have changed our minds about: Part 2. Borderline personality disorder, epistemic trust and the developmental significance of social communication. *Borderline Personality Disorder and Emotion Dysregulation.* 4 (9). Available from: doi: 10.1186/s40479-017-0062-8.

Fotopoulou, A. & Tsakiris, M. (2017). Mentalizing homeostasis: The social origins of interoceptive inference. *Neuropsychoanalysis.* 19, 3-28.

Friston, K.J., Stephan, K.E., Montague, R. & Dolan, R.J. (2014). Computational psychiatry: the brain as a phantastic organ. *Lancet Psychiatry.* 1 (2), 148-158.

Friston, K.J. (2017). Self-evidencing babies: Commentary on "Mentalizing Homeostasis: The social origins of interoceptive inference" by Fotopoulou, A. & Tsakiris, M. *Neuropsychoanalysis.* 19, 43-47.

Gergely, G., Egyed, K. & Kiraly, I. (2007). On pedagogy. *Developmental Science.* 10 (1), 139-146.

Gibbon, L., Nolte, T., Constantinou, M., Montague, R. & Fonagy, P. (in preparation). Modelling axis I and personality disorder symptomatology and its associations with childhood trauma and mentalizing.

Gross, J.J. (1998). The emerging field of emotion regulation: an integrative review. *Review of General Psychology.* 2 (3), 271-299.

Gu, X., Hof, P.R., Friston, K.J. & Fan, J. (2013). Anterior insular cortex and emotional awareness. *Journal of Comparative Neurology.* 521 (15), 3371-3388.

Gunderson, J.G. & Lyons-Ruth, K. (2008). BPD's interpersonal hypersensitivity phenotype: a gene-environment developmental model. *Journal of Personality Disorders.*22 (1), 22-41.

Hauser, T.U. Fiore, V.G., Moutoussis, M. & Dolan, R.J. (2016). Computational psychiatry of ADHD: Neural gain impairments across Marrian levels of analysis. *Trends in Neurosciences.* 39 (2), 63-73.

Holmes, J. and Slade, A., (2017). *Attachment in therapeutic practice.* Sage.

Hula, A., Montague, P.R., & Dayan, P. (2015). Monte Carlo planning method estimates planning horizons during interactive social exchange. *PLoS Computational Biology.* 11: e1004254.

Hula, A., Vilares, I., Dayan, P. and Read Montague, P., 2017. A Model of Risk and Mental State Shifts during Social Interaction. arXiv preprint arXiv:1704.03508.

Kalisch, R., Muller, M.B. & Tuscher, O. (2015). A conceptual framwork for the neurobiological study of resilience. *Behavioural and Brain Sciences.* 38: e92.

Kendler, K.S., Aggen, S.H., Czajkowski, N., Roysamb, E., Tambs, K., Torgersen, S., Neale, M.C. & Reichborn-Kjennerud, T. (2008). The structure of genetic and environmental risk factors for DSM-IV personality disorder: a multivariate twin study. *Archives of General Psychiatry.* 65 (12), 1438-1446.

Koenig, M.A. & Harris, P.L. (2005). The role of social cognition in early trust. *Trends in Cognitive Sciences.* 9 (10), 457-459.

Laceulle, O.M., Vollebergh, W.A.M. & Ormel, J. (2015). The structure of psychopathology in adolescence: replication of a general psychopathology factor in the TRAILS study. *Clinical Psychological Science.* 3 (6), 850-860.

Lahey, B.B., Rathouz, P.J., Keenan, K., Stepp, S.D., Loeber, R. & Hipwell, A.E. (2015). Criterion validity of the general factor of psychopathology in a prospective study of girls. *Journal of Child Psychology and Psychiatry.* 56 (4), 425-422.

Lane, J.D. & Harris, P.L. (2015). The roles of intuition and informants' expertise in children's epistemic trust. *Child Development.* 86 (3), 929-926.

Lazarus, R.S. & Folkman, S. (1984). *Stress, appraisal and coping.* New York, NY: Springer.

Linehan, M.M. (1993). *Cognitive-behavioural treatment of borderline personality disorder.* New York, NY: Guilford Press.

Marr, D. (1982). *Vision: a computational investigation into the human representation and processing of visual information.* San Francisco, CA: W. H. Freeman.

Mascaro, O. & Sperber, D. (2009). The moral, epistemic, and mindreading components of children's vigilance towards deception. *Cognition.* 112 (3), 367-380.

Montague, P.R., Dolan, R.J., Friston, K.J. & Dayan, P. (2012). Computational psychiatry. *Trends in Cognitive Sciences.* 16 (1), 72-80.

Moutoussis, M., Fearon, P., El-Deredy, W., Dolan, R.J. & Friston, K.J. (2014). Bayesian inferences about the self (and others): A review. *Consciousness and Cognition.* 25, 67-76.

Murray, A.L., Eisner, M. & Ribeaud, D. (2016). The development of the general factor of psychopathology 'p factor' through childhood and adolescence. *Journal of Abnormal Child Psychology.* 44 (8), 1573-1586.

Nanni, V., Uher, R. & Danese, A. (2012). Childhood maltreatment predicts unfavorable course of illness and treatment outcome in depression: a meta-analysis. *The American Journal of Psychiatry.* 169 (2), 141-151.

Nolte, T., Bolling, D.Z., Hudac, C.M., Fonagy, P., Mayes, L. & Pelphrey, K.A. (2013). Brain mechanisms underlying the impact of attachment-related stress on social cognition. *Frontiers in Human Neuroscience.* 7 (816), 1-12.

Recanati, F. (1997). Can we believe what we do not understand? *Mind & Language.* 12 (1), 84-100.

Rudrauf, D. (2014). Structure-function relationships behind the phenomenon of cognitive resilience in neurology: insights for neuroscience and medicine. *Advances in Neuroscience.* 2014, 1-28.

Rumelhart, D.E. & MCClelland, J.L. (1986). *Parallel distributed processing.* Cambrdige, MA: MIT Press.

Russell, B. (1940). *An inquiry into meaning and truth.* London, UK: Allen & Unwin.

Scott, K.M., McLaughlin, K.A., Smith, D.A. & Ellis, P.M. (2012). Childhood maltreatment and DSM-IV adult mental disorders: comparison of prospective and retrospective findings. *The British Journal of Psychiatry.* 200 (6), 469-475.

Seth, A.K., 2013. Interoceptive inference, emotion, and the embodied self. *Trends in cognitive sciences,* 17(11), pp.565-573.

Shafto, P., Eaves, B., Navarro, D.J. & Perfors, A. (2012). Epistemic trust: modelling children's reasoning about others' knowledge and intent. *Developmental Science.* 15 (3), 436-447.

Shai, D. & Belsky, J. (2011). When words just won't do: introducing parental embodied mentalizing. *Child Development Perspectives.* 5 (3), 173-180.

Sharp, C., Pane, H., Ha, C., Venta, A., Patel, A.B., Sturek, J. & Fonagy, P. (2011). Theory of mind and emotion regulation difficulties in adolescents with borderline traits. *Journal of the American Academy of Child & Adolescent Psychiatry.* 50 (6), 563-573.

Sharp, C. Fonagy, P. & Allen, J.G. (2012). Posttraumatic stress disorder: a social-cognitive perspective. *Clinical Psychology: Science and Practice.* 19 (3), 229-240.

Sharp, C., Ha. C., Carbone, C., Kim, S., Perry, K., Williams, L. & Fonagy, P. (2013). Hypermentalizing in adolescent inpatients: treatment effects and association with borderline traits. *Journal of Personality Disorders.* 27 (1), 3-18.

Sharp, C., Wright, A.G.C., Folwer, J.C., Frueh, B.C., Allen, J.G., Oldham, J. & Clark, L. (2015). Borderline personality pathology as the 'g' factor of personality disorder. *Journal of Abnormal Psychology.* 124, 387-398.

Shneidman, L. & Woodward, A.L. (2016). Are child-directed interactions the cradle of social learning? *Psychological Bulletin.* 142 (1), 1-17.

Siever, L.J. & Davis, K.L. (1991). A psychobiological perspective on the personality disorders. *American Journal of Psychiatry.* 148 (12), 1647-1658.

Sperber, D. & Wilson, D. (1995). *Relevance: communication and cognition, 2nd edn.* Malden, MA: Blackwell.

Sperber, D., Clement, F., Heintz, C., Mascaro, O., Mercier, H., Origgi, G. & Wilson, G. (2010). Epistemic vigilance. *Mind & Language.* 25 (4), 359-393.

Stephan, K.E. & Mathys, C. (2014). Computational approaches to psychiatry. *Current Opinion in Neurobiology.* 25, 85-92.

Stephan, K.E., Schlagenhauf, F., Huys, Q.J.M., Raman, S., Aponte, E.A., Brodersen, K.H., Rigoux, L., Moran, R.J., Daunizea, J., Dolan, R.J., Friston, K.J. & Heinz, A. (2017). Computational neuroimaging strategies for single patient predictions. *Neuroimage.* 145 (B), 180-199.

Stern, Y. (2009). Cognitive reserve. *Neuropsychologia.* 47 (10), 2015-2028.

Teicher, M. H. & Samson, J.A. (2013). Childhood maltreatment and psychopathology: a case for ecophenotypic variants as clinically and neurobiologically distinct subtypes. *American Journal of Psychiatry.* 170 (10), 1114-1133.

Torgersen, S., Lygren, S., Oien, P.A., Skre, I., Onstad, S., Edvardsen, J., Tambs, K. & Kringlen, E. (2000). A twin study of personality disorders. *Comprehensive Psychiatry: The Journal of Psychopathology.* 41 (6), 416-425.

Torgersen, S. Czajkowski, N., Jacobson, K., Riechborn-Kjennerud, T., Roysamb, E., Neale, M.C. & Kendler, K.S. (2008). Dimensional representations of DSM-IV cluster B personality disorders in a population-based sample of Norwegian twins: a multivariate study. *Psychological Medicine.* 38 (11), 1617-1625.

White, C.N., Gunderson, J.G., Zanarini, M.C. & Hudson, J.I. (2003). Family studies of borderline personality disorder: a review. *Harvard Review of Psychiatry.* 11 (1), 8-19.

Wilson, D.S. (1976). Evolution on the level of communities. *Science.* 192 (4246), 1358-1360.

Wilson, D.S. & Wilson, E.O. (2007). Rethinking the theoretical foundation of sociobiology. *The Quarterly Review of Biology.* 82 (4), 327-348.

Xiang, T., Ray, D., Lohrenz, T., Dayan, P. & Montague, P.R. (2012). Computational phenotyping of two-person interactions reveals differential neural response to depth-of-thought. *PLoS Computational Biology.* 8 (12): e1002841.

Zanarini, M.C., Barison, L.K., Frankenburg, F.R., Reich, D.B. & Hudson, J.I. (2009). Family history study of the familial coaggregation of borderline personality disorder with axis I and nonborderline dramatic cluster axis II disorders. *Journal of Personality Disorders.* 23 (4), 357-369.

Third Part: Practitioner Research and Social Dimensions of Mental Health

Chapter 10

Multidimensional study of the infant/juvenile population with severe and profound Intellectual Developmental Disorders

Anabela Janeco
Centro de Recuperação de Menores D. Manuel Trindade Salgueiro, Portugal

Inês Gomes
Centro de Recuperação de Menores D. Manuel Trindade Salgueiro, Portugal

João Albuquerque
Centro de Recuperação de Menores D. Manuel Trindade Salgueiro, Portugal

Liliana Pacheco
Centro de Recuperação de Menores D. Manuel Trindade Salgueiro, Portugal

Margarida Morais
Centro de Recuperação de Menores D. Manuel Trindade Salgueiro, Portugal

Natália Andrade
Centro de Recuperação de Menores D. Manuel Trindade Salgueiro, Portugal

Patrícia Santos
Centro de Recuperação de Menores D. Manuel Trindade Salgueiro, Portugal

Pedro Amaro
Centro de Recuperação de Menores D. Manuel Trindade Salgueiro, Portugal

Pedro Valente
Centro de Recuperação de Menores D. Manuel Trindade Salgueiro, Portugal

Mental deficiency, here referenced as Intellectual Developmental Disorder (IDD), in accordance with the ICD -11, presents a large prevalence of psychiatric and somatic morbidity that is insufficiently diagnosed. The Double Diagnosis perspective in IDD attempts to clarify this scenario among the global difficulties presented by this population. This article presents the data results obtained through the administration of a set of tools, either selected or built, for the evaluation of a group of people with Severe or Profound IDD, looking for a description that assembles all the intervention areas that usually coexist in supporting this group of people, namely social service, psychology, physical therapy, speech therapy, occupational therapy, psychomotor rehabilitation, nursing and psychiatry. We concluded that the great majority of the studied population suffers serious limitations to their autonomy and life quality. These limitations are of sensorial, expressive and motor nature, associated with the multiple pathologies concerning IDD. To evaluate people with IDD, especially of a Severe and Profound level, even in a designed study and with adapted tools, is always a challenge. The exploration of new tools, its translation, calibration and validation for the Portuguese population, can bring an added value for people with IDD.

Conflict of interest

This work is supported by Plano Nacional da Saúde Mental – Direcção Geral de Saúde, obliging the authors to finalize it. There is no conflict of interest with any other entity.

10.1 Introduction

Intellectual Developmental Disorder (IDD), the chosen designation of the work group of ICD-11 to replace the one of Mental Deficiency in ICD-10, is an entity present across all areas of rehabilitation intervention. This reality has been a source of serious limitations to working with this people because of the difficulty in articulating all the scopes of practice that should be simultaneously present in the intervention plan.

Namely, in what concerns psychiatry, mental health and actually health in general, there is not a policy that enhances the specificity and the difficulties of the approach to these situations. For that reason, there is no specific education on this subject in medical curricula, thus perpetuating an intervention model that rarely meets the needs of people with IDD, of their families, or of the technicians accompanying them in rehabilitation institutions that are usually linked to the social branch.

From amongst these specificities stands out the approach of the Double Diagnosis (Carvalho, 2012) perspective that attempts to clarify, within the global difficulties of the person with IDD, what is a consequence of IDD itself and what is connected with other diagnosis, somatic or psychiatric, that many times are not recognized, as stated before, and thus are included in the IDD frame.

We consider of great importance to produce scientific evidence that awakens the medical community to the reality of Double Diagnosis, as a means of improving the intervention of health professionals, as well as raises the awareness of the people responsible for the education in the different technical areas, to the need of including IDD in their curricula. The higher prevalence of disease in people with IDD is a widely recognized fact. Unfortunately, there are few instruments designed to test this population and the ones existing usually have significant limitations on being applied (age, deficiency degree), on the possibilities of gathering information and on the amount of diagnosis they allow.

The Centro de Recuperação de Menores D. Manuel Trindade Salgueiro (CRM) is part of the health centres net of the Instituto das Irmãs Hospitaleiras do Sagrado Coração de Jesus (IIHSCJ). CRM assists 120 female residents and the admission age ranges from 3 to 17 years old. It is a prerequisite to present a diagnosis of Intellectual Developmental Disorder. Currently, as a more specific clinical intervention in their praxis, CRM gives preference to Double Diagnosis (IDD and other mental health and psychiatric disorders) and Multi-deficiency (IDD and other sensorial, expressive or motor disabilities). Geographically its intervention is of national scope.

10.2 Epidemiology and Grounds for This Study

According to several studies (Mercer, 1973; Tarjan et al. 1973, cited in Albuquerque, 2000), IDD presents a prevalence of 1% when at least two evaluation criteria are used, such as intelligence quotient (IQ) and a scale of adaptative behavior. This value differs significantly from the one we find using only IQ, which will be of 3%. This also explains the variation of results we find among diverse studies.

The prevalence of Severe and Profound IDD is significantly lower than the prevalence of Mild and Moderate IDD (0,4% versus 2,5 to 3%) and its causes are usually clearer (Tejada, 2006). Even more scarce are the studies on the prevalence of mental illness in IDD. Nevertheless, all the existing studies concur on the idea that people with IDD have an increased vulnerability to mental health problems (Cooper, 2007; Moss, 2001; Whitaker & Read, 2006). A prevalence of 30 to 40% of mental illness is found on people with IDD, which corresponds approximately to the quadruple of the prevalence in the general population. This ratio is maintained for each specific pathology we may consider. Hence, for the Schizophrenic Disorders the prevalence is of 3% (O'Hara & Sperlinger, 1998), for Depression between 1,3 and 3,7% (Deb et al., 2001), and Anxiety Disorders occur in circa 40% of the people with IDD (Raghavan, 1997).

On the somatic field there is an emphasis on the neurological diseases (epilepsy, neurofibromatosis, neuromuscular diseases), on endocrinal and metabolic diseases (Hypothyroidism), on cardio-respiratory diseases (congenital cardiopathies, respiratory infections needing frequent hospitalization), on osteo-articular pathology (osteoarthritis and osteoporosis), on chronic pain and on associated motor and sensorial deficits. Moreover, there is a particular vulnerability to multiple substance abuse and addiction, mainly of alcohol (Supporting Complex Needs – A practical guide for supporting staff working with people with a learning disability who have mental health needs, Estia Centre). Concerning the links between motor development and IDD, Frick et al. (2016) make reference to a relation between cognitive and motor development. This correlation between motor skills and intelligence is complex (Barnhill & Kartheiser, 2006), and it is clearer in children with severe global development delay, who have significant intellectual difficulties and grave motor disorders, with very limited functional movements and dependency for daily activities (Winfield et al., 2014).

Motor coordination is a complex process that includes fine and large motor skills that can be rehabilitated or adapted to new needs. Its acquisition and application processes follow the development course. This results in an improvement of speed, planning and execution, precision and quick adaptation to new actions, in an increased ability to inhibit undesired or previous movements, and the monitoring of the success of the task in real time (Barnhill & Kartheiser, 2006). Individuals with severe and profound intellectual deficit have more difficulty in adapting their limited sensorial-motor abilities to new tasks, thus jeopardizing the acquisition of more evolved capacities like the ones (such as handwriting) that need the ability to initiate or stop a motor action, to inhibit independent or simultaneous answers, fluency, sequence and precision in complex movements, synkinesia, and mirror movements, tonus and posture (Barnhill & Kartheiser, 2006). Children with severe global development delay have significant pulmonary dysfunction due to immobility, muscle weakness, skeletal deformities and parenchymal damages caused by aspiration, which are causes of morbidity and mortality. Literature refers respiratory pathology as the main cause of death in individuals with cerebral palsy, neuromuscular disorders and severe development disorders, and also as one of the leading causes of morbidity, with repeated hospitalizations and significant impact in life quality (Winfield et al. 2014). These individuals show a diminished pulmonary capacity (respiratory function below normal, clearance disorders of the airway due to muscular weakness, diminished thorax mobility, diminished extension of the torso and weakness of the respiratory muscles), poor nutritional state, food aspiration, gastro-esophageal reflux, and lack of motor coordination which increases the risk of morbidity and infection (Choi et al., 2016). In addition, the long range side-effects of the pharmacological therapy are more complex in children with Severe IDD (Barnhill & Kartheiser, 2006) and that can increase their difficulties.

In terms of speech therapy, we considered it important to evaluate communication, language and swallowing in the face of the characteristics of this study sample (Infantile and Juvenile Population with Severe and Profound Intellectual Developmental Disorder). Dysphagia is a swallowing disorder of neurological and/or structural causes and may be a consequence of head and neck trauma, of Cerebral Vascular Accident, of degenerative neuromuscular diseases as well as of dementias and encephalopathies. Usually, the severity of the dysphagia accompanies the increasing severity of the disease. Language is composed of complex rules organizing sounds, words and meaningful sentences. In addition to its structure and significance, it requires an intention and a purpose. Language is acquired through an interactive process determined by environment, psycho-social, cognitive and biological factors and involves the manipulation, combination and integration of linguistic forms and its subjacent rules, enabling the abilities of understanding (comprehensive language) and producing (expressive language). According to Bloom & Lahey, language is a complex combination of components categorized into 3 levels: Form, Content and Use. Children with language disorders have difficulties in all components, which is what usually happens in people with IDD.

It is important to distinguish those disorders with the specific language impairment (SLI), language does not evolve within the normal development parameters, and usually, only one or two of language areas are affected. It is an acquired language and/or communication impairment in consequence of cerebral damage. The language of a child with this kind of disorder does not evolve in homogeneous states, both in expression and understanding. Children with communication disorders are not able to transfer information to another person or groups of people. These characteristics are usually evident in children with autism and profound IDD, and it can be associated with delayed language development (DLD), in which the language level coincides with that of a younger child. The possible causes of DLD are: total or partial hearing loss, neurological problems, delayed psycomotor development, genetic inheritance, lack of stimulation, bilingualism, and also autism spectrum disorders. We have illustrated along this chapter the reasons why it is important and difficult to evaluate people with IDD, and also why so many domains should be considered.

10.3 Objectives

The main objectives of this work were:

- Assess the global needs of a population with IDD, evidencing the relevance of the intervention integrated model in IDD,
- Organize the intervention of all technical areas in a structured and coordinated way.

Following the initial objectives, we understood we had the possibility of creating a specific evaluation grid for the infant/juvenile population with Severe and Profound Double Diagnosis, determining which evaluation instruments would be more adequate in a multidisciplinary context.

We also considered relevant to identify instruments worth translating, calibrating and validating for the Portuguese population since the material adapted to the population with Severe and Profound IDD is scarce, and to create guidelines for future work on IDD.

10.3.1 Material and Method

The results here presented correspond to the evaluation of the Severe and Profound IDD population at CRM residents with ages ranging from 3 to 25 years old, with Severe and Profound IDD (n: 33). The Study project was presented to the Comissão de Ética of Instituto Politécnico de Portalegre, a CRM partner, who approved its implementation. The specificity of CRM emerges as a consequence of the characteristics of the technical corpus that includes both medical and nursing staff, which is not common in institutions linked to rehabilitation. It comprises 1 psychiatrist, 1 general and familiar practitioner, 1 pediatrician, 1 dentist, 14 nurses, 1 psychologist, 1 social worker, 1 occupational therapist, 1 psychomotricist, 1 physiotherapist, 1 speech therapist, 2 social educators, 52 caregivers of whom 4 have specific functions of accompanying occupational activities, 2 special education teachers, and 6 Sisters of the IHSCJ congregation that have places of direction and coordination of services as well as care services. It is important to understand that, if such a team can enrich the therapeutic intervention, at the same time this intervention becomes much more complex, with the need to coordinate everyone's work.

Each technical area of expertise decided which evaluation tool would be used, based on theoretical grounds from bibliography research and on possibilities and difficulties found in their practice. Thus, psychiatry chose ChA-PAS, the PAS-ADD (Psychiatric Assessment Schedules for Adults With Developmental Disabilities - Moss, 2001) version for children and adolescents, which evaluates with great precision the probability of a psychiatric disorder being present, considering the Psychosis, Maniac Episode, Depressive Episode, Anxiety Disorder, Obsessive-Compulsive Disorder, Conduct Disorder, Attention Deficit Hyperactivity Disorder (ADHD), and Autism Spectrum Disorder (ASD).

ChA-PAS belongs to a set of 4 instruments called PAS-ADD (http:/www.pas-add.com) designed to improve the detection and diagnose of mental health problems in people with IDD. ChA-PAS has the great advantage of enabling untrained personnel to detect eventual psychiatric disease, thus widening the possibilities of assessing the population with IDD, in different contexts and in every age or disability degree. Nursing chose NOC (Nursing Outcomes Classification), a taxonomic structure that scores from "1" (non desirable) to "5" (desirable), 385 nursing outcomes, organized into 31 classes each divided into 7 domains. Being a technical area of very broad intervention that overlaps almost all other intervention areas we chose to assess Autonomy thus completing the evaluation of the specific therapies.

In making this decision, it was crucial the participation of Universidade de Évora as a study partner, through Escola Superior de Enfermagem São João de Deus, that focused on the specificity of mental health nursing. More specific aspects of general nursing interventions, like insuring the administration of prescriptions, wound dressing and caring, managing physical restrain situations, and so on, are securely monitored by individual intervention plans, and do not document the orientation model of multidisciplinary intervention we advocate. In the Social area, a grid was designed based on the social service record sheet of IIHSCJ, to which we added some items relating to the specific intervention at CRM. Psychology used WISC-III (Wechsler Intelligence Scale for Children), a tool available in the department and already calibrated for the Portuguese population, with a complete profile of cognitive functions, which allows the identification of weaknesses and strengths of the target population and corresponds to the instrument usually used in cognitive evaluations at CRM. At the same time, we avoided an evaluation without scientific value which could occur if we tried to design other instruments for such a complex area.

We admit the limitation in the use of WISC-III, since only 51,5% of the sample could be evaluated with this instrument, and we followed DSM-5

(Diagnostic and Statistic Manual of the American Psychiatric Association) guidelines, that point to a Non-specified IDD diagnose, whenever the use of a standard test is either difficult or impossible. The specific therapy areas, Physical therapy (Physio) Psychomotricity (PM), Speech Therapy (ST), Occupational Therapy (OT), are the ones in which the limitations of the available instruments are more obvious. In fact, patients with Global Delayed Development have neither a specific nor a definite diagnose (Winfield et al., 2014), since its pathology is very complex and evolutive. This high heterogeneity, the scarce scientific research and shortage of adequate instruments to evaluate and gather data enough to characterize the CRM population, motivated the joint elaboration of a grid which tries to gather as many data as possible from the different technical areas and that can be used in different age groups. The inclusion of the subject's perception of her interests and needs stands out in the Occupational Therapist evaluation. The grid is structured into 12 evaluation areas, organized into 4 domains in a total of 273 items.

All of the chosen or created instruments were submitted to the study coordination, in order to avoid overlaps and unnecessary duplications. That meant having to make some alterations, being the most evident the selection made by nursing, as was already mentioned. Having completed this stage, we proceeded to apply each instrument. Each collaborator concluded the evaluation within a definite time frame, between September and November 2015. The results of each technical area were delivered to the study coordination in order to prepare its computer processing and presentation.

All the possible techniques for collecting data were used. We consider observation and interview of the subject (according to her possibilities of involvement), the information of other caregivers linked to basic subsistence activities, to other technical interventions, and to the pedagogic field. Also taken into consideration, was all the available data collected throughout the history of the selected person.

10.3.2 Results

A total of 33 female CRM residents were evaluated, with ages ranging from 3 to 25 years old, 23 of whom presented Severe IDD and 10 Profound IDD, as previously diagnosed in the psychiatric consultation and psychology evaluation. The charts 10.1 to 10.9 illustrate the assessment made at a social level and clearly document the features of resident population concerning this area. We underline that all of the subjects had a previous stay in another care institution. The ages are mostly within the higher half of the considered range which is consistent with the fact that most of them have been institutionalized for more than 5 years. They come from all over the country with higher prevalence from more populous or nearer districts. We also want to point out that the majority of the subjects either has no reference person on the exterior or has only one, and the possibility of later having some kind of support in the residence area is rare. All of the school-aged residents attend school, most of them in the community. The high number of cases in which the family history is unknown corresponds to nonexistent or disrupted families, suggesting the existence of family backgrounds that contribute to the disorder.

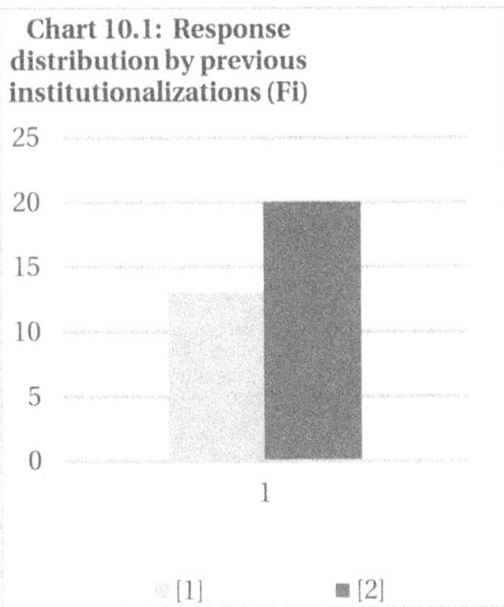

Chart 10.1: Response distribution by previous institutionalizations (Fi)

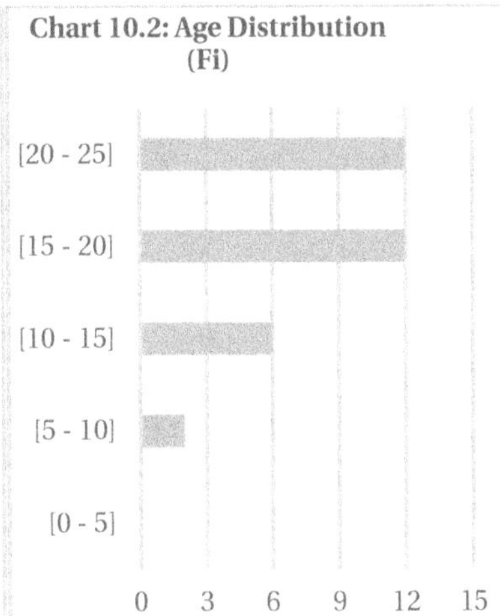

Chart 10.2: Age Distribution (Fi)

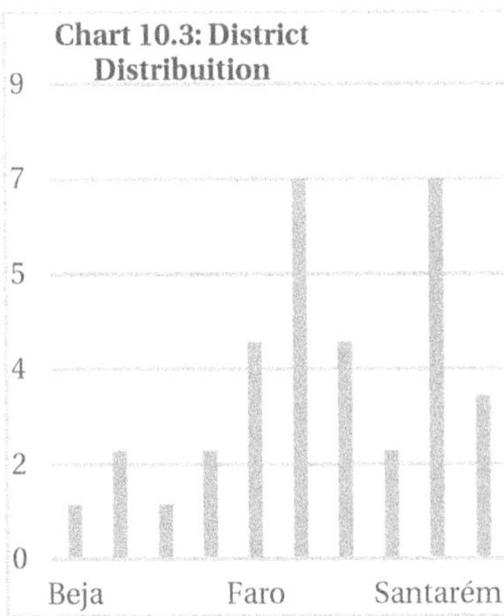

Chart 10.3: District Distribuition

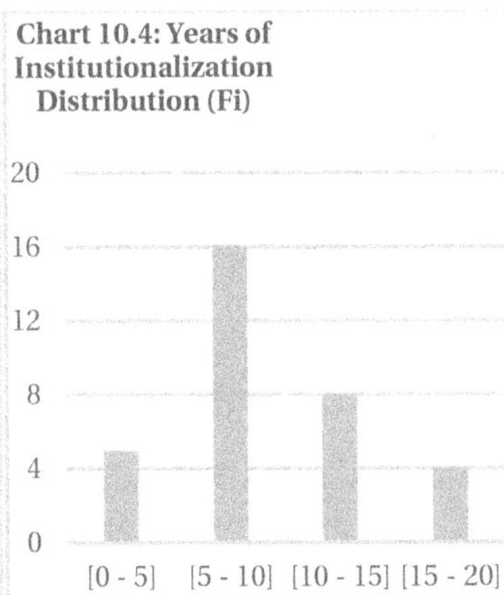

Chart 10.4: Years of Institutionalization Distribution (Fi)

Chart 10.5: Reference Person Distribution (Fi)

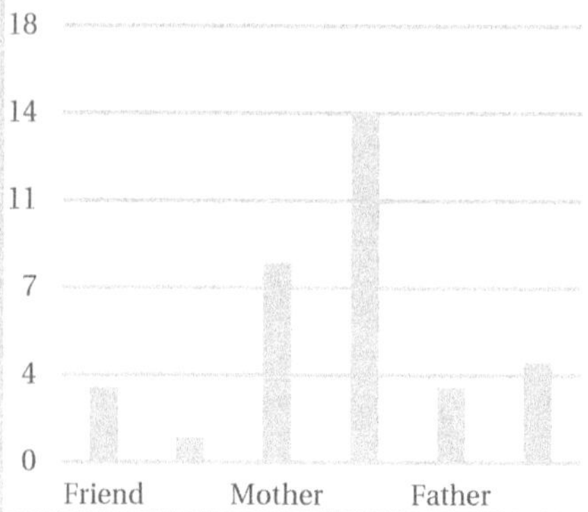

Chart 10.6: Possibility of Institutional Transference to the Residence Area of the Reference Person Distribution (Fi)

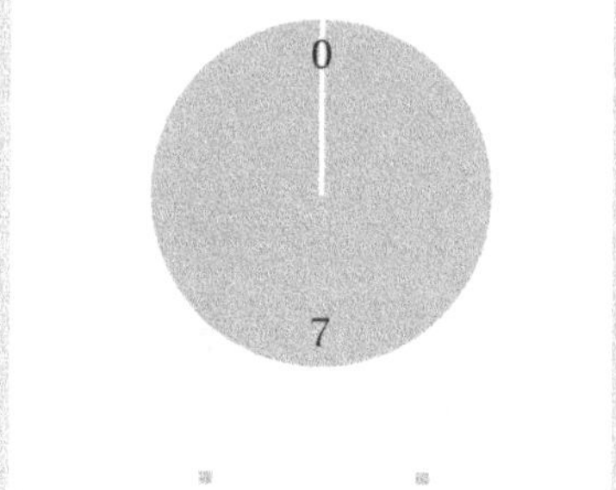

Chart 10.7: School Attendence Distribution (Fi)

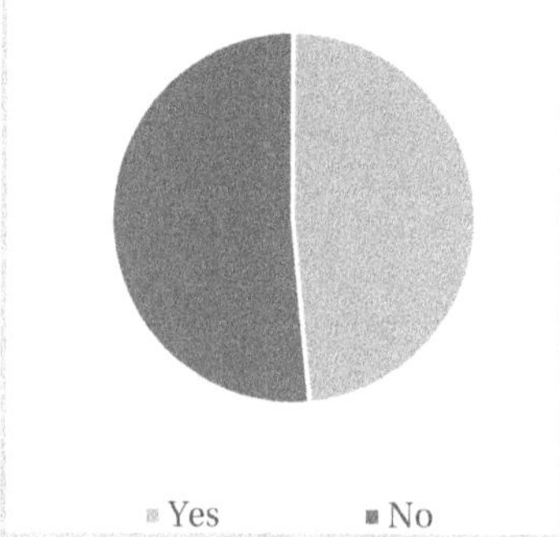

Chart 10.8: Type of Education Distribution (Fi)

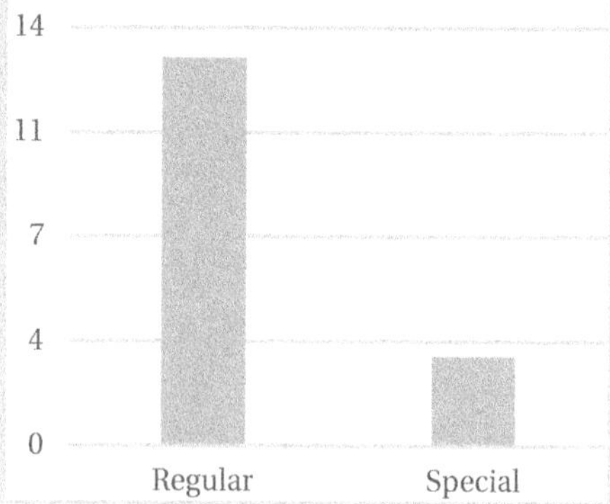

Chart 10.9: Family History Distribution (Fi)

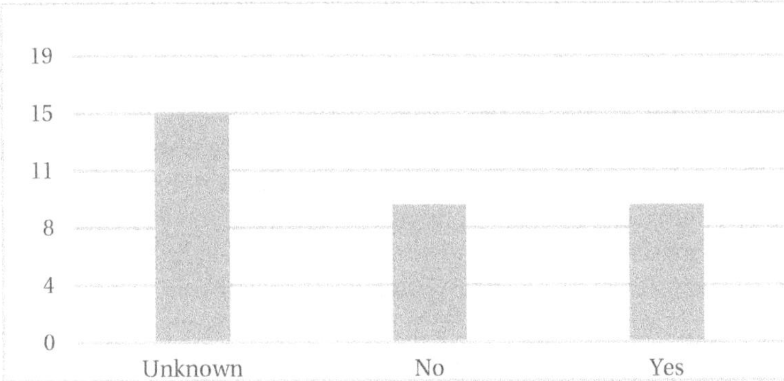

Psychology faced significant limitations when applying the WISC (Chart 10.10) only obtaining numerical results for 9 subjects (29,7% of the sample) on the Verbal Intellectual Quotient (VIQ), for 7 subjects (23,1% of the sample) on the Achievement Intellectual Quotient (AIQ), and 6 subjects on the Full Scale Intellectual Quotient (FSIQ). The majority was evaluated as "non-specified", thus showing how little quantitative value this scale has for this population. Nevertheless, it allows a discriminatory evaluation of the better functioning areas of each person tested. Following these results, we attempted to identify a sub-test that could produce meaningful results for most subjects, but it was still not possible to find an adequate solution since each presented different better functioning cognitive domains. Thus, the use of WISC with a descriptive presentation of data seems to be the best solution for the evaluation of IQ in this population.

Chart 10.11

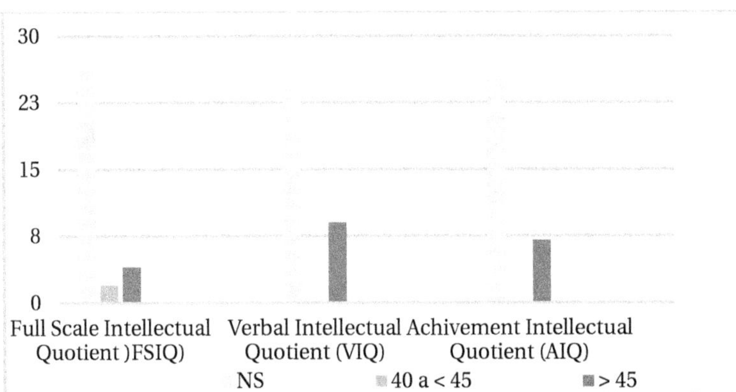

As already stated, Physiotherapy, Occupational Therapy, Psychomotricity and Speech Therapy established a common scale gathering different areas of intervention in which they usually work together. Since the assessment to autonomy is made by the nurses, this domain was not contemplated in the grid conceived by the therapists, thus being reserved to be used in other specific studies in this area or in every day care activities. These are the used domains: 1 – Sensorio-Motor, 2 – Cardio-Respiratory, 3 – Verbal and Non-Verbal Communication, and 4 – Language (Chart 10.11). The number 1 domain is divided into several areas: Global Motricity, Balance, Fine Motricity, Postural Changes, Laterality, and Sensorial. The number 2 domain in two areas: Cardio-Respiratory and Swallowing. Domain number 3 is divided into: Means of Communication, Communicative Interactions, and Communicative Functions. Finally, number 4 domain is divided into: Understanding and Expression (Chart 10.12).

As expected, the basic vital functions are the more maintained and it is evident that the more limited areas are those of Communication and Language. However, as referred in some literature, there is a considerable compromise of the sensorio-motor functions.

There is also some compromise of the basic respiratory and swallowing functions that may acquire significance in the long run due to the sample's age range. The high prevalence of those problems correlates with the fact that causes of dysphagia are neurological, structural, and neuromuscular degenerative diseases, situations that are common in our sample (Padovani AR, Moraes DP, Mangili LD, Andrade CRF, 2007). The evaluation grid proved to be essential on the more detailed and objective characterization of each tested individual, allowing a quick and interdisciplinary plan of intervention.

Chart 10.11: Therapists grid by domains

Domain	% Domain
Sensorio-motor	49,92
Cardio-Respiratory and Swallowing Functions	80,35
Oral and Non-Oral Communication	43,81
Language	43,85

Chart 10.12: Therapists grid by areas

Area	% Area
1 - Global Motricity and Balance	49,05
1 – Fine Motricity	40,09
1 – Postural Changes	59,83
1 - Laterality	45,71
1 - Sensorial	53,23
2 - Swallowing	87,10
2 – Cardio-respiratory Function	76,30
3 – Means of communication	41,04
3 – Communicative Interactions	46,86
3 – Comunicative Functions	46,86
4 - Understanding	45,02
4 – Expression	42,38

Nursing found the results for the different areas that determine autonomy (Chart 10.13). Feeding and intimate hygiene are the areas of more autonomy whereas there is great limitation in the remaining areas.

Chart 10.13: DA (FI)

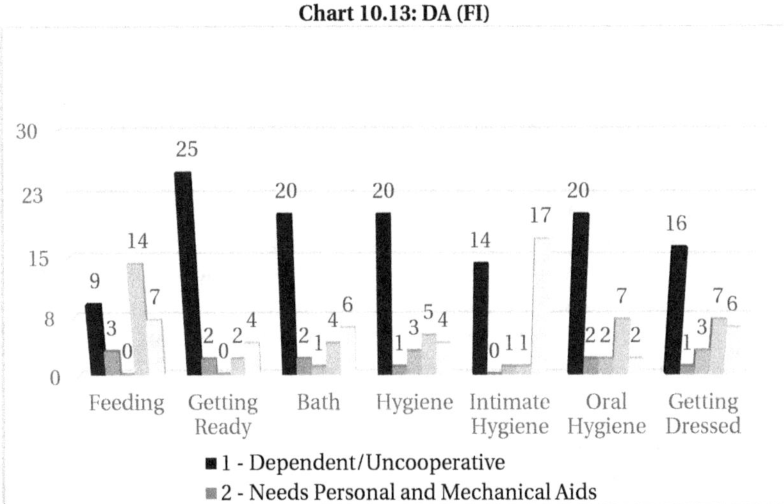

In psychiatry, a predominance of above the average rate was found for the Autism Spectrum Disorder (ASD), and for the Attention Deficit Hyperactivity Disorder (ADHD) diagnosis (chart 10.14), which is in accordance with the experience of the technicians who work in this area as well with data found in literature. Several cases present quotations below the significant threshold for Conduct Disorder, but all of them have significant quotation for ADHD, a diagnosis that includes the identified changes in behavior. Others present under the threshold quotations for psychosis and obsessive disorder, but all of them are encompassed in the cases that punctuate significantly for ASD, which include the identified psychotic and obsessive symptoms. Lastly, although only two people punctuated for Anxiety Disorder (AD), there were 11 more that, besides meaningfully scoring for AD, also punctuated for ADHD and ASD. These are more specific diagnoses that include non-specific AD present in a great number of pathologies, and thus, in these cases, must not be rated as an isolated diagnosis.

Chart 10.14: Results for the ChA-PAS application

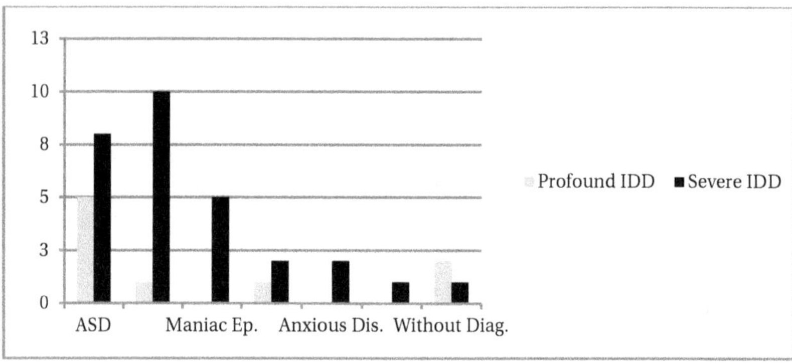

10.4 Discussion and Limitations

We are in the context of a study which orientation follows the model of Investigation-Action, in which the problem being assessed derives from a daily care activity. Usually, in their practice, the technicians are confronted with the need of using non-calibrated evaluation tools, because there are no specific methods or instruments to gather data from this population. That can compromise the reliability of the collected data.

With this work, we open the possibility of easily obtaining a reliable, global, transversal and detailed evaluation of a given person, enabling a more effective intervention plan.

During the time this work took to be completed we faced some evaluation limitations due to the characteristics of this population. Therefore, it was evident the difficulty in assessing symptoms of cognitive nature (for instance, suicidal ideas, self-confidence, delirious ideas) because of the cognitive problems of the people being tested. Also, the personal expression limited the involvement of the person in the study, either because of being bedridden or because of the absence of oral expression or even for not having any expression capacity (the facial expression is stereotyped or severely impaired).

Alongside the expressive and conceptual difficulties, there are also sensorial difficulties that jeopardize the possibility of effective communication with the evaluators, namely blindness and total deafness. Some items, such as sleep and appetite, are shaped by medication and daily routines at the institution. Thus, we clearly identified the need to use the history data, to support our understanding of the person's natural way of functioning. This process has been adopted by all technical areas, along with the use of the new instruments created during the study, and the adapted use of former (previous?) instruments (e.g., the utilization of subscales of the WISC by psychology).

We still have to refer the inherent limitations of the contexts in which the study was carried, with an intense care and technical activity of all the staff, as well as the time limit for concluding the project. Moreover, the dispersion of the assessed items and the involvement of various entities, concur to the complexity of the study deployment well beyond what was initially apparent.

10.5 Conclusion

We may conclude that most of the studied population show severe limitations in their life quality and autonomy associated with the multiple pathologies present in IDD. Those limitations are of sensorial, expressive and motor nature. Even in a study designed to address the specificity of people with Severe and Profound IDD, there were significant difficulties during the evaluation process. It will be possible to conduct a more structured assessment using adapted instruments that will enable a better-oriented intervention. The use of a set of instruments designed in agreement with the structure of an interdisciplinary team allows for a more effective evaluation (less work loss and easier application in the intervention plans).

The prevalence of psychiatric pathology associated with IDD is much higher in the studied population than in the data found in the literature (91% to 30-40%). That may be explained by the institutional context in which the study was carried, since, as stated before, CRM gives priority to people with Double Diagnosis, in whom IDD is associated with another psychiatric diagnosis. The outside social network is either insufficient or inexistent. The cognitive evaluation must be descriptive, by clarifying the areas better functioning and avoiding quantitative evaluations that are of little guidance for the intervention.

The basic vital functions are the more maintained along greater limitations in language and communication. The activities more autonomous are those of intimate hygiene and of feeding. The fact that there is less variety of psychiatric diagnosis within the population with Profound IDD is in line with the literature data. Nevertheless, we still have to question this fact in the face of the described limitations, namely in what concerns the expressive abilities of the subjects even more so because they are more relevant in people with multidificiency.

10.5.1 Future Orientations

Even with adapted instruments, the evaluation of people with IDD, especially of Severe and Profound type, is always a challenge. Therefore, to confirm the data of the present study, further studies are needed, using these or other instruments, better adapted to the characteristics of this population. It is also pertinent in order to an increasing number of technicians acquiring more experience and confidence in the use of these instruments, and to broaden the evidence of multidisciplinary needs in these cases. The exploration of new instruments, its translation, calibration and validation for the Portuguese population can be an inestimable added value for the people with IDD, by helping the people that work with them, regardless of their background education, to detect the presence of concomitant pathology. Carrying out the study, we found new limitations on using the instruments recommending that they are modified in the future. This is the context of a study that does not close in itself but, above all, gives us leads for the future, evolving and stimulating the initiative and development of others. The sensibilization and training of health care professionals for the reality of Double Diagnosis in IDD is crucial, given the dimension of this population. In Portugal a structured education on IDD still does not exist, neither in pre or post-graduate curricula. Only with the production of scientific evidence on this matter will we have arguments to claim the needed curricular changes that include this subject.

Acknowledgments

This study was financially supported by Plano Nacional de Saúde Mental (PNSM) – Direcção Geral da Saúde (DGS) – Ministério da Saúde, to whom we thank this support as well as the incentive to carry out this work. The technical staff at CRM thanks all the CRM community, namely its residents and their families, for their availability and the support they gave to this project at all times.

10.6 References

Albuquerque, M. C. (2000). *A criança deficiente mental ligeira : aspectos comportamentais e familiares.* INR, Lisboa.

Barnhill, L. J. & Kartheiser, P. H. (2006). *Textbook of Neural Repair and Rehabilitation.* Cambridge University Press. Vol. II. Chapter 6. Pag. 89-96.

Carvalho, A, (2012). Conferência "Compromisso Institucional com Testemunho Pessoal". Encontro de Trabalho Saúde Mental na Deficiência Intelectual,2012-Oct-12. *Organização FENACERCI. Fundação Calouste Gulbenkian,* Lisboa.

Choi, J. Y., Rha, D. & Park, E. S. (2016). Change in pulmonary function after incentive spirometer exercise in children with spastic cerebral palsy: a randomized controlled study. *Yonsei Medical Journal.* 57(3). 769-775.

Cooper S.A., Smiley E., Morrison J., Williamson A., Allen L. (2007). Mental ill-health in adults with intellectual disabilities: prevalence and associated factors. *British Journal of Psychiatry,* 190, 27-35

Day, B.L., & Cole, J. (2002). Vestibular-evoked postural responses in the absence of somatosensory information. *Brain,*125: 2081-2088.

Deb, S., Thomas, M. & Bright, C. (2001). Mental disorder in adults with intellectual disability: prevalence of functional psychiatric illness among a community-based population aged between 16 and 64 years. *Journal of Intellectual Disability Research,* 45, 495– 505.

Franki, I., Desloovere, K., De Cat, J., Feys, H., Molenaers, G., Calders, P., Vanderstraeten, G., Himpens, E &, C. Broeck, C. (2016). The evidence-base for basic physical therapy techniques targeting lower limb function in children with cerebral palsy: a systematic review using the international classification of functioning, disability and health as a conceptual framework. *Journal Rehabilitation Medicine.* 44. 385-395.

Frick, A. & Möhring, W. (2016). A matter of balance: motor control is related to children´s spatial and proportional reasoning skills. *Frontiers in Psychology.* 6.

Katz, G. & Lazcano-Ponce, E. (2008). Intellectual disability: definition, etiological factors, classification, diagnosis, treatment and prognosis. *Salud Pública de México.* 50 (12). 132-141.

Mercer, J. (1973). *Labeling the mentally retarded: clinical and social system perspective on mental retardation.* University of California Press.

Moss, S. (2001). "Psychiatric Disorders in Adults with Mental Retardation", *International Review of Research in Mental Retardation,* 24, 211- 243.

O'Hara, J., Sperlinger, A. (1998). *Adults with learning disabilities: a pratical aproach for health professionals.* Wiley.

Ortiz, K. Z. (2006). *Distúrbios neurológicos adquiridos: fala e deglutição.* (1ª edição). Barueri, São Paulo: Editora Manole.

Padovani, A. R., Moraes, D. P., Mangili, L. D., & Andrade, C. R. F. (2007) Protocolo fonoaudiológico de avaliação do risco para disfagia (PARD). *Revista de Sociedade Brasileira de Fonoaudiologia.* [online], vol.12, n.3, pp.199-205.

Raghavan, R. (1997). Anxiety disorders in people with learning disabilities: a review of the literature", *Journal of Learning Disabilities for Nursing, Health and Social Care,*2, 1, 3–9, 199.

Song, G. S., Kim, J. J. & Park, E. C. (2015). The effect of Swiss ball exercise and resistance exercise on balancing ability of scoliosis patients. *Journal Physical Therapy Science.* 27(12). 3879–3882.

Tejada, M.I. (2006). Retraso mental de origen genético. Presentación de la Red GIRMOGEN", *Revista de Neurología,* 42 (Supl 1): S1-6, *2006.*

Winfield, N.R., Barker, N.J., Turner, E.R. Qu, I G.L. (2014). Non-pharmaceutical management of respiratory morbidity in children with severe global developmental delay. *Cochrane Database Syst Rev.* 19 (10).

Whitaker S. & Read S. (2006). The prevalence of psychiatric disorders among people with intellectual disabilities: an analysis of the literature. *Journal of Applied Research in Intellectual Disabilities,* 19, 330-45.

Chapter 11

Assessment of Outcomes at Casa de Alba: Learning from the residents' views

Daniel Guerra
Metalentejo, Portugal

João G. Pereira
Romão de Sousa Foundation, Estremoz, Portugal

Célia Sales
University of Porto, Portugal

In Portugal, service user perspectives to outcome assessment have not yet been seriously considered. We present preliminary results of a small participatory research study involving the patients from Casa de Alba (CA –Residential Therapeutic Community in Mental Health), aiming to explore their views on the outcome measures that form part of the routine assessment at Casa de Alba. A focus group was carried out, where the patients were asked to speak their minds on the advantages and disadvantages of two well-established nomothetic measures (BSI and CORE-OM), and two client-generated outcome measures (CGOMs) (PQ and PSYCHLOPS). Measures were individually rated for its appropriateness until a consensus on the overall score was reached. Suggestions as to possible changes to the assessment protocol being carried out at Casa de Alba were also made, and a discussion of the appropriateness of the procedures ensued. Participants rated nomothetic measures slightly higher than CGOMs; the former was seen as comprehensive, but negatively charged; the latter were described as providing more freedom in choosing topics, but the item-generation procedures were considered difficult. The implications of the findings to outcome instruments in general, and to the assessment process at Casa de Alba, are discussed.

11.1 Introduction

Service user involvement is of cardinal importance because it promotes the empowerment of patients, patient-centered care, and service quality (Mockford, Staniszewska, Griffiths, & Herron-Marx, 2012; Rutter, Manley, Weaver, Crawford, & Fulop, 2004). Service user involvement implies considering the voice of patients regarding decisions that affect care at different levels, such as public policy, service management, treatment planning, and provision and evaluation of outcomes (Alves, Sales, & Ashworth, 2013; Crawford et al., 2011). Nonetheless, decisions regarding which measures are better suited to assess treatment outcomes do not frequently include user perspectives (Perry & Gilbody, 2009). This is even more evident when it comes to therapeutic communities where, to our knowledge, only one study addressing the user perspectives to outcome measures has been carried out in Portugal (Alves, Sales & Ashworth, 2016). At the same time, there is a growing recognition of the need to extend service user involvement to research settings, given that the research priorities of users can be different from those of mental health staff (e.g., Thornicroft, Rose, Huxley, Dale & Wykes, 2002). By using participatory research methods, we aimed to learn from the residents at Casa de Alba, regarding their views and preferences of the routine outcome assessment procedures, while also allowing their perspectives to influence future research priorities.

Outcome assessment and patient monitoring is generally carried out using two main types of tools, namely nomothetic or idiographic instruments. Traditional nomothetic instruments, made up of researcher and clinician defined pre-set items, are considered the gold standard measures for the assessment of treatment outcomes (Ashworth, Evans, & Clement, 2009). Nomothetic measures have the advantage of extensive evidence of their psychometric properties, and the possibility of deriving norms and establishing cut-off scores, which facilitates the clinical interpretation of scores (Evans, Margison & Barkham, 1998; Sales & Alves, 2012). However, it is not clear whether nomothetic measures reflect the perspectives of service users. Even when its development included feedback from service users (e.g. CORE-OM, Evans et al., 2000), clinician concerns regarding symptoms and psychopathology were generally favored over outcomes that are mostly of concern to patients (e.g. medication side-effects) (Perry & Gilbody, 2009; Shumway et al., 2003). On the other hand, idiographic measures (most commonly, Client-Generated Outcome Measures, CGOMs) give the patient the chance to select which topics or domains of his/her concern should be used to measure progress (Fitzpatrick, Davey, Buxton & Jones, 1998). These individualized items are generated by the patients using a standardized structure, resulting in instruments that only contain items that are meaningful. By avoiding the assessment of irrelevant items, CGOMs save time of administration (Wagner & Elliott, 2001) and are more sensitive to change (Ashworth et al., 2004; Ashworth et al., 2009).

Very little is known about the opinion of users regarding outcome instruments, and knowledge is even more scant regarding their preferences for nomothetic or idiographic measures. Crawford et al. (2011) questioned expert groups of service users about their views on the relevance and acceptability of widely used outcome measures. Overall, users favored patient-rated measures over clinician-rated measures and emphasized the importance of the process of collecting data over the content of the measures used (e.g. personal qualities of the interviewer, such as sensitivity to mental health issues). In addition, users were dissatisfied with measures that provided a single score that aggregated functioning with mental health, since users considered that these two dimensions are relatively independent. Lastly, users disagreed with some of the theoretical underlying assumptions (e.g. better social functioning must mean a good relationship with family members). Also, Perry and Gilbody (2009) found that the emphasis given by users to some outcome domains was not in line with the domains considered most important in current outcome measures. When taken together, these studies point to the need of voicing the users' perspectives regarding outcome assessment instruments and procedures.

11.2 The involvement of residents in the outcome monitoring at Casa de Alba

Casa de Alba is a residential therapeutic community with the primary aim of helping individuals and families affected by severe mental health problems. A multidisciplinary team provides support in promoting autonomy and self-management in everyday activities, and in improving quality of life and social integration. These goals are strived for using a relational-integrative framework and therapeutic community theory. Casa de Alba bends towards the democratic type of therapeutic community, integrating contemporary relational thinking whilst detaching itself from radical anti-psychiatry models, characteristic of the 1960´s (Laing, 1960; Pereira and Romão de Sousa, 2014). Everything that happens in Casa de Alba is taken as an opportunity to learn. The relational environment within the house is the main therapeutic tool – this is called the therapeutic milieu. Apart from the therapeutic potential of the day to day activities, there are a number of formal therapies in place. Dyadic psychotherapy, group and multi-family therapy are central. A number of other activities take place over the 5-Day week and also on Saturdays

(Sunday is free of activities): psychomotor therapy, art therapy, horse riding, swimming, gardening, occupational therapy, motivational groups, employment and educational tasks (Pereira, 2015).

At Casa de Alba, the aim is to involve patients in decisions regarding their own treatment and other residents´ treatment. Staff follow participatory research approach, i.e., involving residents in planning and conducting the research process of the topics that affect them (Bergold & Thomas, 2012). This is also the case in what concerns to evaluation tools. This study was a first step in identifying residents' perspectives regarding the advantages and disadvantages of outcome measures in use, so that changes to the outcomes' assessment protocol of Casa de Alba could be made accordingly. It is part of a larger project that aims to listen to the views of residents and staff regularly in order to involve them in the protocol for outcome measuring and progress monitoring. Ultimately, we aim to inform the literature and research in this field. The present paper describes preliminary results concerning the residents' experience and preferences of using nomothetic and individualized outcome tools in routine monitoring in a residential setting.

11.3 Method

11.3.1 Outcome assessment and monitoring at Casa de Alba – pilot protocol

Since its start, Casa de Alba has planned to use routine outcome assessment as part of its quality strategy. The strategy was to start with a pilot protocol to be later discussed and redesigned with residents and the clinical team, following a strategy of ongoing participatory-action research (Project *Quality assessment of Casa de Alba: a patient-centred process*). This pilot protocol included three nomothetic measures (CORE-OM, BSI, and the Global Assessment of Functioning Scale - DSM-IV-TR) administered at intake, at every three months, before discharge, and at follow-up (6 and 18 months); and two CGOMs (PQ and PSYHCLOPS), also administered at intake and before each weekly psychotherapy session, to monitor progress regarding individualized aspects of each patient. The reason for combining these two CGOMs (both measuring psychological distress) was to gain the resident's views regarding the most appropriate for residential care formats.

Casa de Alba also took part in the testing of the Individualized Patient-Progress System (IPPS, Sales et al, 2014), the first feedback system combining standardized measures with CGOMs, granted by the Portuguese Foundation of Science and Technology in collaboration with the CORE IMS (www.coreims.com) and is integrated in the CORE-NET System (Sales & Alves, 2012). Data from PQ and CORE-OM were electronically registered via the IPPS system, which would provide real time graphical representation of progress over the course of treatment.

11.3.2 Brief Symptom Inventory (BSI; Derogatis, 1982; Portuguese version)

The BSI is a widely used and accepted screening tool for general psychopathology that constitutes a brief version of the SCL-90-R (Derogatis & Melisaratos, 1983). It comprises 53 items, rated on a 5-point Likert-Scale, that assess nine primary symptom dimensions of psychopathology: somatization, interpersonal sensitivity, obsessive-compulsive, anxiety, depression, phobic anxiety, hostility, paranoid ideation, and psychoticism. Scorings provide results on each dimension, a global severity index, a positive symptom total, and a positive symptom distress index (Derogatis, 1993). Several studies have attested the good psychometric properties of the BSI, regarding reliability (e.g. Derogatis & Spencer, 1982), internal consistency (e.g. Hayes, 1997; Ruipérez, Ibáñez, Lorente, Moro, & Ortet, 2001), and convergent validity (e.g. with the MMPI; Derogatis & Spencer, 1982).

11.3.3 Clinical Outcomes in Routine Evaluation – Outcome Measure (CORE-OM; Evans et al., 2000; Portuguese version)

CORE-OM is a 34-items measure of psychological distress that assesses four domains: Problems, Functioning, Subjective Well-Being, and Risk (Evans et al., 2000). Its development was carried out in the UK in response to the lack of a general outcome measure suitable for routine settings and included input from mental health practitioners (Barkham et al., 2001; Mellor-Clark et al., 1999). In one of the most cited studies on the psychometric properties of the CORE-OM, the authors concluded that there was good internal consistency, test-retest reliability, and convergent validity. There was also evidence of the instruments' responsiveness.

11.3.4 Psychological Outcome Profiles (PSYCHLOPS; Ashworth et al., 2004; Portuguese version)

PSYCHLOPS is a self-report CGOM that measures psychological distress and that was initially developed for routine use in primary health services (Ashworth et al., 2004). It is one page-long and contains four items that assess three domains: Problems, Functioning, and Well-Being (Ashworth et al., 2009). Apart from the item from the Well-Being domain, all items are generated by the patient using a free-text box where he/she is invited to select the problems that are of greatest concern to him/her, and what has become difficult to perform as a consequence of those problems. Items are rated on a 5-point scale to indicate how much the problem(s) have bothered the patient in the last week, and how much difficulty was felt performing the activities or tasks he/she indicated previously (Functioning item). Good internal consistency, satisfactory test-retest reliability, good responsiveness, and good convergent validity of PSYCHLOPS with a nomothetic measure have been found (Ashworth et al., 2009; Czachowski et al., 2011; Evans et al., 2010).

11.3.5 Personal Questionnaire (PQ; Wagner & Elliott, 2001; Portuguese version)

Like PSYCHLOPS, the PQ is a CGOM where the patients are allowed to generate and rate the items that cause them psychological distress (Wagner & Elliott, 2001). However, unlike PSYCHLOPS, the items in the PQ are generated via a semi-structured interview that is carried out before treatment, where the interviewer helps the patient develop a list of problems that caused him or her to seek help (Sales & Alves, 2016). The final items are stated in the patient's own words and are rated using a 7-point scale (Wagner & Elliott, 2001). There is no limit regarding the number and content of the items. Evidence stemming from three countries show that PQ is a measure with has good internal consistency, moderate test-retest reliability, and good convergent validity (Elliott, Wagner, Sales, Rodger, Alves and Café, 2016).

11.3.6 Participants and procedure

All residents staying at Casa de Alba in November, 2015, were invited to participate in the study. Four residents refused to take part in the study, without providing any further explanation. The three service users that made up the group were all inpatients, taking part in programs that lasted between 9 and 18 months. Two participants were male. Ages ranged from 35 to 44 years old (m = 40.3). All participants suffered from severe disturbance; two of the participants had a diagnosis of Schizophrenia, and one was diagnosed with mixed Personality Disorder.

11.3.7 Data collection

Approximately one week before the research team arrived, written information was given to the staff at Casa de Alba. The staff then proceeded to discuss this information with all residents in a group meeting, to ensure that everyone was clear on the purpose of the research, and on the voluntary nature of participation. The invitation to take part in the study was made after this discussion. A preliminary and informal inquiry regarding the intention to participate was carried out, with four residents agreeing to take part in the study. One of the residents ended up choosing not to take part in the study, and so the group was carried out with three residents. On the day of the meeting, the first and third authors of this study had lunch with all residents and staff, in order to make the meeting less formal and to reduce the residents' anxiety regarding participation. The meeting took place after lunch at Casa de Alba premises.

The meeting started with a group re-reading of the purpose of the study, and participants were reminded of the voluntary nature of participation and thanked for choosing to take part in the study. After making sure that there were no questions regarding the study, participants were invited to freely express their opinions on the instruments used at Casa de Alba (PSYCHLOPS, BSI, PQ, and CORE-OM). This was done sequentially (i.e. participants were asked to state their perspectives on each individual instrument before moving on to the next one). After discussing each instrument, participants were asked to provide a rating on a 0-10 points scale regarding the usefulness and appropriateness of the instruments (e.g., considering any negative aspects regarding difficulty in understanding the content, excessive items, phrasing, and so on). A low score meant the instrument was not very useful or appropriate and had several negative aspects, while a high score meant the instrument was highly appropriate for measuring progress and had few negative aspects. After reviewing all the instruments, the average rating for each instrument was obtained and shared with the participants. The combined averages of the nomothetic and idiographic groups were also obtained and shared with the participants. The purpose of this exercise was (1) to assess whether the participants agreed that the average was a fair representation of the usefulness and appropriateness of each instrument, or if the group considered that changes should be made in order to reach a consensus, and (2) to outline the differences obtained between the ratings of the nomothetic and idiographic instruments, so that participants could speak their minds as to the fairness of any differences obtained. Finally, participants were invited to propose recommendations regarding the outcome assessment protocol at Casa de Alba. After completing these steps, participants were asked whether they wished to make any further contributions and were thanked for their time. A new focus group was settled, where results obtained would be shared with the participants, and any changes made to the assessment protocol at Casa de Alba as a result of the data collected would be reported. During the group, notes were taken regarding the most significant and meaningful statements by the patients (one of the residents opposed the recording of the session, and so we had to rely on note taking).

11.4 Results

Table 11.1
Overall mean for each measure, before and after group discussion, and sample of patient statements.

Outcome Measure	Overall Mean Before Group Discussion	Overall Mean After Group Discussion	Sample Patient Statements
PSYCHLOPS	6,66	7	"Understanding what the main problem is is hard, but it's a good thing." "Me distinguishing the most troubling problem is difficult, here [at Casa de Alba] the problem may not be present or be something else."
CORE - OM	8	8	"This one is not as heavy [as the BSI]. Even though it has negative symptoms it is written with more neutral words, I don't feel it is as heavy." "I feel like I am being assessed, I feel the look of someone else, on this this one and on the BSI."
PQ	6	7	"The freedom [to generate and add problems] isn't necessarily positive because I have not changed anything [the problems]." "I can't relate to this instrument because the problems I included in the first one [PSYCHLOPS] are the same."
BSI	7	6,5	"It´s good in practical terms, but it´s bad in emotional terms. It´s very heavy." "I think it encompasses a lot, but at the same time it is negative and heavy. Everything [the content] is negative. At the same time, I can understand the need to assess everything."

Residents gave an overall positive mean rating of the instruments (7.13 out of 10). After nominal consensus, the CORE-OM was the measure considered most useful and appropriate (scoring 8), followed by PQ and PSYCHLOPS (scoring 7 points) and BSI (6.5) (see Table 11.1.).

The experiences of answering CORE-OM and BSI, which have pre-set items and only require that residents rate its intensity, were clearly distinct from answering PQ and PSYCHLOPS, that involved residents creating their own items. Nomothetic tools were seen as designed to reach a diagnosis, to formally assess all the clinically relevant symptoms ("it is all there", BSI), and participants said they trusted the accuracy of the clinical assessment provided. However, answering it, especially BSI, was emotionally challenging because the items were focused on symptoms and negative experiences ("it is heavy", BSI). The content of the items (i.e. what the question was about) contributed as much to this negative emotional experience as the clinical and "cold" phrasing of the items (i.e. how the question was written). Moreover, the rating scale on the BSI was criticized for not allowing precise answers (i.e. more options should be made available). Less perspectives were expressed on idiographic instruments. Participants saw the possibility of identifying idiosyncratic problems as an advantage. However, the item generation process was seen as difficult because identifying the problems that need to be worked on in therapy is a very difficult task when one is suffering. For instance, a specific difficulty was pointed out in the creation of the PQ: after identifying the major problems in an interview, respondents are asked to rank order the problems, indicating the one that bothers him or her the most, and so on; having to order the identified problems considering their importance was reported as a difficult task.

Regarding the process of ongoing assessment of outcomes, residents reported insufficient knowledge on how data was used. Residents felt that there was a disconnection between the assessment and the sessions because their answers to the instruments were not debated in session, nor was progress jointly reviewed with their therapists. Because someone other than their therapist administered the tools, all residents wondered if and how results were discussed among professionals and used for clinical decision making, and they manifested interest in knowing more about this process. Moreover, one resident experienced distress with the process of weekly monitoring, because he thought that not presenting progresses would mean staying longer at Casa de Alba. Importantly, residents indicated they had never shared these issues with their therapists. Finally, participants also felt that answering both PSYCHLOPS and PQ every week was redundant because the items generated on both measures were identical.

11.5 Discussion

In this study, we present the views of the residents at Casa the Alba regarding the instruments used to assess psychological change and the procedures used to monitor clinical progress. Globally, participants rated the instruments positively, manifesting interest in openly discussing its meaning with their therapist as a part of the therapeutic work.

CORE-OM was rated the highest, BSI rated the lowest, and the CGOMs received intermediate levels of preference. These results suggest that the users' experiences when filling in outcome measures are influenced by other factors besides their nomothetic or idiographic nature. Nomothetic instruments are often accused of not being sensitive to the specific concerns of patients and of mostly reflecting the views of professionals as to what items should measure progress (Evans et al., 2010). CGOMs, on the other hand, are commonly presented as measures that can be adjusted to fit the patient's idiosyncratic problems; measures with personal meaning and therefore preferred by patients. However, residents at the Casa de Alba did not report this experience of freedom in item definition as a positively distinctive aspect of CGOMs. On the contrary, even though CGOMs allowed for the inclusion of highly specific problems, generating, selecting and ranking the items was difficult. Difficulties in item generation had previously been reported by residents in a community for substance misuse treatment, where some patients in a severe condition at intake reported that they could not even feel or understand what was wrong with them (Alves et al., 2016). To what extent are CGOM less suitable for individuals presenting a severely disturbed condition? Could it be the case that some of the difficulties that residents experienced when answering CGOMs were due to the nature of their disturbance? The severe nature of the residents' mental disorder at Casa de Alba (most had been through long-term hospitalizations) may mean that the less structured format of the CGOMs, as well as the level of introspection required to generate the items, is more demanding. However, previous research that used the PQ in a psychiatric day care unit did not find such difficulties (Sales, Neves & Alves, 2017). Further research to understand the views of patients regarding item generation in CGOM, especially in populations with problems under the psychosis spectrum, is required.

Results also suggest that users' preferences may be linked to *what* the instrument is measuring, the dimensions under evaluation. BSI focuses more on psychopathological issues, which is a domain more sensitive that residents found harder to talk about, while the remain measures focus on general distress, and the CORE-OM includes positive dimensions, such as wellbeing. Moreover, emotional reactivity could be related to *how* the items are phrased. However, perhaps more than the characteristics of the items or the format of the measures, the meaning of the assessment and how it influences the course of therapy were central issues for residents. Monitoring outcomes may mean being systematically confronted with the failure to improve when rating the same items with the same intensity week after week. Residents viewed assessment as part of the treatment and held unspoken expectations regarding how their therapist was informed about their answers, and how it influenced the course of care.

These testimonies show the complexity of variables present when incorporating outcomes assessment into the clinical routine, and the need to involve users and therapists in the definition of the procedures. Participatory action-research, as the present study, can be a valuable approach, resulting in more feasible, creative and useful solutions. In this specific case, the results of the focus group were shared with the residents in a second visit for validation and for eventual comments. The contribution of one resident regarding the importance of choosing the right measures to monitor change was enriching. From her point of view, the intensity of the PQ and the PSYCHLOPS items could not change from one week to the other since they corresponded to the problems that affect them the most and that will cause them a high degree of distress, perhaps for their entire lifetime. However, this resident observed that she was having small improvements, small gains that were not captured by any of these tools (e.g. waking up on time, taking a shower, etc.). This revelation suggested that perhaps an individualized measure based on objectives, such as the Goal Attainment Scaling (Kiresuk & Sherman, 1968) could be more appropriate to monitor progress in residential units with the characteristics of Casa de Alba. The results of the focus group were also shared with the clinical team. In light with the dissatisfaction voiced by participants on how therapy sessions did not include the discussion of results obtained by the instruments, the assessment process was adjusted: the purpose of data collection was clarified with each resident and the forms started to be taken into therapy to allow for discussion. The recommendation to discuss expectations regarding progress with residents was also made to the clinical team, who later explored with the residents new ways of transforming the ongoing assessment of outcomes into a useful and meaningful experience.

There are some limitations to this study that should be addressed. The small number of cases may not be a problem *per se* as in most qualitative research the aim was not to generalize but rather to provide a rich, contextualized understanding of specific aspects of human experience through the intensive study of particular situations. Even if generalization was not an aim, a number of reasons could be pointed out to explain why only three (out of seven residents approached) accepted to take part: firstly, the fact that the group was conducted on the same day that the research team were introduced to the residents could have prevented less willing residents to participate. It would have been beneficial if a prior encounter had been arranged, where the research team introduced themselves to the residents and provided a detailed explanation of the study objectives. Secondly, it could also be the case that the low number of participants was related to the lack of clarity surrounding the purpose and usefulness of monitoring progress and measuring outcome. Residents could also feel their opinion on these instruments was not important, which would warrant action towards increasing the value attributed to the users' perspectives. Given that all participants were suffering from severe mental health disturbance, it could also be the case that the mental state of the residents could have biased the results. The less structured nature of the CGOMs, as well as the level of introspection required to generate the items, is more demanding, and might not be the most suitable for patients experiencing severe disturbance.

Another limitation of this study was the uncertainty regarding the interpretation of the preference ratings since it was difficult to discriminate what was being assessed: the instrument itself, the administration format, or both? For instance, it might be unreasonable to compare the rating of appropriateness of CGOMs directly against the nomothetic measures in the first place, since the latter were used every three months, while the former was used weekly. Also, PQ and PSYCHLOPS were criticized for being redundant in this weekly administration, which was not surprising given that both intend to assess psychological distress and do so in a very similar manner.

From a research point of view, two main conclusions can be drawn from these findings. Firstly, the assumption that patients will invariably prefer instruments that can adjust to their specific problems is unwarranted. Secondly, the presentation of nomothetic and idiographic instruments as opposing alternatives may not reflect the users' perspectives. Patients do not see one type of instruments as better suited to assess progress than the other; nomothetic and idiographic instruments complement each other and allow for different contributions to patient monitoring and outcome assessment. These findings support the hybrid approach presented by some authors (e.g. Sales & Alves, 2012), where positivist and constructivist assumptions are brought together to improve outcome assessment. In this study, the complementary nature that users perceive between nomothetic and idiographic instruments indicate that a hybrid approach to outcome assessment could be more than a theoretical proposition; it seems to reflect the users' perspectives. This is also in line with current views that present clinical trials methodology and practice-based research as complementary; by extension, the same could be argued in relation to nomothetic and idiographic measures (Barkham, 2016). In a way, these findings help to settle the dispute between which type of instruments are better suited for patient monitoring and outcome assessment because they suggest that users view nomothetic and idiographic instruments as complementary. The question remains as to how nomothetic and idiographic instruments should be combined and integrated into treatment. In this matter, users have much to say.

11.6 References

Alves, P., Sales, C., & Ashworth, M. (2013). Enhancing the patient involvement in outcomes: a study protocol of personalised outcome measurement in the treatment of substance misuse. *BMC Psychiatry,* 13, 337.

Alves, P., Sales, C., & Ashworth, M. (2016). "It is not just about the alcohol": service users' views about individualised and standardised clinical assessment in a therapeutic community for alcohol dependence. *Substance Abuse Treatment, Prevention, and Policy, 11,* 25.

Ashworth, M., Evans, C., & Clement, S. (2009) Measuring psychological outcomes after cognitive behaviour therapy in primary care: a comparison between a new patient-generated measure "PSYCHLOPS" (Psychological Outcome Profiles) and "HADS" (Hospital Anxiety and Depression Scale). *Journal of Mental Health, 18,* 169-177.

Ashworth, M., Shepherd, M., Christey, J., Matthews, V., Wright, K., Parmentier, H., Robinson, S., & Godfrey, E. (2004). A client-generated psychometric instrument: the development of 'PSYCHLOPS'. *Counselling and Psychotherapy Research, 4,* 27-31.

Barkham, M. (2016). Patient-Centered Assessment in Psychotherapy: Toward a Greater Bandwidth of Evidence. *Clinical Psychology: Science and Practice, 23,* 284-287.

Barkham, M., Margison, F., Leach, C., Lucock, M., Mellor-Clark, J., Evans, C., Benson, L., Connell, J., Audin, K., & McGrath, G. (2001). Service Profiling and Outcomes Benchmarking Using the CORE-OM: Toward Practice-Based Evidence in the Psychological Therapies. *Journal of Consulting and Clinical Psychology, 60,* 184-196.

Bergold, J. & Thomas, S. (2012). Participatory Research Methods: A Methodological Approach in Motion [110 paragraphs]. *Forum Qualitative Sozialforschung / Forum: Qualitative Social Research, 13.* Art. 30, http://nbn-resolving.de/urn:nbn:de:0114-fqs1201302.

Crawford, M.J., Robotham, D., Thana, L., Patterson, S., Weaver, T., Barber, R., Wykes, T., & Rose, D. (2011). Selecting outcome measures in mental health: the views of service users. *Journal of Mental Health, 20,* 336-346.

Czachowski, S., Seed, P., Schofield, P., Ashworth, M. (2011). Measuring Psychological Change during Cognitive Behaviour Therapy in Primary Care: A Polish Study Using 'PSYCHLOPS' (Psychological Outcome Profiles). *Plos One, 6,* 1-6.

Derogatis, L.R. (1982). *BSI: Brief Symptom Inventory.* Minneapolis, MN: National Computers Systems.

Derogatis, L. R. (1993). *Brief Symptom Inventory: Administration scoring and procedures manual* (3rd ed.). Minneapolis, MN: National Computer Systems.

Derogatis, L. R., & Melisaratos, N. (1983). The Brief Symptom Inventory: An introductory report. *Psychological Medicine, 13,* 596–605.

Derogatis, L. R., & Spencer, M. S. (1982). *The Brief Symptom Inventory (BSI): Administration, scoring, and procedures manual-I.* Baltimore, MD: Johns Hopkins University School of Medicine, Clinical Psychometrics Research Unit.

Elliott, R., Wagner, J., Sales, C., Rodgers, B., Alves, P., & Café, M. (2016). Psychometrics of the Personal Questionnaire: A Client-Generated Outcome Measure. *Psychological Assessment, 28,* 263-278.

Evans, C., Ashworth, M., & Peters, M. (2010). Are problems prevalent and stable in non-clinical populations? Problems and test-retest stability of a patient-generated measure, PSYCHLOPS (Psychological Outcome Profiles), in a non-clinical student sample. *British Journal of Guidance &Counseling, 38,* 431-439.

Evans, C., Margison, F., &Barkham, M. (1998). The contribution of reliable and clinically significant change methods to evidence-based mental health. *Evidence Based Mental Health, 1,* 70-72.

Evans, C., Mellor-Clark, J., Margison, F., Barkham, M., Audin, K., Connell, J., & McGrath, G. (2000). CORE: Clinical Outcomes in Routine Evaluation. *Journal of Mental health, 9,* 247-255.

Fitzpatrick R., Davey C., Buxton M.J. & Jones D.R. (1998) Evaluating patient-based outcome measures for use in clinical trials. *Health Technology Assessment, 2,*1–74.

Hayes, J. A. (1997). What does the Brief Symptom Inventory measure in college and university counselingcenter clients? *Journal of Counseling Psychology, 44,* 360–367.

Kiresuk, T., & Sherman, R. (1968). Goal attainment scaling: a general method of evaluating comprehensive mental health programmes. *Community Mental Health, 4,* 443-453.

Laing, R.D. (1960) *The Divided Self.* London: Tavistock.

Mellor-Clark, J., Barkham, M., Connel, J., & Evans, C. (1999). Practice-based evidence and need for a standardized evaluation system: Informing the design of the CORE system. *European Journal of Psychotherapy, Counselling and Health, 3,* 357-374.

Mockford, C., Staniszewska, S., Griffiths, F., & Herron-Marx, S. (2012). The impact of patient and public involvement on UK NHS health care: a systematic review. *International Journal for Quality in Health Care, 24,* 28-38.

Pereira, J. G. (2015) An Integrative-Relational Framework of Intervention at Casa de Alba: therapeutic community for severe mental health problems. *British Journal of Psychotherapy Integration, 11*(2), 37-52.

Pereira, J. G., & Romão de Sousa, J. (2014) Casa de Alba: therapeutic community for severe mental health problems. *European Journal of Psychotherapy and Counselling, 16,* 61-68.

Perry, A., &Gilbody, S. (2009). User-defined outcomes in mental health: A qualitative study and consensus development exercise. *Journal of Mental Health, 18,* 415-423.

Ruipérez, M. Ã., Ibáñez, M. I., Lorente, E., Moro, M., &Ortet, G. (2001). Psychometric properties of the Spanish version of the BSI. Contributions to the relationship between personality and psychopathology. *European Journal of Psychological Assessment, 17,* 241–250.

Rutter, D., Manley, C., Weaver, T., Crawford, M.J., &Fulop, (2004). Patients or partners? Case studies of user involvement in the planning and delivery of adult mental health services in London. *Social Science & Medicine, 10,* 1973-1984.

Sales, C. & Alves, P. (2012). Individualized Patient-Progress Systems: why we need to move towards a personalized evaluation of psychological treatments. *Canadian Psychology, 53,* 115-121.

Sales, C., & Alves, P. (2016). Patient-centered assessment in psychotherapy: a review of individualized tools. *Clinical Psychology: Science and Practice, 23,* 265-283.

Sales, C.M.D., Alves, P.G., Evans, C., & R. Elliott, onbehalfof IPHA Group (2014). Individualized Patient Progress System (IPPS): A decade of international collaborative networking. *Counselling & Psychotherapy Research, 14*(3), 181–191.

Sales, C., Neves, I., & Alves, P. (2017). Capturing and missing the patient's story through outcome measures: a thematic comparison of patient-generated items in PSYCHLOPS with CORE-OM and PHQ-9. *Health Expectations, 00,* 1-5.

Shumway, M., Saunders, T., Shern, D., Pines, E., Downs, A., Burbine, T., et al. (2003). Preferences for schizophrenia treatment outcomes among public policy makers, consumers, families, and providers. *Psychiatric Services, 54,* 1124–1128.

Thornicroft, G., Rose, D., Huxley, P., Dale, G., & Wykes, T. (2002). What are the research priorities of mental health service users? *Journal of Mental Health, 11,* 1-5.

Wagner, J., & Elliott, R. (2001). *The Simplified Personal Questionnaire.* Manuscript submitted for publication. Department of Psychology, University of Toledo, Toledo, OH.

Chapter 12

Burnout in teaching: the importance of personal and professional variables

Adelinda A. Candeias
Department of Psychology, University of Évora, Portugal

Inês P. Calisto
Center for Research in Education and Psychology, University of Évora, Portugal

Liberata Borralho
Department of Educational Sciences, University of Évora, Portugal

António Portelada
Department of Educational Sciences, University of Évora, Portugal

In this chapter we present a characterization of the levels of burnout in teachers of Portuguese public schools in function of personal and professional variables. A total of 9,269 teachers participated in the study, which makes up about 7% of the pre-school and school teachers from all regions of Portugal. Participants completed the Maslach Burnout Inventory (MBI; Maslach, Jackson & Leiter, 1996), which allows a three-dimensional characterization of burnout: personal accomplishment, emotional exhaustion, and depersonalization. A differential burnout pattern was identified based on gender, years of professional experience and professional qualifications. Female teachers experience greater emotional exhaustion, while male teachers feel more depersonalized. Teachers with more years of working experience are more exhausted. Teachers with higher professional qualifications have higher levels of personal and professional accomplishment. This pattern suggests the need for an intervention promoting the health and wellbeing of teachers, oriented towards the training and professional valorization and the prevention of the emotional exhaustion that happens with the career advancement and years of experience.

12.1 Introduction

Teaching is one of the professions most commonly associated with high levels of stress and burnout, and it is classified by the International Labor Organization as a profession of physical and mental risk (Jesus, 2000; Picado, 2005). Since in the past teachers did not display levels of stress higher than those observed in other professional groups, it is appropriate to try to understand the reasons underlying this situation at the present time. In addition to factors related to the current lifestyle, it is also necessary to distinguish between factors found at macro and micro levels (Jesus, 2001).

At the macro level, socio-political factors, such as the excessive number of students, lead to students' lack of motivation and indiscipline, causing stress and burnout and excessive demands on the teachers and, consequently, making them disregard their training and the quality of their work, hindering them from responding adequately to the demands placed on them (Jesus, 2001). The accelerated technological development has also contributed to teachers' stress, as it induces a greater rate of knowledge expiration, and impairs the social image transmitted through the media with its tendency to emphasize the negative aspects of the situations. At a more micro level, the main cause of stress is students' indiscipline, with disruptive situations being more and more frequent and serious, thus making unfeasible all the planning of the classes and damaging the quality of the teaching and learning process intended by the teachers (Jesus, 2001).

Currently, teaching, regardless of the level of education or the type of school, public or private, is becoming a profession subject to numerous contextual psychosocial stressors, with emphasis on mental and emotional overloads (Carlotto, 2011). This job is characterized by high levels of concentration, precision and diversified attention. Emotionally, teachers are required to interrelate with students, parents and guardians, colleagues and technical staff, relationships that can often become conflicting (Salanova, Martínez & Lorente 2005).

Several investigations have established the relationship between stress and the appearance of health problems in teachers. Regarding health issues related to the teaching profession, Scheuch, Haufe and Seibt (2015) mention that there are several international studies that describe the most common health problems among teachers, with emphasis on psychic disorders (depression, stress, and burnout being the most common), vocal problems and musculoskeletal disorders. Stress is not just a concept that vaguely relates to some uncomfortable situation. There is currently evidence that occupational stress, when it is intense, unpredictable, uncontrollable and repetitive or prolonged, has negative consequences at the individual and organizational levels, representing high costs for societies.

One of the possible consequences of prolonged or chronic stress in the teaching profession is the professional burnout syndrome or "teacher syndrome". The term burnout in the teaching profession is used to describe the "degenerative cycle of teacher effectiveness". Exhaustion arises as a consequence of teacher's malaise, which is designated as the set of permanent negative consequences that affect the teacher caused by the joint action of the psychological and social conditions in which teaching is exercised. It is known today that the burnout syndrome can generate psychological and body disorders and mood and conduct fluctuations. Maslach and Jackson (1981) consider burnout as a primarily emotional response, emphasizing work and institutional factors as conditioning and antecedents of the syndrome. It is defined as a psychological disturbance characterized by three dimensions: emotional exhaustion, depersonalization and decreased personal and professional achievement (Schaufeli, Malasch & Marek, 1993). Emotional exhaustion is characterized by a lack of energy, enthusiasm, and a feeling of emotional exhaustion, the main cause being overwork. Depersonalization occurs when, in this case, the teacher begins to treat students and peers distantly and impersonally. Finally, the decrease in personal and professional achievement is characterized by a tendency of the teacher to evaluate themselves negatively, feeling dissatisfied with their professional development and experiencing a decline in their sense of competence and their capacity for social interaction (Maslach, Schaufeli & Leiter, 2001).

The experiences of stress and burnout in the teacher should, therefore, be understood as a threat to their wellbeing, self-esteem and personal values, which may lead to the development of negative feelings, revealed in their professional practice by the diminishing quality of the activities performed in a school context. (Patrão, Rita & Marôco, 2012, p.180).

For example, it may lead to the teacher's falling-out from work, alienation, cynicism, apathy, absenteeism, and ultimately the decision to abandon the profession. The teachers wear off and, in extreme cases, cannot endure it and "give up".

It is important to study occupational malaise since increased knowledge will allow a more effective intervention to prevent it. The study of burnout is an important element in the context of the prevention of occupational risks and the analysis of working conditions on account of the different European rulings that recognize that psychopathologies (mental disorders) may have an occupational-type etiology, although they do not appear in the usual occupational disease lists of European countries. In this sense, teachers are a professional category especially exposed to psychosocial risks. In fact, several investigations (Carlotto, 2002; Gomes et al., 2012) coincide in considering that the level of burnout is higher in teachers than in other professional groups. Without the willingness to relate to the students, the role of the teacher risks being mischaracterized with the automatic and standardized fulfillment of tasks, and consequent loss of richness of the teaching and learning process. According to Carlotto (2002), the burnout syndrome affects many teachers nowadays, manifesting itself through physical and emotional exhaustion, anxiety, frequent irritations, insomnia, in addition to other more serious conditions, such as alcohol abuse and self-medication. Thus, stress and burnout in the teaching profession, in addition to deteriorating the teachers' work relationship with their students and leading to a break in the commitments that the teacher is able to assume, can reduce the quality of teaching and learning, which will affect students.

The depreciation of the traditional role of the teacher we have witnessed along these last two decades in Portugal and, simultaneously, the increase of the social requirements of the role of the teacher, have reaffirmed the pertinence of the study of the phenomenon of stress in this professional group. Therefore, more studies are needed that address this issue and help teachers dealing with a phenomenon that is characteristic of today's society, where time runs swiftly, the tasks to be performed are multiple and complex, and the conflicts between the different roles teachers are obliged to play are constant. In Portugal, since the end of the 1980s and throughout the 1990s (after the approval of the Basic Education Act of 1986 - Law no. 46/86 of October 14th), the references to teacher's malaise in scientific publications, union reports, and public debates have increased (Lopes, 2001). Several studies carried out in the last two decades established the existence of professional burnout, high stress prevalence, reduced working capacity, health problems and professional dissatisfaction among Portuguese teachers in the 2nd and 3rd cycle of basic education and in secondary education and support an in-depth analysis of these problems (Jesus, 2002). The need for further studies has been reinforced by the recent changes in fundamental aspects of the Career's Statute for Early Childhood Educators and Primary and Secondary School Teachers. The coming into force of these changes created new contexts and constraints (e.g. worsening teachers' working hours at school), with repercussions on the personal, family and professional life of these teachers. According to Gomes, Montenegro and Peixoto (2010), the modification of the legal norms decided at the political level can be a source of tension for professionals and organizations, highlighting in their study that 689 Portuguese teachers of 3rd cycle and secondary education observed significant values of occupational stress (close to 40%), and burnout (10% in emotional exhaustion, 3% in low personal achievement and 1% in depersonalization) and several physical health problems and values of professional dissatisfaction close to 20%.

In the case of Portugal, it is important to mention that there is some research on stress and burnout in teachers in general. The studies that do exist, however, show similar conclusions to international ones (Carlotto, 2011; Gomes et al., 2012; Jesus & Conboy, 2001; Jesus, 1996, 2000, 2002, 2006; Martins, 2008; Patrão et al., 2012; Ramos, 2012), namely that Portuguese teachers present high levels of stress and burnout. These data associated with the evidences that show the high impact of burnout on teachers' health and their negative consequences on the quality of learning and teaching (Klusmann, Richter & Lüdtke, 2016; Ramos, 2012; Rumschlag, 2017; Scheuch et al., 2015), justify why in Portugal there is an urgent need to characterize burnout problems, could thus contribute to sensitize government entities to adopt policies that promote the health and wellbeing of teachers.

Based on these evidences and the urgent need to better understand and characterize the situation of Portuguese teachers in terms of stress, burnout and wellbeing, we develop the project: Stress, Burnout and Wellbeing among Portuguese teachers, with the support of the University of Évora and the University of Leipzig (Candeias et al. 2016). This project aims at characterizing the levels of stress, burnout and wellbeing in terms of personal and professional variables.

The study that we are presenting in this chapter is part of this project and intends to characterize the levels of burnout among teachers of Portuguese public schools in function of personal and professional variables.

12.2 Method

12.2.1 Participants

A total of 9269 teachers (Table 12.1.) participated in the study, of whom 7187 (77.5%) are female and 2082 (22.5%) are male. The teachers' age ranged between 21 and 69, with a mean of 50 years of age. In the current sample the level of education most verified is the graduation degree (61.2%).

Table 12.1. *Sample description*

Gender	Degree	Months of service (Median)	Age (Median)	N (%)
Female	Bachelor	312	54	143 (2%)
	Graduation	336	50	4453 (62%)
	Specialization	313	48	766 (10.7%)
	Master Degree	288	48	1541 (21.4%)
	PhD	205	50	284 (4%)
Male	Bachelor	373	57	44 (2.1%)
	Graduation	288	50	1218 (58.5%)
	Specialization	276	49	205 (9.8%)
	Master Degree	229	48	473 (22.7%)
	PhD	217	50	142 (6.8%)

12.2.2 Instruments

In order to analyze the levels of burnout, we use the Maslash Burnout Inventory (MBI), adapted by Martins (2008), to the Portuguese population. The MBI consists of 22 items whose form of response is on a Likert scale of 1 (never) to 7 (every day), and the participants must respond according to the last three months of professional activity. The instrument assesses emotional exhaustion (e.g., I feel tired when I get up to go to work*)*, depersonalization (e.g., I am worried that this work makes me colder emotionally*)* and personal and professional performance (e.g., I feel that I can do useful things in my work).

12.2.3 Procedures

In this research, we proceed in accordance with the Guidelines for Improving the Quality of Web Surveys, based on the Checklist for Reporting Results of Internet E-Surveys (CHERRIES) (Eysenbach, 2012). We began with a test of the usability and functionality of the electronic questionnaires in Limesurvey (version 2.05) of the Portuguese version of the MBI (in paper version). For this test, we use a platform to survey the electronic version with students of master Degree in Education that work as teachers (N=35). The electronic survey was headed by an introduction with the identification of the research team and its organization contacts, the main aims of the research, the assurances of anonymity, confidentiality and data usage only for scientific purposes. After this procedure the electronic survey inside the whole project of research that supports this study, was submitted to the endorsement from the Ministry of Education. After the approval, we distribute the invitation through all the groups of Portuguese public schools, using the mailing list of the public schools made accessible by the Ministry of Education. Each School Director received, by e-mail, the request to disseminate the link of the electronic survey between February 2016 and February 2017. Teachers sent it back to an e-mail box explicitly generated for this, till the end of February 2017. To create a data base we analyse each one of the surveys received in the Limesurvey platform and downloaded and converted them into SPSS (Statistical Package for the Social Sciences) Statistics a data file (version 22).

The next step was data analysis with software for data processing SPSS - Statistics Data and Document.

12.2.4 Results

We begin data analysis with descriptive statistics of burnout based on professional qualifications, age, months of professional experience and gender among Portuguese teachers. Through univariate analysis of variance for professional qualifications (Table 12.2.) it was possible to verify a significant difference for the dimensions emotional exhaustion and personal and professional accomplishment. There is a significant difference in the emotional exhaustion dimension between licensed teachers and doctorates ($p = .000$) and between graduates and masters ($p = .002$), with graduates having the highest levels of emotional exhaustion, followed by the masters, whereas the doctorates present smaller values by comparison of means. Regarding the sense of accomplishment, it is the teachers with doctoral degrees who are more satisfied at a personal and professional level ($p = .000$).

Table 12.2. *Burnout levels according to professional qualifications*

	G1: Bachelor	G:2 Graduation	G3: Specialization	G4: Master Degree	G5: PhD	F	*p*	Post-doc
Emotional exhaustion	36.18 (12.12)	36.28 (11.11)	35.50 (11.43)	35.20 (11.10)	33.66 (10.31)	8.21	.000	1 < 2 1 > 3 1 > 4 1 < 5
Personal accomplishment	22. 68 (10.89)	23.29 (9.96)	21.71 (9.35)	23.34 (10.31)	29. 47 (11.64)	45.84	.000	1 < 2 1 > 3 1 < 4 1 < 5
Depersonalization	13.76 (7.12)	13.80 (6.89)	14.18 (7.03)	14.19 (7.11)	14.08 (6.80)	1.56	.182	1 < 2 1 < 3 1 < 4 1 < 5

In relation to age, no significant differences were found for any of the dimensions (Table 12.3.); however, it can be seen that teachers aged between 44 and 50 years show a greater trend towards higher values of emotional exhaustion and also depersonalization.

Table 12.3. *Burnout levels according to age*

	G1: 21 - 43	G:2 44 - 50	G3: 51 – 55	G4: 56 - 69	F	*p*	Post-doc
Emotional exhaustion	35.33 (10.34)	36.14 (10.94)	35.98 (11.39)	35.90 (12.02)	2.47	.060	1 < 2 1 < 3 1 < 4
Personal accomplishment	23. 28 (10.18)	23.71 (10.13)	23.24 (10.16)	23.32 (10.24)	1.18	.315	1 < 2 1 > 3 1 < 4
Depersonalization	13.76 (6.94)	14.08 (6.90)	13.73 (6.78)	14.19 (7.18)	2.41	.064	1 < 2 1 > 3 1 < 4

Regarding the years of professional experience (Table 12.4.), there is a difference between the groups (p = .000) at the level of emotional exhaustion; those with a work experience of less than 216 months present lower values of exhaustion. Concerning personal and professional achievement (p = .235) and depersonalization (p = .282), no significant differences were observed.

Table 12.4. ***Burnout levels according to professional experience***

	G1: 0 -216	G2: 217 - 288	G3: 289 – 360	G4: 361 - 388	F	*p*	Post-doc
Emotional exhaustion	34.80 (10.44)	36.25 (10.70)	36.06 (11.44)	36.28 (11.91)	9.37	.000	1 < 2 1 < 3 1 < 4
Personal accomplishment	23. 44 (10.42)	23.73 (9.95)	23.32 (10.13)	23.12 (10.20)	1.42	.235	1 < 2 1 > 3 1 > 4
Depersonalizatio n	13.75 (6.94)	14.04 (6.82)	13.87 (6.95)	14.11 (7.11)	1.27	.282	1 < 2 1 < 3 1 < 4

Finally, in relation to gender (Table 12.5.), it can be seen that women are more emotionally exhausted (p = .000), while men show higher levels of depersonalization.

Table 12.5. ***Levels of burnout according to gender***

	G1: Women	G:2 Men	F	*P*
Emotional exhaustion	36.69 (10.91)	32.92 (11.44)	188.60	.000
Personal accomplishment	23.31 (10.07)	23.73 (10.53)	2.82	.093
Depersonalization	13.61 (6.72)	15.07 (7.59)	71.27	.000

12.3 Discussion and Conclusion

Teaching is a profession that requires scientific and pedagogical knowledge to help and prepare the teacher to carry out with the diversity of students and school requirements of success and quality. Success and quality represent a continuous challenge to personal and professional competences of teachers that could represent sources of stress and burnout and strong threats to the teacher's wellbeing.

The main aim of the present chapter was to characterize the levels of burnout among teachers of Portuguese public schools in function of personal and professional variables.

It was found that teachers with lower professional qualifications and years of professional experience present higher emotional exhaustion and the group of specialized, masters and PhD present lower values. These results are similar to those presented by Mukundan and Khandehroo (2009) and Majali et al. (2015).

Regarding the gender differences in the various dimensions of burnout, we found higher results in the depersonalization dimension for men as some studies show previously (Martins, 2008; Esteras, Chorot & Sandin, 2014; Rumschlag, 2017; Scheuch et al., 2015), although some authors mention that there is still no affirmative consensus about the gender that predominates in the various dimensions of burnout (Norlund et al., 2010). Other authors (Briga, 2013; Chang, 2013) they suggest that women express greater emotional exhaustion, which also occurred with the Portuguese participants of this study.

Regarding age, no significant differences were found for any of the dimensions, which is in agreement with studies that obtained a similar result (Colakoglu & Yilmaz, 2014). Regarding the years of professional experience, it is found that early-career teachers less express the symptoms of burnout, similar to other studies (Goddard and O'Brien, 2004; Dias & Queirós, 2010; Kokkinos, 2007). Teachers who had more experience and more time in the profession felt more exhausted and depersonalized, while those who had less time showed higher levels of achievement.

12.4 Implications

Burnout affects the health of teachers, which has increasingly become an object of interest in the field of education, as proposed by, for example, Klusmann, Richter and Lüdtke, 2016; Rumschlag, 2017. Such interest is justified when there is evidence that the deterioration of teachers' health has had negative consequences not only at the personal level but also at the organizational level, impairing the quality of teachers' professional performance and, consequently, the school's mission as an organization for maximizing school achievement and wellbeing.

In general, the main research studies about school achievement seek answers that focalize students' process of learning, trying to solve the problem of school failure and the generalized difficulties of education as an organizational system. In another approach, we suggest research studies focused on teachers' wellbeing, namely about the teaching process as well as in the way they feel and experience their professional life, in order to enhance students' success through promotion of the health and wellbeing of teachers as a key factor for school achievement and wellbeing of students. Healthy teachers improve healthy schools and healthy students.

In this sense, the measures to be adopted should not only be related to medical care of the consequences of burnout, but especially in the development of prevention and health promotion policies, as a guarantee of quality and success of education and the wellbeing of teachers and students. The primary intention is to encourage changing policies that promote healthy educational environments, that is, to transform the school into a place of involvement, motivation, and learning, for both students, teachers, and other school staff. In short, a school that promotes health and wellbeing for all.

Change must be geared, firstly, towards initial and continuing training, leading to a continuous supervision of teachers, at the level of their personal and social skills, which promote emotional self-regulation, health and wellbeing, as well as an effective management of curricular aspects in general; secondly it should improve prevention programs of stress and burnout management. Such programs, in accordance with the results presented here suggested, should be discriminated by gender, age, years of professional experience and level of professional qualification. Different characteristics of teachers represent different needs of support and consequently diverse needs of support and levels of preventive intervention. Next steps of these studies should improve a deeper look at the personal, professional and psychological variables that influence emotional exhaustion, depersonalization and personal accomplishment, in order to identify the variables or group of variables that explain variability in such dimensions of burnout. This work will be useful to create profiles of needs and to organize prevention programs, differentiated by gender, years of professional experience and professional qualifications; in one word accurate programs for teachers' needs of wellbeing and quality of life and consequently for improving learning for the wellbeing among students.

12.5 References

Briga, D. (2013) *Atitudes face à Mudança e Burnout - Um estudo com docentes da Universidade do Minho.* PhD Thesis. Universidade do Minho, Bragança.

Candeias, A., Calisto, I., Borralho, L. & Vilia, P. (2016). Stress, burnout, wellbeing in teaching profession: Portuguese studies. In, Candeias, A.A., Galindo, E., Pires, H., Reschke, K. & Witruk, E. (eds.) (2016). *Psychological Issues in Education and Health. Proceedings of 1st Scientific Meeting University of Evora and University of Leipzig* (35-41). Aachen: ShakerVerlag.

Carlotto, M. (2002) A Síndrome de Burnout e o Trabalho Docente. *Psicologia em Estudo,* 7(1), 21-29. doi: 10.1590/S1413-73722002000100005

Carlotto, M. (2011) Síndrome de Burnout em Professores: Prevalência e Fatores Associados. *Psicologia: Teoria e Pesquisa,* 27 (4), 403-410. Available from: doi: 10.1590/S0102-37722011000400003

Chang, M. L. (2013) Toward a theoretical model to understand teacher emotions and teacher burnout in the context of student misbehavior: Appraisal, regulation and coping. *Motivation and Emotion,* 37 (4), 799-817. Available from: doi: 10.1007/s11031-012-9335-0

Colakoglu, F. & Yilmaz, T. (2014) Burnout levels of physical education teachers according to personal factors. *Procedia Social and Behavioral Sciences,* 152, 409-414. Available from: doi: 10.1016/j.sbspro.2014.09.221

Dias, S. & Queirós, C. (2010) A influência dos traços de personalidade no burnout dos professores. *IN*: Nogueira, C., Silva, I., Lima, L., Almeida, A. T., Cabecinhas, R., Gomes, R., Machado, C., Maia, A., Sampaio, A. and Taveira M. C. (eds.). *Actas do VII Simpósio Nacional de Investigação em Psicologia,* 1066-1080. Available from: http://www.actassnip2010.com

Eysenbach G (2012). Correction: Improving the Quality of Web Surveys: the Checklist for Reporting Results of Internet E-Surveys (CHERRIES). *Journal of Medical Internet Research,* 14(1):e8. Available from: doi:10.2196/jmir.2042.

Esteras, J., Chorot, P. & Sandín, B. (2014) Predicción del burnout en los docentes: papel de los factores organizacionales, personales y sociodemográficos. *Revista de Psicopatologia y Psicología Clinica.* 10 (2), 79-92. Available from: doi: 10.5944/rppc.vol.19.num.2.2014.13059

Goddard, R. & O'Brien, P. (2004) Are Beginning Teachers with a Second Degree at a Higher Risk of Early Career Burnout. *Australian Journal of Teacher Education,* 29(1), 1-10. Available from: doi: 10.14221/ajte.2004v29n1.4

Gomes, R., Montenegro N. & Peixoto, A. (2010) Stress ocupacional no ensino: um estudo com professores dos 3ºciclo e ensino secundário. *Psicologia & Sociedade,* 22 (3), *587-597.* Available from: doi:10.1590/S0102-71822010000300019.

Gomes, A., Peixoto, A. & Silva, R. (2012) Stress ocupacional e alteração do Estatuto da Carreira Docente português. *Educação e Pesquisa*, 38 (2), 357-372. Available from: doi: 10.1590/S1517-97022012005000008

Jesus, S. N. (1996) *A motivação para a profissão docente. Contributo para a clarificação de situações de mal-estar e para a fundamentação de estratégias de formação de professores.* Aveiro: Estante Editora

Jesus, S. (2000) *Motivação e formação de professores.* Coimbra: Quarteto Editora.

Jesus, S. (2001) *Como prevenir e resolver o stress dos professores e indisciplina dos alunos? Cadernos do CRIAP- 4.* Porto: ASA.

Jesus, S. & Conboy, J. (2001). A stress management course to prevent teacher distress. *International. Journal of Educational Management, 15* (3),131-137. https://doi.org/10.1108/09513540110384484

Jesus, S. (2002) *Perspectivas para o bem-estar docente: Uma lição de síntese.* Porto: ASA Editores II.

Jesus, S. N. (2006). Avaliação do desempenho e bem-estar docente. Psicologia, Educação e Cultura, 10 (1), 7-22.

Klusmann, U., Richter, D. & Lüdtke, O. (2016). Teachers' emotional exhaustion is negatively related to students' achievement: Evidence from a large-scale assessment study. *Journal of Educational Psychology, 108*(8), 1193-1203. http://dx.doi.org/10.1037/edu0000125

Kokkinos, C. M. (2007) Job stressors, personality and burnout in primary school teachers. *British Journal of Educational Psychology*, 77(1), 229–243. Available from: doi: 10.1348/000709905X90344

Lopes, A. (2001) *Mal-estar na docência? Visões, razões e soluções.* Porto: ASA Editores II.

Majali, A., Adwan, S., Shaheen, H. & Momani, R. (2015) The relationship between emotional intelligence and burnout among special education teachers in Jordan. *International Journal of Adult and Non Formal Education*, 3 (4), 94-100. Available from: www.globalscienceresearchjournals.org

Maslach C, Jackson SE. (1981). The measure-ment of experienced burnout. *Journal of Occupational Behavior* 99–113.

Maslach, C., Jackson, S. E. & Leiter, M. P. (1996) *Maslach Burnout Inventory: Manual* 3rd (ed.). Mountain View, CA: Consulting Psychologists Press.

Maslach, C., Schaufeli, W. & Leiter, M. (2001) Job burnout. *Annual Review of Psychology, 52*, 397-422. Available from: doi: 10.1146/annurev.psych.52.1.397

Martins, J. (2008) *Burnout na profissão docente.* Master Thesis. Faculdade de Psicologia e Educação do Porto, Porto.

Mukundan, J. & Khandehroo, K. (2009) Burnout in Relation to Gender, Educational Attainment, and Experience among Malaysian ELT Practitioners. *The Journal of Human Resource and Adult Learning*, 5(2), 93-98. Available from: psasir.upm.edu.my

Norlund, S., Reuterwall, C., Hoog, J., Lindah, B., Janler, U. & Birgander, L. (2010) Burnout, working conditions and gender- results from the northern Sweden Monica STUDY. *BMC Public Health*, 10. Available from: doi: 10.1186/1471-2458-10-326

Patrão, I., Rita, J. & Maroco, J. (2012) Avaliação do burnout em professores: Contributo para o estudo de adaptação do CBP-R. *Psychology, Community and Health*, 12, 179 – 188. Available from: doi:10.5964/pch.v1i2.29

Picado, L. (2005) *Ansiedade na Profissão Docente*. Mangualde: Pedago.

Ramos, A. (2012). *Stresse percebido, burnout, estratégias de coping e estilos de vida em docentes portugueses* (Doctoral thesis). UTL-FMH, Lisbon, Portugal.

Rebelo, M. (2011) *Concepções e práticas de professores do 2º e 3º ciclo do Ensino Básico face à inclusão de crianças com Necessidades Educativas Especiais*. Master Thesis. Escola Superior de Educação de Lisboa, Lisboa.

Rumschlag, K. (2017). Teacher burnout: A quantitative analysis of emotional exhaustion, personal, accomplishment, and depersonalization. *International Management Review*, 13 (1), 22-36.

Salanova, M., Martínez, M. & Lorente, L. (2005) ¿Cómo se relacionan los obstáculos y facilitadores organizacionales con el burnout docente?: Un estúdio longitudinal. *Revista de Psicología del Trabajo y de las Organizaciones*, 21 (1-2), 37-54. Available from : www.redalyc.org/articulo.oa?id=231317039003

Santos, S. (2011) *Burnout e diferenças de género em professores*. Master Thesis. Universidade Lusófona de Humanidades e Tecnologias, Lisboa.

Schaufeli, W., Maslach, C., & Marek, T. (1993) Professional Burnout: Recent Developments. *Theory and Research*. Washington, DC: Taylor & Francis

Scheuch, K., Haufe, E. & Seibt R. (2015) Teachers' health. *Dtsch Ärzteblatt International*, 112 (20), 347–356. Available from: doi: 10.3238/arztebl.2015.0347

Silva, J. (2012) *Clima escolar, satisfação, stresse profissional e colaboração entre docentes: Que relações?*. PhD Thesis. Universidade Aberta, Lisboa.

Kickbusch, I. (2012). *Aprender para o Bem-Estar: Uma Prioridade Política para as Crianças e os Jovens na Europa. Um Processo para a Mudança*. (Trad. Portuguesa). Lisboa: Fundação Calouste Gulbenkian.

Chapter 13

Parenting and/or mental health?

Dora Pereira
University of Madeira, Department of Psychology, Portugal

In child protection, parenting and mental health issues are frequently related; questions about continuity of care and family reunification are answered mostly by assessing parents and implications of their behaviour for child development. However, many times parenting capacity and mental health assessments are not clearly distinguished, and psychiatric diagnoses are improperly taken as parenting incapacity evidence. This chapter aims to re-locate mental health assessments on parenting capacity evaluation main frame, and to present the Parenting Capacities Assessment Guide as a tool that could help professionals in clinical judgements about parenting behaviour. The Functional Model of Parenting Behaviour as a new conceptual approach and the results of the exploratory study developed in Alentejo (Portugal) about Guide's reliability are presented. To conclude it will be argued that to understand relations between parenting and mental health and to foster the optimal children development and parents' wellbeing, professionals should go beyond psychopathology diagnosis and focus on actual and future parenting functioning.

13.1 Introduction

The impact of parent´s mental health on children development has been approached from perspectives as mentalization, attachment, parentification or risk (Chatelle & de Becker, 2016; Duncan & Reder, 2003; Holzer, Jaugey, Sofia, & Corthésy, 2013; Keren & Tyano, 2014; Muzik et al., 2016). Child protection and mental health professionals acknowledge that psychopathology can have detrimental effects on parenting (Cleaver, Unell & Aldgate, 2011) and they must explain "how" that risk could increase or decrease in order to make decisions that assure the best development conditions to children and protect them from dangerous situations (Deutsch & Clyman, 2016; Holzer et al., 2013). In daily practice, parents' mental health assessments have a major influence in child protection service planning; Salvaggio (2015) refers as an outcome of her qualitative study that recommendations of mental health assessments are integrated into service planning and child's permanency is decided based on a mother's compliance with them.

The relation between parenting and psychopathology is more than just check if diagnostic criteria are present or not. It means to understand their interaction through each parent developmental process, identifying what factors could minimize detrimental effects on parenting and maximize children and parents' wellbeing. In a short-term perspective, those factors will be directly related to risk assessments and decisions about immediate protective measures (Boursnell, 2014). In middle or long term, they should be used to plan interventions with children and families. With these aims, assessment processes cannot be similar to those intended to inform decisions about psychopharmacological intervention in a clinical context. Besides that, taking a psychiatric diagnosis as evidence of parenting (in)capacity (Risley-Curtiss, Stromwall, Hunt, & Teska, 2004) would origin biased conclusions as it does not explain what can trigger some behaviours, how they impact parenting and child development and how they can change across time.

In this chapter, the main focus will be on the relation between parenting and mental health, more precisely in what refers to assessment processes in child protection context. Mental health assessments will be read from parenting capacity evaluation mainframe, and the Parenting Capacities Assessment Guide will be presented as a tool that could help professionals in clinical judgements about parenting behaviour.

13.2 Parenting

To understand parenting three main frames will be used, developmental, ecologic and systemic. From Bronfenbrenner's ecological model (1986), human development is influenced (simultaneously) by different ecological levels (microsystem, mesosystem, exosystem and macrosystem); microsystem is the context where occur very important proximal processes like the enduring interaction between child and parents that have a major influence on individual development. That means child development is more directly influenced by parenting and less directly by elements from other levels, like members of the social network, as mental health and child protection professionals. A systemic perspective moves away from linear causality and cause-effect relations to circular causality, understanding interactions between the different system elements, looking for the "pattern that connects" (Bateson, 1972) in "a multicausal reciprocal framework" (Feiring & Lewis, 1987, p. 4). Therefore, it makes no sense to ask if a mental health disorder causes some behaviours but instead what are the circumstances, relations and contexts under which they occur and how they change through time. In practice, the need to develop good communication and coordinated work between professionals becomes emphasized; parenting cannot be excluded from mental health intervention and neither parents' mental health from child protection work (Göpfert, McClelland, & Wilson, 2010). The specificity of professional functions should not be a limit to a systemic practice: even if they do not work directly in child protection or in mental health services, professionals should consider how the different contexts interact and influence the aims of their interventions.

13.2.1 Functional Model of Parenting Behaviour and mental health

Conceptual approaches to parenting behaviour are at the origin of differences between professionals about the planning of assessment and intervention processes. The description of parenting practices in one moment does not say anything about how it can be in the future; besides, as parenting is a developmental process (Pereira & Alarcão, 2014), it would not be correct to conclude that in the future it would have the same features. Such description could be useful to justify immediate protective measures but will not be enough to plan future interventions aimed to promote change in parental functioning; it is needed a comprehensive model that locates parenting in a developmental timeline, considering past, present and future.

The Functional Model of Parenting Behaviour (Pereira & Alarcão, 2014) approaches parenting functioning from three main elements hierarchically organized: parenting capacity, mediating processes and parenting skills.

Figure 13.1.: Functional Model of Parenting Behavior

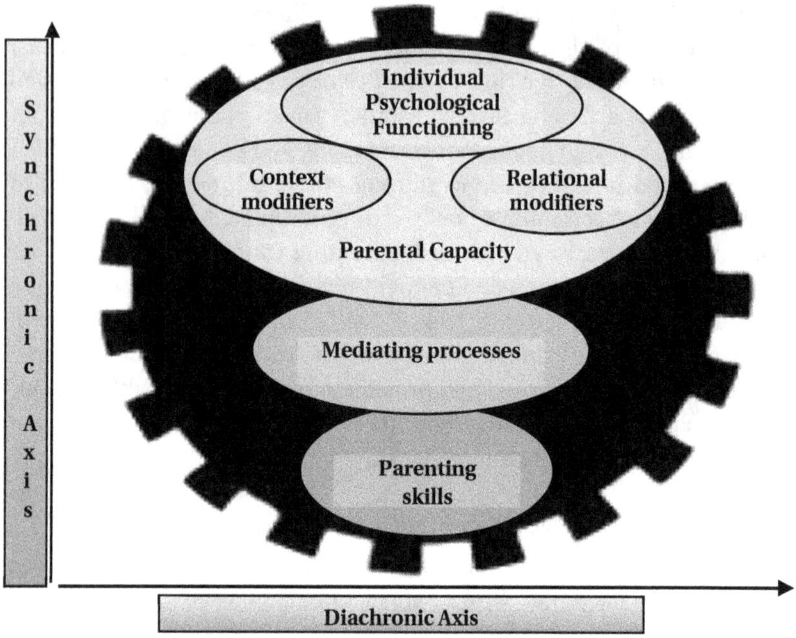

Parenting capacity is the potential that parents have (now and in the future) to address child's needs at his/her different stages of development (Steinhauer et al.,1993). Such potential results from the features of parental psychological functioning interacting with contextual (the specific features of contexts in which parents are included) and relational (their social network, including the specific child in order to which parenting is considered) modifiers. Parents' psychological functioning include their working models, personal beliefs, narratives and mental capacity – the capacity to make specific decisions as defined by the Mental Capacity Act (2005). Göpfert et al. (2010) claim that mental capacity is tested by whether a person can understand, retain, weigh up information in order to make a decision and communicate what has decided. Mental disorders may have a detrimental effect on capacity to acknowledge and address child's needs, because parents are, at the same time, striving to cope with symptoms. In a recent longitudinal study (during 6 years) with children of mothers diagnosed with chronic depressive disorder, Apter-Levi et al. (2016), concluded that the disorder limited mother's capacity to take care of children in a regular and sensitive way. Keren and Tyano (2014) state that as humour disorders (depressive and bipolar disorders), psychotic spectrum disorders (schizophrenia) and borderline personality disorder are the most studied about their influence on parenting, post-traumatic stress disorder or obsessive-compulsive disorder have been less researched. Besides the particular results of the different studies, it should be acknowledged that even if mental capacity is impaired it does not mean that parenting capacity also is. Parenting capacity could be reinforced by context and social network; it does not depend only on psychological functioning.

Contextual modifiers and mental health are interrelated. On the one hand, parents with a mental health disorder refer they have to cope with many different stress factors, like financial problems, difficulties at work, medication effects, the symptoms, access to support services while they receive medical care (e.g. a nanny to take care of the child during weekly psychotherapy sessions), fear and anxiety related to the possibility of being separated from the child by protective services, and stigma related to mental illness (Rampou, Havenga, & Madumo, 2015). On the other hand, as a mental disorder becomes chronic and stress factors increase, it is common that parent's coping resources decrease, closing a vicious circle that impairs parenting capacity.

Relational modifiers are also very important (de Becker, 2011), because positive relations between parents and other family members, neighbours, friends, and professionals from different support services may foster parents' resilience and stress coping skills, and also assure child care in crisis times. Members of parent's social network that are aware of their difficulties and assume a supportive and containing role may be a protective factor to child development and reinforce parental capacity. However, if they are unsupportive or they do not acknowledge parent's difficulties they could become a risk factor. The child is also considered a relational modifier of parental capacity because their specific needs and developmental features require different parenting and coping skills from parents. From this perspective, assessments (of parental capacity) must be developed in order to a specific child.

In summary: "the impact of mental illness on parenting capacity can be complex to determine, but critical to understand" (Deutsch & Clyman, 2016, p. 37). Considering parental capacity as a result of psychological functioning, contextual modifiers and relational modifiers, it becomes clear that a psychiatric diagnosis does not mean parental incapacity and that parenting capacity is not the same as mental capacity.

Parenting skills refer to how parents practice their role at the present moment, with their cognitive, emotional and behavioural resources (Barudy & Dantagnan, 2010). Parents must adjust their skills as the child develops according to his/her specific needs (Pereira, 2014): some skills, (e.g. baby hygiene) are used just in some specific developmental phases; others must be adjusted as the child develops (e.g. promotion of learning opportunities); and new others must be learned and used (e.g. to negotiate with an adolescent). Hornstein et al. (2006) claim that mothers with psychotic spectrum disorders are more distant, less sensitive, more intrusive to her babies and self-centered than mothers with mood disorders. However, most parents with mental health disorders are really concerned with his/her child's wellbeing (Duncan & Reder, 2003) requiring professionals to make very difficult decisions such as to separate a child from his/her parents to ensure protection when all want to stay together.

The process to turn capacity in skills is mediated by cognitive and affective processes, as responsibility recognition and impulse control. To recognize responsibility implies that parents with mental health disorders can acknowledge and reflect about the way their behaviour impact parenting and child development; that could be different according to the egosyntonic or egodystonic features of the disorder. As an example, parents with anxiety disorders may consider they are protecting their children keeping them home and avoiding they could contact with - what they consider as - dangerous stimulus (Van Brunt, Zedginidze, & Light, 2016). If egosyntonic features are present, it could be more difficult for the parents to recognize that such attitude may have a negative impact on child's development, and to get involved in a change process through psychiatric or psychotherapeutic intervention. Impulse control will be determinant to deal with contextual stress and address child's needs, using successfully their coping skills. Difficulties in impulse control could also be an obstacle to learn and use new parenting skills.

As represented in Figure 13.1., parental functioning is framed by two axes: synchronic and diachronic. That means that parental behaviour is being assessed at a specific moment in time when a set of conclusions can be reached in the present circumstances (synchronic axis). But, it must be considered within a developmental pathway, the one that leads to the actual parental behaviour and that will drive parents – for sure – through changes in it (diachronic axis). Parental behaviour has a dynamic nature (as all human behaviours), which prevents from taking as definitive the statements made about it in some moment. A mental disorder does not course independently of individual's life course (Benjet, Azar, & Kuersten-Hogan, 2003); therefore, parents' mental health and its influence in parenting should be approached from a developmental and constructivist stance.

Based on a longitudinal research during 20 years (from 18 to 38 years of the participants) focusing structural features of psychopathology as dimensionality, persistence over time, co-occurrence and sequential comorbidity of the mental disorders in a group of 1007 participants (members of the Dunedin Multidisciplinary Health and Development Study), Caspi et al. (2014) suggest that the structure of mental disorders could be summarized in four main dimensions: internalization (related to depression and anxiety and female gender), externalization (related to anti-social disorders, substance abuse and male gender) thought disorder (related to psychosis and to a higher severity) and *p* factor (psychopathology); *p* factor is described as a general dimension related to a developmental progression, that is the individual's propensity to develop some of the more common psychopathologies. Higher scores on *p* were related to higher severity, duration and extent of sequential comorbidity of mental disorders, and also to higher impairment on adult functioning and to childhood maltreatment. Based on results authors suggest that a high number of individuals could develop psychopathology in a brief and isolated moment, and a smaller group of those could progress to a persistent internalization or externalization disorder. An even smaller group will progress through a high *p* value, associated to psychotic disorders. If future studies come to confirm this structural hypothesis, as Caspi et al. (2014) well point out, it could be very useful to understand and make a sound prognosis about how mental health impacts parenting. It will be very important that researchers could clarify if *p* values progression through life course development could be related to a progressive detrimental effect on parenting from parenting skills to parenting capacity.

Applying the Parenting Functional Model as conceptual frame, a mental disorder with its symptoms and course over time, could have three types of influence: 1) on parenting capacity (and mediating processes and parenting skills) if it limits parents possibility to learn and/or use new skills required by child's needs; 2) on mediating processes (and parenting skills) if it influences processes that prevent the use of learned skills or parents involvement in processes leading to their acquisition; or 3) on parenting skills, if it limits the use of acquired skills. The "same" diagnosis could impact parenting in different functioning elements, and that impact could change over time: as a disorder becomes chronic or show a more severe course with relapses or recurrent episodes (Manderscheid et al., 2010) more parenting functioning elements could be progressively impaired, from parenting skills to parenting capacity. In child protection work it will be very important that parenting assessments could specify the quality of impact for two main reasons: first, this information will be decisive to justify if future interventions with parents are viable in a time respecting child developmental needs; second, to clarify if psychotropic medication can influence, and how, parenting functioning elements (e.g. influence on learning and use of new parenting skills).

13.2.2. Child protection practice: integrating knowledge

Mental health assessments have different goals and should have different features in these two practice fields: in mental health practice clinical assessments aim to reach a diagnosis and plan psychiatric or psychotherapeutic intervention according to it, focused on the individual wellbeing; in child protection assessments are focused on a specific role of the adult – parenting – and aim to inform future interventions with families (Göpfert et al., 2010), in order to promote the wellbeing of the child in first place. That means that besides a comprehensive picture of adult's psychological functioning, assessments in child protection must consider the specific features of parent-child interaction.

Different barriers have been pointed to assessments' accuracy (Risley-Curtiss et al., 2004) as, the urgency of decisions and its implications for child's development, lack of resources to assure longer and deeper assessment processes (based on different information sources and different evaluation methods), or poor professional training about this issue. Besides those, two additional barriers are considered important. One is the overvaluing of observed parenting skills: to centre parenting assessment mostly on this parenting functioning element, makes professionals adopt a reactive position that changes with the quality of the last observed behaviours. It becomes more difficult to understand *how* they change or not - the "pattern that connects". Frequent variations in parenting functioning are the main issues in the cases of parents with mental health disorders (Duncan & Reder, 2003). In these situations, parents show adequate parenting skills in some periods alternating with others when it does not happen, and judgments about parenting quality may change with them; the reports refer very rarely what factors contribute for variability and what are its consequences for parents and for child development, namely on attachment processes.

The second is to maintain a linear perspective, a main threat to parenting assessments' quality because professionals focus their work mostly in one of the many factors that influence parenting (e.g. mental disorders; social and economic conditions; child behavioural characteristics), valuing it as *the* cause, and do not approach how the different factors interact with each other. It is important to note that "The

Family Model" (Falkov, 2017) proposes a systemic approach to integrate mental health and family functioning based on the identification of risk and protective factors that will be the focus of future interventions. However, it does not differentiate the elements of parenting functioning that could help improve the adjustment of intervention to the specificities of children and families.

13.3 So, is there a different way to do it?

To understand parenting and plan intervention in child protection requires: (1) a functional description of parenting capacity, mediating processes and parenting skills in order to a specific child; (2) information gathering in order to promote change in parenting – not only to judge if parents are adequate or not. Concerning the impact of a mental disorder in parenting, it will be decisive to answer these specific questions: a) Do parents recognize the impact of their behaviour in children? b) How is his/her social network functioning? Does it reinforce parenting capacity? c) How do parents cope with mental health disorder, namely with risk for children? d) How parenting functional elements are or could be affected?

To assess complex interactions requires a flexible and dynamic approach, able to find a pattern through the difference gathered from multiples sources of information - parents, family members, neighbours, friends, professionals – and methodologies – home visits, interviews, psychological assessment, observation of interactions and contact with other professionals. Each dyad parent-child has different features, and scientific evidence points to patterns that could be observable under specific circumstances. As such, adequate parenting won't be reducible to a quantitative value, a norm, with which parents could be compared. From a systemic and developmental perspective, that would be a biased process as it would not be considering equifinality (similar outcomes reached through different pathways) or multifinality (different outcomes reached through a similar pathway) (Feiring & Lewis, 1987). The "pattern" corresponds to a case formulation, the level of inference of clinical judgment (Tippins & Wittmann, 2005), that gives meaning to the information gathered during assessment, applying the actual scientific evidence. With clinical judgment, professionals could justify inferences related to the prognosis for parenting behaviour change and the "how's" of change process.

The nonexistence of validated assessment tools with these characteristics in the Portuguese context have been contributing to decisions about parenting assessments in child protection based almost exclusively on professionals' life and training individual experiences

(Pereira & Alarcão, 2010). Assessment and intervention with parents having a mental disorder becomes very heterogeneous processes, and in some instances with professionals devaluing or overvaluing signs and symptoms of psychopathology, and postponing referral to health services (Duncan & Reder, 2003). Even professionals from the same academic and working field (e.g., mental health) could reach very different conclusions about the same situation.

The Parenting Capacities Assessment Guide (De Rancourt, Paquette, Paquette, & Rainville, 2006) is an assessment tool, in order to which have been developed the exploratory studies of validation to the Portuguese context (Pereira, 2014). It aims to offer professionals a common frame to assess parenting behaviour that could help them to organize the information gathered, to make and communicate clinical inferences (Tippins & Wittmann, 2005) relevant to child protection work.

13.3.1 Parenting Capacities Assessment Guide

This qualitative tool integrates the three inference levels (observation, clinical judgment and practical implications for change) referred by Tippins and Wittmann (2005) in four main assessment areas – context, child's health and development, parents-child relation and parents - corresponding to nine specific dimensions (Table 13.1.). Each dimension has three sections: the first includes questions focusing observations and clinical judgments made during the assessment process about specific contents; the second – based on the first set - includes questions (summary questions) related to their practical implications for parenting and child development; in the last one the professional will use specific criteria to classify the dimension as a *main resource*, a *secondary resource*, a *secondary difficulty* or a *main difficulty*. The nine classifications form the parenting capacity profile. Considering the parenting functional model this profile enables assessors to identify in which parenting element are located the resources and difficulties, making service planning more clearly related to case formulation. The Guide also includes the Prognosis Scale, that allows the professional to identify the set of factors related to a high risk of chronicity or to a high potential to change parenting behaviour. The tool is used through the assessment process, helping the professional to gather relevant information to decision making whilst having in mind what is missing, what is inconsistent and what is regular. As isolated events or observations are not enough to justify conclusions about patterns, parenting assessments should last not less than a month and a half, in order to gather enough information as consistent as possible.

Table 13.1.: Parenting Capacities Assessment Guide (Areas and Dimensions)

Areas	Dimensions
Context	**D1: Context**: What are the major strengths and weaknesses on the different daily life areas, such as household, neighbours, finances, employment, couple relation, imigration?
Child	**D2: Child's health and development**: How good is child's health and development?
Parents-child relation	**D3: Attachment:** How secure is child-parents attachment?
	D4: Parenting skills: How good are the parenting skills?
Parents	**D5: Impulse control**: How good are parents' impulse control?
	D6: Responsibility recognition: Does the parent recognizes his/her responsibility on the present situation?
	D7: Personal factors that affect parenting capacities: How does the parent's personal factors (such as mental health problems) affect the parental capacities?
	D8: Social support network: How does the parent' relates with his/her social network?
	D9: Access to services: How is/was the parent's relation with clinical services?

The nine dimensions included in the Guide should be considered as a system, because just all the set will reflect the ecological nature of parenting on the assessment process, making possible to reach conclusions about parenting functioning that consider the three main frames referred. A partial use of the Guide would just allow professionals to describe specific dimensions, and not to make inferences about parenting functioning. Giving an example, if one of the dimensions is classified as a *main difficulty* it would not be correct to conclude that the parent does not have enough parental capacity to take care of his/her child. Conclusions should be focused on the need and/or feasibility of an intervention process aimed to change parenting functioning in order to maximize the developmental potential of the child and promote parent's personal development.

13.3.2 The 7th dimension: personal factors that affect parenting capacity

The 7th dimension of the Parenting Capacities Assessment Guide includes a set of contents that are usually considered in assessments made by mental health professionals, adding the questions considered relevant to child protection workers. Considering all of them, the communication between child protection and mental health' professionals could become more fluid.

More specifically, in the first place, this dimension requires that the assessor reports the frequency with which parents show positive characteristics (as being organized, good communicator, able to express his /her needs or being flexible in his /her positions). This positive stance (also present in the other dimensions) helps to reach conclusions not just based on problems and deficits. After that, the assessor should register (based on the information gathered from parents and their formal and informal social network) with which frequency are observed signs of depression, anxiety, perfectionism, aggression (self or other directed), substance abuse, fantasies, hallucinations or persecutory ideas, difficulties in judgment, neurological symptoms, cognitive impairment, inappropriate sexual behaviour to the child and also how parents think about themselves and interpersonal relations. The summary questions are focused on: a) the influence of the signals reported in parenting functioning, more precisely if it is detrimental, and in positive case, if that occurs chronically, within a specific set of circumstances or in isolated brief episodes; b) the need of inpatient treatment, and in positive case, it's frequency and duration; c) the consequences of these problems on child and other family members; the assessor should also register how the child understand parents 'problems and the kind of impact it has on his/her behaviour (e.g., parentification); d) the reference of other adults in the family able to compensate parents difficulties, especially in critical periods.

At last, what are the meaning of the different classifications in the 7th dimension? Classifying it as a *main resource* means that do not exist important difficulties in mental health and that parents have positive personal characteristics; as *secondary resource* means that difficulties do exist *but* parents have (and use) enough resources to cope with them and there are no observable negative effects on child development; as *secondary difficulty* means that there exist important difficulties parents are not coping with, but it´s feasible an intervention with that goal because they are willing to, have enough capacity to benefit from it, and their social network have enough and adequate resources to offer in order to promote that change process (e.g., adequate services); as *main difficulty* it means that there exist important difficulties, that are not being object of some kind of effort to change them, and that intervention is not feasible in short term because parents are not willing to, or do not have enough capacity to benefit from it, or the social network does not have enough or adequate resources to offer. This is very important since it clearly points that parenting capacity may be – or not – reinforced by social network.

A future intervention should be based on a prognosis pointing to the hypothesis of a high potential for behavioural change. Future interventions should correspond to parenting functioning comprehensive assessment and consider as the main goal to promote change in the higher element (capacity, mediating processes or parenting skills) presenting difficulties[1].

13.3.3 Validation studies

The Parenting Capacities Assessment Guide has been the object of exploratory validation studies for Portuguese context, namely ecological validity (Pereira & Alarcão, 2015) reliability (Pereira & Alarcão, 2014) and predictive validity studies (Pereira, 2014). On reliability studies the Guide was applied by three different raters to 10 cases of children in residential care following a court order (corresponding to 13 parents assessed) in Alentejo (south of Portugal); mean-values of inter-rater agreement higher than 90% point that the Guide is a potentially reliable tool for parenting capacities assessment in child protection work field. Validation studies are continuing, with parents from different regions and in different parenting conditions (e.g. children living with parents). The Functional Model of Parenting Behaviour is also being researched.

13.4 Concluding remarks

To evaluate parenting when signs and symptoms of psychopathology are present is not a matter of presence or absence of a set of diagnostic criteria. Implies a comprehensive process based on systemic, ecological and developmental frames, able to integrate its complexity Assessment tools as the Parenting Capacities Assessment Guide may be a critical factor to improve the efficacy of child protection intervention, clearly linking theory and practice. Although validation studies are not complete, the

[1] see Fig. 13.1 to locate assessment dimensions (D) on The Functional Model of Parenting Behaviour

results reached at the moment are encouraging. To work with parents with a mental health disorder is a very demanding task technically, emotionally and ethically. To adopt a preventive perspective - and not just reactive - could decrease possible biases related to the urgency of decisions and the emotional impact of them on professionals, parents and children.

Parenting and mental health are not mutually exclusive terms. Many professionals contribute to making it a "and "(not "or") question, deepening joint practices from referral and assessment until the success of their interventions.

13.5 References

Apter-Levi, Y., Pratt, M., Vakart, A., Feldman, M., Zagoory-Sharon, O. & Feldma, R. (2016). Maternal depression across the first years of life compromises child psychosical adjustement; relations to child HPA-functionig. *Psychoneuroendocrinology* 64, 47-56. Doi: http://dx.doi.org/10.1016/j.psyneuen.2015.11.006

Barudy, J., & Dantagnan, M. (2010). *Los desafíos invisibles de ser madre o padre: Manual de evaluación de las competencias y la resilencia parental.* Barcelona: Gedisa.

Bateson, G. (1972). *Steps to an ecology of mind: Collected essays in anthropology, psychiatry, evolution, and epistemology.* Chicago: University of Chicago Press.

Benjet, C., Azar, S. T., & Kuersten-Hogan, R. (2003). Evaluating the parental fitness of psychiatrically diagnosed individuals: Advocating a functional-contextual analysis of parenting. *Journal of Family Psychology, 17*(2), 238-251. doi: 10.1037/0893-3200.17.2.238

Boursnell, M. (2014). Assessing the capacity of parents with mental illness: Parents with mental illness and risk. *International Social Work, 57*(2), 92-108. doi: 10.1177/0020872812445197

Bronfenbrenner, U. (1986). Ecology of the family as a context for human development: Research perspectives. *Developmental Psychology, 22*(6), 723-742.

Caspi, A., Houts, R., Belsky, D.W., Goldman-Mellor, S., Harrington, H.L., Israel, S., Meier, M.H., Ramrakha, S., Shalev, I., Poulton, R., Moffitt, T.E. (2014). The p factor: One general psychopathology factor in the structure of psychiatric disorders? *Clinical psychological science : a journal of the Association for Psychological Science, 2*(2), 119-137. doi: 10.1177/2167702613497473

Chatelle, N., & de Becker, E. (2016). Troubles psychiatriques et parentalité : Entre soins psychiatriques pour l'adulte et protection de l'enfance. *Annales Médico-psychologiques, revue psychiatrique, 174*(8), 627-633. doi: https://doi.org/10.1016/j.amp.2016.04.012

Cleaver, H., Unell, I., & Aldgate, J. (2011). *Children's Needs – Parenting Capacity: Child abuse, parental mental illness, learning disability, substance misuse and domestic violence* (2nd ed.). London: The Stationery Office.

De Becker, E. (2011). Avoir un parent malade mental: repères diagnostiques et thérapeutiques. *Neuropsiquiatrie de l´ énfance et de l´adolescence,* 59 (6), 348-355. Doi:10.1016/j.neuernf.2011.04.003

De Rancourt, L., Paquette, F., Paquette, D., & Rainville, S. (2006) *Guide d'évaluation des capacités parentales: Adaptation du guide de Steinhauer: 0 à 5 ans* (3éd.). Montréal: Centres Jeunesse de Montréal.

Deutsch, R. M., & Clyman, J. (2016). Impact of mental illness on parenting capacity in a child custody matter. *Family Court Review, 54*(1), 29-38. doi: 10.1111/fcre.12201

Duncan, S., & Reder, P. (2003). How do mental health problems affect parenting? In P. Reder, S. Duncan & C. Lucey (Eds.), *Studies in the Assessment of Parenting* (pp. 201-216). Hove and New York: Brunner-Routledge.

Falkov, A. (2017). The Family Model. *Journal of Parent and Family Mental Health, 2*(2), 1-3.

Feiring, C., & Lewis, M. (1987). *Equifinality and Multifinality: Diversity in Development from Infancy into Childhood.* Paper presented at the Biennial Meeting of the Society for Research in Child Development, Baltimore.

Göpfert, M., McClelland, N., & Wilson, J. (2010). Maternal mental health: An ethical base for good practice. In D. Kohen (Ed.), *Oxford Textbook of Women and Mental Health* (pp. 59-71). Oxford: Oxford University Press.

Holzer, L., Jaugey, L., Sofia, C., & Corthésy, Y. (2013). Les enfants de parents souffrant de psychose. *Annales Médico-psychologiques, revue psychiatrique, 171*(9), 618-622. doi: http://dx.doi.org/10.1016/j.amp.2011.05.011

Hornstein, C., Trautmann-Villalba, P., Hohm, E., Rave, E., Wortmann-Fleischer, S., & Schwarz, M. (2006). Maternal bond and mother–child interaction in severe postpartum psychiatric disorders: Is there a link?. *Archives of Women's Mental Health, 9*(5), 279-284. doi: 10.1007/s00737-006-0148-6

Keren, M., & Tyano, S. (2014). Psychopathology and its early impact on parenting behaviors in mothers: The interface between adult and infant psychiatry. *The Israel Journal of Psychiatry and Related Sciences, 52*(2), 92-98.

Manderscheid, R. W., Ryff, C. D., Freeman, E. J., McKnight-Eily, L. R., Dhingra, S., & Strine, T. W. (2010). Evolving Definitions of Mental Illness and Wellness. *Preventing Chronic Disease, 7*(1), A19.

Mental Capacity Act (2005). London: HMSO.

Muzik, M., Morelen, D., Hruschak, J., Rosenblum, K. L., Bocknek, E., & Beeghly, M. (2016). Psychopathology and parenting: an examination of perceived and observed parenting in mothers with depression and PTSD. *Journal of Affective Disorders, 207,* 242-250. doi: 10.1016/j.jad.2016.08.035

Pereira, D. & Alarcão, M. (2010). Avaliação da parentalidade no quadro da proteção à infância. *Temas em Psicologia,* 18 (2), 499-517.

Pereira, D. (2014). Parentalidade e proteção à infância: Um guia para a avaliação da capacidade parental. (Doctoral dissertation). Retrieved from https://estudogeral.sib.uc.pt/.

Pereira, D., & Alarcão, M. (2014). Guia de Avaliação das Capacidades Parentais: Estudo exploratório da fiabilidade em profissionais da proteção à infância. In M. M. Calheiros & M. V. Garrido (Eds.), *Crianças em Risco e Perigo* (Vol. 4, pp. 171-193.). Lisboa: Ed. Sílabo.

Pereira, D., & Alarcão, M. (2015). Guia de Avaliação das Capacidades Parentais: Estudo de validade ecológica. *Psicologia: Teoria e Pesquisa, 31*(2), 203-212. doi: 10.1590/0102-37722015021900203212

Rampou, A. M., Havenga, Y., & Madumo, M. (2015). Parenting experiences of mothers living with a chronic mental illness. *Health SA Gesondheid, 20*(1), 118-127. doi: http://dx.doi.org/10.1016/j.hsag.2015.04.004

Risley-Curtiss, C., Stromwall, L., Hunt, D. T., & Teska, J. (2004). Identifying and reducing barriers to reunification for seriously mentally ill parents involved in child welfare cases. *Families in Society: The Journal of Contemporary Social Services, 85*(1), 107-118. doi: 10.1606/1044-3894.240

Salvaggio, M. (2015). *Mothers' mental health evaluations in foster care practice: A social constructionist, qualitative data-mining study* (Doctoral dissertation). Retrieved from: http://academicworks.cuny.edu/gc_etds/1120

Steinhauer, P., Leitenberger, M., Manglicas, E., Pauker, J., Smith, R., & Gonçalves, L. (1993). *Guide d'evaluation de la compétence parentale.* Toronto: L'Institut pour la Prévention de l'enfance maltraitée.

Tippins, T. M., & Wittmann, J. P. (2005). Empirical and ethical problems with custody recommendations: A call for clinical humility and judicial vigilance. *Family Court Review, 43*, 193-222.

Van Brunt, B., Zedginidze, A. A., & Light, P. A. (2016). The unfit parent: six myths concerning dangerousness and mental illness. *Family Court Review, 54*(1), 18-28. doi: 10.1111/fcre.12200

Chapter 14

Boredom: humanising or dehumanising treatment

Josefa Ros Velasco

Department of Romance Languages and Literatures, Harvard University

Boredom has been understood as a pathological personality trait and a psychological and psychiatric disorder from the beginning of the twentieth century. Since then, specialists in psychology and psychiatry have done their utmost to offer suitable treatments in order to avoid or cope with boredom. Such attempts consisted of therapies for ages, but also took the form of medical treatments. In keeping with the trend established in recent years, since the *DSM-V* was published, many everyday annoyances are under the category of mental disorders, including boredom. This phenomenon is actually dangerous, – particularly as it is applied to children – because there is a lack of knowledge and agreement in this regard. On the contrary, boredom is also understood as an adaptive emotion whose role is to make people aware of a contextual situation that should be changed. In this sense, many thinkers have explained that boredom is an essential part of every human being and, consequently, it should be neither avoided nor treated in clinical terms. In this approach, we will carry out a criticism of this current clinical trend by supporting the idea that it is promoting a process of dehumanization of the so-called 'patients'.

14.1 Introduction

At the very beginning of the last century, the metaphor of 'boredom as a disease' began to be taken at face value in the context of psychological and psychiatric research from industrial psychological tests of efficiency in the workplace (Malkovsky et al., 2012). Since then, the study of boredom has been mainly carried out from the mental health field (Ros Velasco, 2017), from which it has been spread the idea that boredom concerns the individual development, appearing inside the subject as a mental health issue and involving serious psychosocial consequences (Eastwood et al., 2012; German and Latkin, 2012; Bergler, 1945). It has been in recent years that boredom has become a matter of scientific and clinical interest, being analysed from its neurological and cognitive conditions (Eastwood et al., 2012; Danckert and Allman, 2005).

To date, it is estimated that boredom arises from a discontinued relationship between the patient and the environment, caused by pathologies of boredom such as the 'pathology of chronic boredom' (Eastwood et al., 2012; Fisher, 1987; Gimbel, 1975; Todman, 2003; Toohey, 2011), among others. Therefore, much of the literature on boredom focuses on the proposal of boredom-fighting approaches such as the avoidance or confrontation therapies (Nett, Goetz and Daniels, 2010; Leung, 2008; Chaney and Chang, 2005), the 'aways' strategy (Goffman, 1959), and the occupational therapy (Conrad, 1997), among many other antidotes (Retana, 2011). Meanwhile, psychiatrists consider that boredom deserves to be treated as a common psychotic disorder (as depression and stress, Butler et al., 2011) due to its neurotic implications (Bergler, 1945). Not only is boredom associated with the range of mental diseases, but also with dysfunctional behaviors stemming from the efforts of the patients themselves to relieve its symptoms. In this sense, the study of boredom is motivated by the need to predict deviant behaviors and avoid and confront them.

Another almost unknown point of view, on the contrary, describes boredom as an adaptive emotion whose role is to alert people to such external conditions in which boredom is taking place (Blumenberg, 2011; Toohey, 2011; Plutchik, 1980; Damasio, 1999; Damasio, 2003). In this regard, boredom is an emotion that makes us aware that something in our context is not working as it should and we have to change it. Accordingly, boredom is not a pathology, but one of the essential parts of our daily lives.

As a result, we can rethink if psychology and psychiatry disciplines may be leading the understanding of boredom to an extent that neglects the importance of socio-cultural context, without taking into account the alternative points of view – as the anthropological one – which consider boredom as a functional emotion, shaping a process of dehumanization. A criticism may be carried out on the fact that specialists are addressing boredom as a pathological personality trait and prescribing all types of therapies and drugs to bored people, without knowing exactly what cognitive and neurological implications come into play. This is part of a deeper criticism of the current trend of treating and medicalising common and daily negative emotions, turning them into mental diseases, marked by the birth of the latest version of the *Diagnostic and Statistical Manual of Mental Disorders* (DSM-V).

In view of the above, in this chapter we will first briefly discuss the understanding of boredom as a mental pathology, which is at the cutting edge today and, at the same time, the starting point of our criticism. Secondly, we will more extensively approach the point of view of those thinkers who see boredom as an adaptive and functional emotion, as this is a position to which scholars have not paid sufficient attention. Lastly, we will conclude our contribution by evaluating the present trend of medicalising boredom and other common everyday annoyances.

14.2 Understanding boredom as a mental health disorder

At present, boredom is understood as a pain caused by a situation of conflict between the subjects' internal need of mental activity and their inability to stimulate themselves (Campbell, 1996; Rule, 1998). Boredom is a consequence of a disconnection between the patient and their environment which remains in spite of changing external conditions. This is caused by pathologies of boredom such as 'the pathology of chronic boredom' (Eastwood et al., 2012; Fisher, 1987; Gimbel, 1975; Todman, 2003; Toohey, 2011), 'the disorder of a propensity towards boredom' (Brissett and Snow, 1993; Conrad, 1997; Eastwood et al., 2012; Farmer and Sundberg, 1986; Fenichel, 1951; Fisher, 1991; Greenson, 1953; LePera, 2011; Martin, Saldo and Stew, 2006; McCormick et al., 2005; O'Hanlon, 1981; Tilburg and Igou, 2011), or 'the problem of subjective perception of boredom' (Fiske and Maddi, 1961; Pekrun et al., 2010; Perkins and Hill, 1985; Slaby, 2010; Watt and Blanchard, 1994). Moreover, it is seen as a corollary of other pathologies, such as 'the inability to clarify one's own desires' (Antón, 2012; Eastwood et al., 2012; Fenichel, 1951; Healy, 1984; Parreño Roldán, 2013; Seib and Vodanovich, 1998; Watt and Blanchard, 1994), 'the distorted perception of time' (Danckert and Allman, 2005; Eastwood et al., 2012; Musharbash, 2007; Revers, 1967; Vodanovich and Watt, 1999), 'the excess or lack of self-awareness' (Chaney and Chang, 2005; Eisnitz, 1974; Gana et al., 2000; Kreisel, 2006; Seib and Vodanovich, 1998; Sommers and Vodanovich, 2000; Meyer Spacks, 1995; Weinberger and Muller, 1975; Wink and Donahue, 1999; Leong and Schneller, 1993; Watt and Blanchard, 1994), 'attention deficit disorder' (Cheyne et al., 2006; Damrad-Frye and Laird, 1989; Danckert and Allman, 2005; Eastwood et al., 2012; Farmer and Sundberg, 1986; Kass et al., 2001; LePera, 2011;

Malkovsky et al., 2012; O'Hanlon, 1981; Pekrun et al., 2010; Seib and Vodanovich, 1998), or many of them at the same time.

As a result, a large part of the extant literature on boredom, as a pathological personality trait, is geared towards providing treatments. Specialists in psychoanalysis and existential philosophy (Martin, Sadlo and Stew, 2006) have learned to deal with boredom and teach others how to get to grips with its symptoms so that they prevent or tackle them (Goffman, 1959; Conrad, 1997). Our time, more than any other, has led to the proliferation of innumerable antidotes to the 'evil of boredom' (Retana, 2011), through the implementation of confrontation or avoidance strategies (Nett, Goetz and Daniels, 2010). On the other hand, psychiatrists have considered boredom as a common pathology like depression or stress (Butler et al., 2011), since it plays a definitive role within psychotic disorders as a result of its neurotic implications (Bergler, 1945). According to some authors, the treatment of boredom requires the genetic location of the disease in order to intervene in it at a neurological level (Bergler, 1945).

What actually concerns researchers on boredom, thus understood, are the consequences arising from its suffering (van Tilbur and Igou, 2011; Martin, Sadlo and Stew, 2006; Vodanovich and Watt, 1999). Not only is boredom associated with the range of mental disorders, but also with dysfunctional behaviours stemming from the efforts of the patients themselves to relieve its symptoms. In this sense, boredom is considered responsible for causing depressions, anxieties, addictions to drugs, sex and gambling, reckless driving, states of despair and loneliness, hostile and aggressive behaviours, sleep disorders, criminal actions, school deviations, suicidal tendencies, low self esteem, lack of social affiliation and eating disorders, among others (see, e.g. Martin, Sadlo and Stew, 2006; van Tilburg and Igou, 2011; Vodanovich and Watt, 1999, for complete lists of behaviour deviations caused by boredom). Consistently, the study of boredom is motivated, in most cases, by the need to learn more about the phenomenon to predict harmful or deviant behaviours and avoid or treat them.

Beyond this understanding of boredom as a pathological personality trait, there is another different and even contrary point of view on boredom which emphasises its capacity to cause responses that are neither necessarily counterproductive nor depend on the personality of the bored, but on the contextual situation. This is the point of view of those who understand boredom as an adaptive emotion, from which we will cast doubt on the fact that boredom must be treated or medicalised as if it were a mental disorder.

14.3 An alternative understanding of boredom as an adaptive emotion

Little attention has been paid by scholars to the understanding of boredom as a causal reality that does not necessarily produce dysfunctional responses, but creative. As an experience that triggers anxiety and agitation (van Tilburg and Igou, 2011), and perhaps even something like nervousness (Fenichel, 1951), boredom, detached from the complexity of the mental disease, leads to innovation. Boredom makes us feel dissatisfied with certain situations because "when we are bored there is something of the context that we are rejecting" (Retana, 2011, p. 283) and, thereupon, drives us to seek new opportunities as a reaction to the environment. It is a symptom that instigates moments of experimentation (Parreño Roldán, 2013; Moravia, 1999), including a critical element: an expression of deep dissatisfaction (Svendsen, 1999).

Leaving behind truly pathological personalities and socio-cultural contexts in which the search for novelty is impracticable – for instead, those marked by predictability, in which there is a pre-planned response for each need; sometimes even a medical one – the potential of boredom is inherently positive. This is because boredom prevents stillness by giving way to something else. As Rossi claims (1986), "your symptom or problem is your friend! Your symptom is a signal that a creative change is needed in your life" (quoted in Rule, 1998, p. 330). In short, boredom meets benign purposes in healthy contexts and individuals because of its consequences translated into compensatory activities.

Unfortunately, the benefits of the driving force of boredom have been less visible in the literature. Nevertheless, some authors have stated that boredom may be understood as an opportunity to learn from oneself (Rule, 1998), a mechanism of anticipation, a starting point of great works (Butler et al., 2011), and a valuable art (Sloterdijk, 2009 quoted in Butler et al., 2011). Thinkers like Siegfried Kracauer (1995) considered to be bored is the only suitable option because boredom provides a kind of guarantee that one is still in control of their own existence. According to Russell (1930), we must admit that a certain degree of boredom is essential to living a happy life and is something that should be taught to young people.

In this regard, another group of thinkers has suggested that boredom plays a key role in our individual and social development by pointing out that the driving force of boredom meets an adaptive function. Professor Peter Toohey is the main representative of this new trend as he stated the driving force of boredom meets an adaptive function in a true Darwinian sense (2011), as inspired by Robert Plutchik and Henry Kellerman's premises on emotions (1980). Plutchik and Kellerman understood that negative emotions help individuals to cope with survival problems and keep them away from harmful environments and situations. According to Toohey (2011), we can see boredom as one of that adaptive emotions since it protects us from danger (Plutchik and Kellerman, 1980) by highlighting our relationship with the environment and forcing us to develop a strategy to explore new opportunities (Todman, 2007). As Plutchik and Kellerman, the neurologist Antonio Damasio suggested (1999; 2003) that negative emotions directly influence self-regulation by foreseeing the dangers, being an advantage and opportunity for individuals. Boredom may be one of these emotions that might actually have been selected in our evolutionary race due to its role to remove us from toxic situations and lead the search for new experiences.

By the same token, the German philosopher Hans Blumenberg (2011) pointed out that boredom meets an adaptive role in stimulating the change that keeps our interest. Blumenberg did not hesitate to claim that if we do not suffer from boredom when we are fully adapted, forcing us to be in constant motion, we would be unable to adapt to future changes. This is the reason why, according to him, boredom was selected: to fulfill a cardinal adaptive function, to open the door to future adaptations. Blumenberg (2011) also argued that one of our biggest problems today is that we are neglecting the importance of negative emotions such as boredom. These are understood as if an error of the evolution were, perhaps because in our society health is defined as "a state of complete physical, mental and social well-being" by WHO (Blumenberg, 2011, p. 516). As Blumenberg stated (2011), "everything becomes a disease and, therefore, the need for treatment becomes a permanent state" (p. 516). Today, boredom is nothing more than a nuisance one can turn into "a way to earn a living" (Blumenberg, 2011, p. 519).

The study of the driving force of boredom has been carried out from psychology and psychiatry disciplines that misrepresent its power to move people to action in a positive sense, focusing only on the pathological personality of the bored and neglecting the possibility that such boredom might be a symptom or a consequence of an unhealthy environment. This alternative understanding of boredom opens the way for rethinking how boredom came to be generally understood in pathological terms, and also for casting doubts on the medical practices concerning boredom and other everyday annoyances.

14.4 The philosophical-anthropological criticism of the clinical treatment of boredom

Boredom is omnipresent in our modern society precisely because of the abundance of preplanned responses to cope with it. Although the philosophical and anthropological disciplines are attempting to understand boredom as a symptom of our context, the mental health ones are understanding it as the disease itself. Thus, the core of the problem totally lays on the individual and, as a result, the cure for boredom is found through the prescription of therapies and drugs.

However, at the beginning of the twenty-first century, there is a lack of agreement by specialists on what is boredom, its cognitive implications and neurological base (Danckert and Allman, 2005; Eastwood, 2012; Merrifield, 2014), despite the growing literature on this complex issue. Researchers describe boredom as a pathology and propose their remedies while at the same time agree that the study of boredom is in its infancy and has not been paid the attention deserved (O'Halon, 1981; Damrad-Frye and Laird, 1989; Vodanovich and Watt, 1999; Fisher, 1991; Vodanovich, 2003; Martin et al., 2006; Pekrun et al., 2010), and acknowledge that the lack of assessment and measuring instruments makes it impossible for specialists to reach a true understanding of the matter (Vogel-Walcutt et al., 2012).

It is important to be cautious and maintain a permanent critical attitude when we elevate everyday annoyances to the status of mental pathologies. We just have to bear in mind that, according to De Chenne and Moddy (1987), mental health professionals are not simply phenomenologists but they take action in a way from theory to therapy.

Boredom is considered a widespread pathology (Ramey, 1974) that may be treated with therapies such as the occupational therapy (Overheul, 2001) or the Mitsubishi Outlander therapy (Ato, 2013). Beyond the lack of knowledge about the phenomenon, we acknowledge that such therapies are not always harmful as some of them are mainly taking into account the environment surrounding the bored. In such cases, the therapies may even help the bored to find the way towards a change. However, when boredom is considered as a pathological personality trait the proposed therapies may not be as harmless, especially when consisting of drug therapies.

About a half century ago, Barmack started to explore the possibility of treating boredom with benzedrine sulphate and ephedrine hydrochloride (Barmack, 1937; 1938; 1939a; 1939b; 1939c). Now that boredom is associated with mental disorders such as depression or stress, specialists talk about dopamine (Branković, 2015). Nowadays, manuals on pharmacology point to methylxanthine – a group of naturally occurring agents present in caffeine, theophylline and theobromine – (Bennet and Morris, 2007; Hancock and Mckim, 2013) as a drug capable of reversing performance decrements caused by boredom and fatigue.

There are strong limitations to diagnose and treat boredom with drugs, as almost nothing is known about its cognitive and neural implications. Many researchers are attempting to locate the neural basis of boredom and are tentatively suggesting the insula plays a key role (Toohey, 2011; Weissman, 2006). However, by the time, how boredom neurologically affects people and how drugs will affect the bored – including side effects – remain unknown.

What we are saying about boredom could certainly be extended to other modern phenomena such as anxiety or stress. We have to acknowledge that there will be cases in which the continued suffering of such a state of discomfort, together with the pathological inability of the subject to deal with it, will require a solution involving drugs. However, many transitory discomforts such as boredom have been diagnosed as mental disorders too lightly without a sound basis. In this regard, we agree with Foucault (1974) when he stated that, since the twentieth century, medicine has extended beyond patients' needs.

It is not surprising that this trend has aroused certain concern and criticism on the part of both, those who think that boredom, like many other negative emotions, plays a key role in our development, and some researchers and specialists in psychology and psychiatry who claim not to elevate everyday annoyances to the status of mental disorders. Many criticisms, including that of Blumenberg (2011), point out that the *Diagnostic and Statistical Manual of Mental Disorders (DSM)* played a key role in this trend. The fifth edition, especially, promotes life medicalisation: "it expands the number of pathologies appearing on the *DSM-IV* [...] causing more transitional behaviors, even normal behaviors, become disease" (Janin, 2013, p. 34). Strictly speaking, this manual pathologises human suffering: "it promotes medicalisation to avoid, for instead, mourning, which becomes a disease according to the *DSM-V*" (Rojas, 2013, p. 46). Allen Frances himself – who edited the previous version of this manual – is promoting this criticism since 2014, when he published his book *Saving Normal*. As far as he is concerned (2014), the *DSM-IV* started to be used wrong and as a result "in the United States, 20% takes psychiatric drugs daily, and a fourth of the population has been diagnosed of a mental disease". Frances (2014) warns that if specialists follow the recommendations of the *DSM-V* "81% between 11-21 years old may be diagnosed of a mental disease".

The reason why we have come to such a critical situation, according to Frances (2014), is that psychiatrists are selfish and unable to look beyond their own perspectives. Psychiatrists have turned daily annoyances into diseases by making wrong decisions and overstating the importance of their fieldwork (Frances, 2014). As Blumenberg, Frances claims that in a culture in which people take pills to alleviate their problems, our ability to cope with stress and also self-security is drastically reduced: "if this behavior becomes widespread, society as a whole will be weakened" (Frances, 2014). The situation is further complicated if we talk about children and adolescents. Campolongo et al. (2015) sounded the alarm in this regard: "in our clinical practice with children we see how the different problems of daily life have started being medicalised: motivation, academic failure, hyperactivity, lack of attention, boredom, sadness, and so on" (p. 141).

We are not going to discuss the economic interests underlying this trend, as they were already announced by Blumenberg and are well known to all (Frances, 2014; Stolkiner, 2013; Illich, 1975; Foucault, 1974). We just concur with the words of Frances (2014) when he states that everybody knows that the real beneficiary of the situation is the pharmaceutical industry. Given the magnitude of the issue, revealed in this last paragraph, now more than ever people have to take a critical and responsible view and demand further information when diagnosed.

14.5 Conclusion

Since the end of the last century, boredom is mainly defined as a pathological personality trait by psychologists and psychiatrists, who propose all types of therapies and treatments to cope with it. What is going on with regard to boredom may be applied to other everyday annoyances, which are being elevated to the state of mental disorders. We examined an alternative understanding of boredom and other negative emotions that highlight the role of such states of discomfort in promoting the search for change from an environmental point of view. Taking this alternative point of view into account, we carried out a criticism of the current trend of turning everyday annoyances into mental disorders. We can safely assume that it is only through criticism that we will be able to break with such a dangerous trend that medicalises the most common upsets, including boredom.

Acknowledgements

The author would like to express her gratitude to the Department of Romance Languages and Literatures at Harvard University and the Real Colegio Complutense at Harvard.

14.6 References

Antón, M. C. (2012) El aburrimiento. Una mirada desde el psicoanálisis. *Perspectivas en Psicología.* 9, 104-109.

Ato, M. C. (2013) Combate el aburrimiento vital con la 'Terapia Outlander' de Mitsubishi. *Ipmark: Información de publicidad y marketing.* 793, 48.

Barmack, J. E. (1937) Boredom and other factors in the physiology of mental effort: An exploratory study. *Archives of Psychology.* 218, 1-83.

Barmack, J. E. (1938) The effect of benzedrine sulphate upon the report of boredom and other factors. *Journal of Psychology.* 5, 125-133.

Barmack, J. E. (1939a) A definition of boredom: A reply to Mr. Berman. *American Journal of Psychology.* 52, 467-471.

Barmack, J. E. (1939b) Studies on the psychophysiology of boredom: Part I. The effects of 15mgs of benzedrine sulphate and 5mgs of ephedrine hydrochloride on blood pressure, report on boredom, and other factors. *Journal of Experimental Psychology.* 25, 494-505.

Barmack, J. E. (1939c) Studies on the psychophysiology of boredom: Part II. The effect of lowered room temperature and an added incentive on blood pressure, report of boredom and other factors. *Journal of Experimental Psychology.* 25, 634-642.

Bennett, P. & Morris, J. (2007) Drug Abuse. In: Bennett, P. & Morris, J. *Clinical Pharmacology.* Oxford: Elsevier Health Sciences, 142-174.

Bergler, E. (1945) On the disease-entity boredom ("alyosis") and its psychopathology. *Psychiatric Quarterly.* 19 (1), 38-51.

Blumenberg, H. (2011) *Descripción del ser humano.* Mexico: FCE.

Branković, S. (2015) Boredom, dopamine, and the thrill of psychosis: psychiatry in a new key. *Psychiatria Danubina.* 27 (2), 126-137.

Brissett, D. & Snow, R. P. (1993) Boredom: Where the future isn't. *Symbolic Interaction.* 16 (3), 237-256.

Butler, N., Olaison, L., Śliwa, M., Sørensen, B. M. & Spoelstra, S. (2011) Work, play and boredom. *Ephemera.* 11 (4), 329-335.

Campbell, R. J. (1996) *Psychiatric Dictionary.* New York: Oxford University Press.

Campolongo, D., Martin, A., Tammaro, M. S. & Torregiani, M. (2015) Niños medicalizados. VII Congreso Internacional de Investigación y Práctica Profesional en Psicología – XXII Jornadas de Investigación – Décimo Encuentro de Investigadores en Psicología del MERCOSUR. University of Buenos Aires, 141-144. Available from: https://www.aacademica.org/000-015/714.pdf

Chaney, M. P. & Chang, C. Y. (2005) A Trio of Turmoil for Internet Sexually Addicted Men Who Have Sex with Men: Boredom Proneness, Social

Connectedness, and Dissociation. *Sexual Addiction & Compulsivity.* 12, 3-18.

Cheyne, J. A., Carriere, J. S. & Smilek, D. (2006) Absent-mindedness: Lapses of conscious awareness and everyday cognitive failures. *Consciousness and cognition.* 15 (3), 578-592.

Conrad, P. (1997) It's Boring: Notes on the Meanings of Boredom in Everyday Life. *Qualitative Sociology.* 20 (4), 465-475.

Damasio, A. R. (1999) *The Feeling of What Happens: Body and Emotion in the Making of Consciousness.* New York: Harcourt Brace & Co.

Damasio, A. R. (2003) *Looking for Spinoza: Joy, Sorrow, and the Feeling Brain.* North Eugene: Harvest.

Damrad-Frye, R. & Laird, J. D. (1989) The experience of Boredom: The role of the Self-Perception of Attention. *Journal of Personality and Social Psychology.* 57 (2), 315-320.

Danckert, J. A. & Allman, A. A. (2005) Time flies when you're having fun: Temporal estimation and the experience of boredom. *Brain and Cognition.* 59 (3), 236-245.

De Chenne, T. K. & Moddy A. J. (1988) Boredom: Theory and therapy. *The Psychotherapy Patient.* 3, 17-29.

Eastwood, J. D., Frischen, A., Fenske, M. J. & Smilek, D. (2012) The Unengaged Mind: Defining Boredom in Terms of Attention. *Perspectives on Psychological Science.* 7 (5), 482-495.

Eisnitz, A. J. (1974) On the metapsychology of narcissistic pathology. *American Psychoanalysis Association.* 22, 279-291.

Farmer, R. & Sundberg, N. D. (1986) Boredom Proneness—The Development and Correlates of a New Scale. *Journal of Personality Assessment.* 50 (1), 4-17.

Fenichel, O. (1951) On the psychology of boredom. In: Rapaport, D. *Organization and pathology of thought: Selected sources.* New York: Columbia University Press, 349-361.

Fenichel, O. (1953) On the psychology of boredom. In: Fenichel, O. *The collected papers of Otto Fenichel.* Vol 1. New York: W. W. Norton, 292-302.

Fisher, C. D. (1987) *Boredom: construct, causes and consequences.* Texas: A&M University, College Station, Department of Management.

Fisher, C. D. (1991) Boredom at work: A neglected concept. *Human Relations.* 46 (3), 395-417.

Fiske, D. W. & Maddi, S. R. (1961) *Functions of varied experience.* Oxford: Dorsey.

Foucault, M. (1974) La crisis de la medicina o la crisis de la antimedicina. Curso de medicina social del Instituto de Medicina Social, Centro Biomédico de la Universidad Estatal de Río de Janeiro, Brasil. Available from: http://hist.library.paho.org/Spanish/EMS/4451.pdf.

Frances, A. (2014) Saving Normal: *An Insider's Revolt against Out-of-Control Psychiatric Diagnosis, DSM-5, Big Pharma, and the Medicalization of Ordinary Life.* New York: Harper Collins.

Frances, A. & Ayuso, M. (2014) Allen Frances. Director del DSM IV. El confidencial. Available from: https://www.elconfidencial.com/alma-corazon-vida/2014-09-09/la-industria-farmaceutica-esta-causando-mas-muertes-que-los-carteles-de-la-droga_188426/.

Frances, A. & Barbosa, J. (2014) Entrevista a Allen Frances. El país. Available from: https://elpais.com/sociedad/2014/09/26/actualidad/1411730295_336861.html.

Gana, K., Deletang, B. & Metais, L. (2000) Is boredom proneness associated with introspectiveness? *Social Behavior and Personality: an international journal.* 28 (5), 499-504.

German, D. & Latkin, C. A. (2012) Boredom, depressive symptoms, and HIV risk behaviors among urban injection drug users. *AIDS.* 16 (8), 2244-2250.

Gimbel, L. (1975) The pathology of boredom and sensory deprivation. *The Canadian Journal of Psychiatric Nursing.* 16 (5), 12.

Goffman, E. (1959) *The Presentation of Self in Everyday Life.* London: Allen Lane The Penguin Press.

Greenson, R. R. (1953) On boredom. *Journal of the American Psychoanalytic Association.* 1, 7-21.

Hancock, S. & Mckim, W. (2013) Caffeine and the Methylxanthines. In: Hancock, S. & Mckim, W. *Drugs and Behavior: An Introduction to Behavioral Pharmacology.* New Jersey: Pearson, 209-246.

Healy, S. D. (1984) *Boredom, Self and Culture.* New Jersey: Fairleigh Dickinson University Press.

Illich, I. (1975) *Némesis médica. La expropiación de la salud.* Barcelona: Barral.

Janin, B. (2013) Intervenciones subjetivas frente al sufrimiento psíquico. In: Dueñas, G. *La patologización de la infancia (II), Problemas e intervenciones en la clínica.* Buenos Aires: Noveduc, 15-33.

Kass, S. J., Vodanovich, S. J. & Callender, A. (2001) State-trait boredom: Relationship to absenteeism tenure, and job satisfaction. *Journal of business and psychology.* 16 (2), 317-327.

Kracauer, S. (1995) *The Mass Ornament: Weimar Essays.* Cambridge: Harvard University Press.

Kreisel, D. K. (2006) Wolf Children and Automata: Bestiality and Boredom at Home and Abroad. *Representations.* 96 (1), 21-47.

Leong, F. T. L. & Schneller, G. R. (1993) Boredom Proneness: Temperamental and Cognitive Components. *Personality and Individual Differences.* 14 (1), 233-239.

Leung, L. (2998). Leisureboredom, sensation seeking, self-esteem, and addition. In: Konijn, E., Utz., S., Barnes, S. B. & Tanis, M. *Mediated interpersonal communication.* New York, Routledge, 359-381.

LePera, N. (2011) Relationships between boredom proneness, mindfulness, anxiety, depression, and substance use. *The New School Psychology Bulletin.* 8 (2), 15-25.

Malkovsky, E., Merrifield, C., Goldberg, Y. & Danckert, J. (2012) Exploring the relationship between boredom and sustained attention. *Experimental Brain Research.* 221 (1), 59-67.

Martin, M., Saldo, G. & Stew, G. (2006) The phenomenon of boredom. *Qualitative Research in Psychology.* 3, 193-211.

McCormick, B. P., Fundeiburk, J. A., Lee, Y. & Hale-Foiight, M. (2005) Activity Characteristics and Emotional Experience: Predicting Boredom and Anxiety in the Daily Life of Community Mental Health Clients. *Journal of Leisure Research.* 37 (2), 236-253.

Merrifield, C. (2014) *Toward a Model of Boredom: Investigating the Psychophysiological, Cognitive and Neural Correlates of Boredom.* Waterloo: University of Waterloo. (Unpublished Doctoral dissertation).

Meyer Spacks, P. (1995) *Boredom. The literary history of a state of mind.* Chicago: The University of Chicago Press.

Moravia, A. (1999). *La noia.* Barcelona: Planeta.

Musharbash, Y. (2007) Boredom, time, and modernity: An example from Aboriginal Australia. *American Anthropologist.* 109 (2), 307-317.

Nett, U. E., Goetz, T. & Daniels, M. (2010) What to do when feeling bored? Students' strategies for coping with boredom. *Learning and Individual Differences.* 20 (6), 626-638.

O'Halon, J. F. (1981) Boredom: Practical consequences and a theory. *Acta Psychologic.* 49, 53-82.

Overheul, V. (2001) A cure for boredom. *Occupational health & safety.* 70 (9), 192.

Parreño Roldán, C. (2013) Aburrimiento y Espacio: Experiencia, Modernidad e Historia. *Revista de la Escuela de Arquitectura de la Universidad de Costa Rica.* 2 (3), 1-15.

Pekrun, R., Goetz, T., Daniels, M., Stupnisky, R. H. & Raymond, P. (2010) Boredom in Achievement Settings: Exploring Control–Value Antecedents. *Journal of Educational Psychology.* 102 (3), 531-549.

Perkins, R. E. & Hill, A. B. (1985) Cognitive and affective aspects of boredom. *British Journal of Psychology.* 76, 221-234.

Plutchik, R. & Kellerman, H. (1980) Emotion: Theory, research, and experience. *Vol. 1. Theories of emotion.* New York: Academic.

Ramey, E. R. (1974) Boredom: The most prevalent American disease. *Harper's Magazine. 249,* 12-22.

Retana, C. (2011) Consideraciones acerca del aburrimiento como emoción moral. *Káñina, Revista de Artes y Letras, Universidad de Costa Rica.* 35 (2), 179-190.

Revers, W. J. (1967) Perspectivas antropológicas del aburrimiento. *CONVIVIUM.* 23, 38-47.

Rojas, M. C. (2013) El mercado de la felicidad, cuestionando el DSM-V. In: Dueñas, G. *La patologización de la infancia (II), Problemas e intervenciones en la clínica.* Buenos Aires: Noveduc, 45-55.

Ros Velasco, J. (2017) Boredom: A Comprehensive Study of the State of Affairs. *Thémata.* 56, 171-198.

Rossi, E. (1986) *The psychobiology of mind-body healing.* New York: W. W. Norton.

Rule, W. R. (1998) Unsqueezing the Soul: Expanding Choices by Reframing and Redirecting Boredom. *Journal of Contemporary Psychotherapy.* 28 (3), 327-336.

Russell, B. (1930) Boredom and Excitement. In: Russell, B. *The Conquest of Happiness.* London: Allen & Unwin, 57-68.

Seib, H. M. & Vodanovich, S. J. (1998) Cognitive correlates of boredom proneness: The role of private self-consciousness and absorption. *The Journal of Psychology.* 132 (6), 642-652.

Slaby, J. (2010) The other side of existence: Heidegger on boredom. In: Flach, S., Margulies, D. & Söffner, J. *Habitus in Habitat II— Other Sides of Cognition.* Bern: Peter Lang, 101-120.

Sloterdijk, P. (2009) Inspiration. *Ephemera.* 9 (3), 242-251.

Sommers, J. & Vodanovich, S. J. (2000) Boredom Proneness: Its Relationship to Psychological- and Physical-Health Symptoms. *Journal of Clinical Psychology.* 56 (1), 149-155.

Stolkiner, A. (2013) *Medicalización de la vida, sufrimiento subjetivo y prácticas en salud mental.* Buenos Aires: Psicolibro.

Svendsen, L. (1999) *Kjedsomhetens filosofi.* Oslo: Universitetsforlaget.

Todman, M. (2003) Boredom and psychotic disorders: Cognitive and motivational issues. *Psychiatry.* 66 (2), 146-167.

Toldman, M. (2007). Psychopathology and Boredom: A Neglected Association. In: Fantil, K. A. *Psychological Science: Research, Theory and Future Directions.* Athens, Atiner Press, 1-12.

Toohey, P. (1988) Some Ancient Notions of Boredom. *Illinois Classical Studies.* 13 (1), 151-164.

Toohey, P. (2011) *Boredom. A Lively History.* New Haven: Yale University Press.

Van Tilburg, W. A. P. & Igou, E. R. (2011). On boredom: Lack of challenge and meaning as distinct boredom experiences. *Motivation and Emotion.* 36 (2), 181-194.

Vodanovich, S. J. (2003) Psychometric measures of boredom: a review of the literature. *The Journal of Psychology.* 137, 569-601.

Vodanovich, S. J. & Watt, J. D. (1999) The relationship between time structure and boredom proneness: An investigation within two cultures. *The Journal of Social Psychology.* 139 (2), 143-152.

Vogel-Walcutt, J. J., Fiorella, L., Carper, T. & Schatz, S. (2012) The Definition, Assessment, and Mitigation of State Boredom Within Educational Settings: A Comprehensive Review. *Educational Psychology Review.* 24 (1), 89-111.

Watt, J. D. & Blanchard, M. J. (1994) Boredom proneness and the need for cognition. *Journal of Research in Personality.* 28 (1), 44-51.

Weinberger, J. L. & Muller, J. J. (1975) The American Icarus Revisited: Phallic Narcissism and Boredom. A Reply to the Discussion by Alan J. Eisnitz. *International Journal of Psycho-Analysis.* 56, 379-380.

Weissman, D. H., Roberts, K. C., Visscher, K. M. & Woldorff, M. G. (2006) The neural bases of momentary lapses in attention. *Nature Neuroscience.* 9, 971-978.

Wink, P. & Donahue, K. (1997) The relation between two types of narcissism and boredom. *Journal of Research in Personality.* 31 (1), 136-140.

Part Four: Phenomenology, Existentialism and Psychoanalysis

Chapter 15

Dialectic of the first and third person accounts in mental health – therapeutic implications

Victor Amorim Rodrigues
ISPA University Institute, Lisbon, Portugal

Twentieth first century psychiatry, and most of contemporary psychology, relies on a third-person account with an emphasis on objectivity and quantification. On the other hand, phenomenology brings the first person perspective to the clinical practice giving voice to the suffering and lived experience of the patients. Despite its name, the study of subjectivity is not in and of itself subjective by virtue of the rigor of the phenomenological method. I contrast third-person psychiatric and psychological reports with first-person reports of lived experience recorded of clinical sessions. In the development of the first-person approach, I propose a hermeneutical critical method relevant for clinical practice that proved more perspicuous than third person approaches in the clinical case presented and discussed.

15.1 Twilight of the clinical idols in the Twentieth First century

Mainstream psychotherapy is caught in the grip of a powerful dilemma: either it completely surrenders itself to 18th century comtean positivism, with its reliance on quantification, objectivity, measurability, in a similar way biological psychiatry and cognitive psychology have done, or it becomes unable to distinguish herself from crackpot "alternative therapies" with quasi-magical foundations, like some forms of body energetic therapy and integral therapy. Let us have a close reading of what each of these alternatives entail and evaluate whether we should accept any of them.

Biological and empirical research have completely dominated psychiatry in its originary mission of understanding the *psyche* in its characteristic subjectivity. To illustrate this, I will propose to make the following exercise, not to be confused with a systematic analysis or review of the literature: taking a look at one of the main psychiatry top ranked journals, randomly selected, and analyze what one finds in the latest issue. If one takes a look at the latest issue of the *American Journal of Psychiatry*, one will find three main articles, each of which identifies, in my view, themes of contemporary psychotherapy: an undue emphasis for empirical studies of psychotherapy outcomes, neurobabble and a medicalizing tendency.

In the first main article of the latest issue of *American Journal of Psychiatry* "PTSD Psychotherapy Outcome Predicted by Brain Activation During Emotional Reactivity and Regulation" (Fonzo et al., 2017b), the authors propose to identify the functional traits of PTSD patients which have a high rate of successfully undergoing psychotherapy via a study of brain systems associated with emotional reactivity. One can immediately detect two main themes of contemporary psychiatry: the allure that neuroscience has over it and the necessity of being evidence-based and empirically attested.

Evidence-based psychotherapists, in line with the previous aforementioned study, try to show empirically that their therapy works. But, alas, the manifold problems of trying to study the effectiveness of psychotherapy are well known. First, what are the criteria of success and on what basis do we select those criteria? Second, how do we know the changes are due to psychotherapy and not some other factors (placebo, selection effects, chance)? Third, how long do the alleged benefits last (Bloch, 1999)? Cognitive-behavioural therapists are very proud and quick to point out that their therapies are based on solid and extensive empirical research establishing an unfavorable contrast with other therapies, such as psychoanalytic and existential ones. Let us look at some examples of such claims: 'CBT [Cognitive behavioural Therapy] should be considered a first-line psychosocial treatment of choice.' (Tolin, 2010); 'The empirically supportive psychotherapies are still not widely practiced. As a result, many patients do not have access to adequate treatment'. (Hollon et al., 2002); 'The disconnect between what clinicians do and what science has discovered is an unconscionable embarrassment' (Mischel, 2008).

These claims are problematic for the previously cited well-known reasons, e.g. the criteria for success are conceptually dependent on the paradigm one chooses. In standard psychodynamic psychotherapy, it counts as an improvement if the patient accesses a previous unconscious feeling or memory, while a pure practitioner of behavioural therapy would count as an improvement only a particular change in behaviour originated from a traumatic situation. It's also very unclear whether the standard measure in many empirical studies on the effectiveness of psychotherapy (Hunsley, J., & Lee, 2007) - perceived increased in well-being the patient reports – does not prejudge the issue in favour of CBT. For example, for psychoanalysts the well-being perceived by the patient cannot be the only criteria of therapeutically success, e.g., the unconscious recollection of a traumatic repressed past event in psychodynamic therapy could be traumatic in several ways, even if the original symptoms disappear. Other kinds of therapy could also doubt this excessive reliance on the patient self-judgement which can be clouded by bad faith and other species of self-deception.

Moreover, the claim that CBT is more empirically supported than other therapies (Tolin, 2010) has come under serious attack in recent years (Baardseth et al., 2013; Sundquist et al., 2015; Johnsen & Friborg, 2016). To take one example, in a meta-analysis conducted by Tom Johnsen and Oddgeir Friborg (2016), titled "The Effects of Cognitive Behavioral Therapy as an Anti-Depressive Treatment Is Falling." after studying seventy cognitive and behavioural therapy outcome studies, the authors conclude "the effects of CBT have declined linearly and steadily since its introduction, as measured by patients' self-reports, clinicians' ratings, and rates of remission" (p. 747). In other words, as empirical evidence accumulates it was discovered CBT is not as effective as it was originally thought it would be from the first seminal trials.

These new findings (Baardseth et al., 2013; Sundquist et al., 2015; Johnsen & Friborg, 2016; Jones et al., 2012) have come with recent scepticism about many of the previously considered "established results" in psychology, medicine and other experimental sciences. Researchers consider there is a *replication crisis* (Nosek, 2015), where many previously established results, like ego depletion (Carter et al., 2015) and social priming effects (Doyen et al., 2012), don't replicate under analysis, which extends to empirical studies in psychotherapy. The replication crisis was not primarily caused by blatant falsification, as in the case of infamous Dutch psychologist Diederik Stapel, who falsified dozens of studies in scientific journals. The problem with many empirical studies, on which third person psychotherapies rely on, is that researchers use inadequate samples, that negative results tend not to be published, that most studies are affected by researcher bias, use outdated statistical methods or have a poor experimental setup. One prominent Stanford medical researcher, John Ionnidis, has defended in print that due to these poor methodological problems and biased incorrect use of statistics most published research findings are false (Ionnidis, 2005).

The second main article "Selective Effects of Psychotherapy on Frontopolar Cortical Function in PTSD" (Fonzo et al., 2017a) studies how therapy affects brain function related to emotions in PTSD patients. They studied the effects of therapy in the brain through a brain scan four weeks after the last session of therapy. One point to make is that the control group was put on a waiting list, instead of using "sham therapy" or simple dialogue. So, the effects may be simply due to ordinary dialogue and an extra human connection, not any psychotherapeutically theory, technique or expertise. Furthermore, although the study is very interesting, four weeks doesn't show a lasting benefit for such an expensive and demanding course of treatment as therapy. Moreover, it's interesting to note that the researchers didn't ask how the subjects subjectively felt – or how well the therapists think the subjects improved -, compared to the control group, and see if that difference corresponds to what they see in the brain scans. After all, just because one detects a difference in brain scans after therapy, that does not automatically mean there is a significant difference in behaviour or mental well-being.

Neurology has seducted and hijacked contemporary psychiatry and psychology. In the twenty first century, it's attempting to do the same with psychotherapy (Broder, 2005). We live in a neuromaniac society, where the newly established fields of neuroeconomics (Glimcher, 2013), neuroethics (Clausen & Levi, 2015), neuromarketing (Renvoisé & Morin, 2007) and even neurotheology (Newberg, 2010) and neuropolitics (Connolly, 2002) are being seriously proposed, at least to the general public if not in academia. It was a matter of time before neuropsychoterapy came on to the scene (Grawe, 2007). As critical thinkers we all owe ourselves the question whether a legitimate enterprise, such as neuroscience certainly is, isn't being misguidedly hijacked by other disciplines in search for of scientific credibility to increase their standing both in academia and society at large.

In my opinion, neuroimaging studies hold a large part of the seductive nature of neuroscience. It's easy to be razzle dazzled by the flashy multi-coloured pics to realize that the bright patches of colour visible on brain scans don't depict brain activity. It may seem to psychotherapists that the future where they will see their patient's brains on a live movie screen has finally begun, but this is factually inaccurate and scientifically uninformed.

PET and fMRI only measure neuronal activity very indirectly. What PET measures is a radioactive tracer isotope, which tracks glucose and water molecules, which correlates in turn to blood flow. In its turn, fMRI measures changes in blood oxygen levels. Oxygen is taken by the haemoglobin in red blood cells to the most active brain regions because neurons need oxygen, so the blood flow to that area increases. The magnetic properties of haemoglobin differ, depending on the amount of oxygen they transport and are detected by a scanner. Then, scientists take a control task and the task they are studying and compare the blood flow in various locations of brain areas. After mathematical complex stages of analysis, the researchers determine whether or not the difference between the two tasks is statistically significant or not (Filippi, 2015).

Simplifying, the coloured pictures don't represent the regions that are active during the intended task, but the brain regions which, after a complicated process of analysis, are deemed to be, at least, statistically slightly more active than in the control task. Since the appropriate statistical methods are subject of hot dispute in the neuroimaging community, there's a lot of room for spurious "breakthroughs". The case where one researcher put a dead salmon on a fMRI and generated a picture depicting brain activity to illustrate this point is well known (Bennett et al., 2009).

Another worry comes from the use of very small samples in neuroimaging studies because the technology and its use are very expensive. Small samples are especially unreliable because irrelevant confounder variables, as simple as different breathing rates and taking a cup of coffee or a soda, may be surreptitiously introduced and drastically change the brain patterns detected without having any observable way of being detected and excluded on that basis (Button et al., 2013).

There's also a semantical mistake in trying to couch what we know from psychotherapy to neuroscientific language and pretending we have achieved a deeper insight. Instead of saying "the patient is scared", one could say the "patient's lateral amygdala neurons are having an enhanced synaptic response which is then transmitted via a pathway, involving the prelimbic cortex, to the nucleus of the amygdala", but it's clear no extra insight was gained in this translation to the language of neuroscience. The idea that just because one can describe a situation in terms of an alleged more basic science is a mistake. To give one example, just because organisms are made out of subatomic particles, it doesn't mean biologists should stop using their vocabulary of "mating", "reproductive fitness", "natural selection", "male sex", etc., in order to use the vocabulary of the quantum physicist of "quark", "entanglement", and so on. This does not imply there's something magical about biology, just that the vocabulary of a higher science can be more perspicuous and insightful than a more basic one.

Let me hasten to point out that I'm not criticizing neurology and brain science themselves, just overhasty applications of that research to others contexts, in particular, clinical contexts where complex lived experiences are being shared and our neuroscience, despite much progress in recent years, still struggles to explain walking, sleeping and other basic functions.

The third main article 'A Randomized Controlled Trial of Medication and Cognitive-Behavioral Therapy for Hypochondriasis' (Fallon et al., 2017) investigated through an RCT whether hypochondriacs self-reported improvement increased through the use of a new drug (fluoxetine) and CBT compared to placebo. One short point to make about this study is that difference between the rate of improvement from the use of fluoxetine alone and the use of said drug with joint therapy was very modest (5%), one might be forgiven to think that therapy has very little usefulness for such a modest effect and that our main focus should be on the drugs.

The critics of excessive medication in psychiatry, in particular SSRIs recently, have a long and distinguished history both within the profession, starting with Thomas Szasz, and from the Humanities, with Michael Foucault. The traditional worries are twofold: Firstly, the clinical trials under which psychiatric drugs are tested are funded by and large by pharmaceutical companies, which have a clear conflict of interests and whose research has a clear bias as attested by a recent study in the famous *Journal of Clinical Epidemiology* (Ebrahim et al., 2015) which finds that many authors of meta-analysis, not just RCTs, have clear ties to the pharmaceutical companies and don't report negative secondary effects. Secondly, pharmacotherapy has always side-effects that are costly to the mental health of the patient. Even if the dosage for each individual patient is correct, which is not always easy to do, there is always a potential cost to the mental health to be weighed against the potential benefits. Many patients report feeling "numb inside" or dysphoria, lack of sex drive, and a generalized lack of interest and un-connectedness to the world when they are on certain medications (Moncrieff et al., 2013).

On my view, one ought to also take into account a third more fundamental worry. Psychopharmaceutic drugs by themselves can never address the problem of meaning and purpose of life that many patients come to look for in psychotherapy. In my practice, many patients tell me that other psychiatrists don't listen to them as attentively as they should and just prescribe drugs as an end all of therapy. But patients come to therapy, not only for relief of symptoms, but very often because they feel in a state of *angst* and existential vertigo. Therapy is essential to this search for meaning, as Viktor Frankl said long ago (Frankl, 1946). And, if we restrict ourselves to simple pharmaceutical recipes and manualized therapies that don't take into account the lived history and subjective body of the individual patient, we miss that dimension entirely.

As a reaction to the scientistic *naivité* I have been criticising, several alternative minority and New Age therapies have come appeared like Ken Wilber's Integral Approach (Wilber, 2000) and energetic therapy (Wardell et al., 2014). The obvious problem with these is that they have no connection with objective reality and are free-floating constructions of their practitioners. Their existence should be regarded as a symptom of the deficiencies of the mainstream approaches.

Of course, I am not suggesting that we stop practicing good empirical research or give up psychopharmacology. But, on the face of these challenging problems, one might be forgiven for looking elsewhere for complementary alternatives.

15.2 The phenomenological and existential approach as an alternative psyche-logia

15.2.1 The phenomenological approach

Nothing is as practical as a good theory

Kurt Lewin

Phenomenology is a philosophical movement whose beginnings date to the twentieth century with the publication of *Logische Untersuchungen.* Originally, phenomenology was conceived as a rigorous eidetic science - that studied the essence of things through the way they reveal themselves - by Edmund Husserl and then taken up in different ways by philosophers like Martin Heidegger, Jean-Paul Sartre, Merlau-Ponty and many others since then. The main task of phenomenology is to describe the world of human experience. That is, to describe the world as experienced from the first-person perspective, and in so doing discover and analyze the essential structure of that human experience. From this philosophical tradition, several psychologists have joined together in the field known as phenomenological psychology (Spinelli, 2005; Langridge, 2007; Giorgi; 2009; Tassone, 2012).

Take one simple example to make things clearer: seeing an oak tree. Vision is a type of perception. Drawing on his teacher Brentano, Husserl argues that perception is intentional. Intentionality is a fancy word for directedness or being about something. Intentional mental states are always about something, like desiring, wishing, imagining, fearing. When one desires, one desires something; when one wishes one wishes something; when one imagines, one imagines something or other; when one fears, one fears something. Perception of an oak tree is object-directed. In a nutshell, perception is always directed at something.

According to Husserl, all intentional states can be divided into immanent content, act and object. In this case: the immanent content is constituted by the visual qualitative sensations each of us privately has; the subjective act is the perception or visual act which has a meaning the sensations by themselves lack; the object is the oak tree. The act can be further analyzed into the intentional matter and the intentional quality. The intentional matter of our perception, in this case, the sense or appearance of the oak tree as an oak tree. We always see something under a mode of presentation. Maybe a native aborigine, which has a different relation to nature in comparison with westerns, could see an oak tree as a spirit of the forest. Our intentional matter would be different. But the intentional quality of the act could be the same, namely the visual mode. The intentional matter provides the perceptive act with a meaning, but the intentional quality gives it a mode. If I imagine the oak tree I previously saw, the intentional matter will be the same, but the intentional quality will be different (Zahavi, 2003). I will not go on, but phenomenologists continue this rich analysis and would include the temporal, spatial, bodily, historical and intersubjective aspects of perception. The project of phenomenological psychology is to extend this kind analysis to all the topics of psychological research: emotions, moods, imagination, memory, fantasy, delusion, perception, cognition, language, motor skills, etc.

Husserl devised a rigorous method to reach the essence of these experiences, which includes several steps that must be followed by all phenomenological researchers, including phenomenological psychologists:

- 1st Step: *Époché*: The researcher gives up his *natural attitude* toward the phenomena and brackets every empirical information and acquired knowledge he may have about the object or experience being considered.
- 2nd Step: Phenomenological Reduction: The researcher focuses his attention on the object or experience exactly as it appears.
- 3rd Step: Eidetic Analysis: Through imaginative variation, imagining the phenomena in every imaginable circumstance, the researcher reaches the essence or invariant structure of the experience or object he is considering.

The results are then put forward to other researchers so they can be discussed and criticized. Phenomenology is not an isolated enterprise and sometimes researchers make mistakes at each of the steps, either by describing the experiences with terms from the natural or social sciences, by mistaking and joining different aspects of an experience together or different experiences, or by being insufficiently imaginative and mistaking contingent characteristics of experiences for necessary ones. One could argue, going beyond Husserl, the final step is the intersubjective validation of results (Giorgi & Sousa, 2010).

Since my practice is influenced not only by Husserl but mainly Heidegger, I find it necessary to distinguish between Husserl's transcendental eidetic phenomenology and Heidegger's hermeneutic ontological phenomenology (Paisana, 1996).

Hermeneutics is a discipline of interpretation that has its origins in nineteenth-century biblical criticism where the objective was to understand the several layers of meaning of biblical texts. From this narrow enterprise, the philosopher Schleiermacher wanted to develop a more general discipline of interpretation, not just concerned with the Bible, or even with textual exegesis, but any kind of interpretation – life events, works of art, juridical interpretation (Palmer, 1969).

Wilhelm Dilthey developed Schleiermacher hermeneutics through his distinction between understanding and explanation. Explanation is the mode of the natural sciences. They explain phenomena by postulating causal relations, as when a scientist explains the earthquake by the movement of the tectonic plaques. In contrast, when dealing with the human element, Dilthey thinks explanation leaves the essential out. The social sciences and humanities should seek not primarily to explain, but to understand. When a parent understands his son or daughter's decision not to go to university, he does not seek to explain this through hidden causes, but rather to understand the motives and context which lead to this decision. With this distinction in mind, Dilthey argues that a hermeneutics of life needs to understand the world-view – the set or norms, assumptions and values - of those around us (Makkreel, 1993).

On this hermeneutical note, Heidegger in *Sein und Zeit* re-introduces in philosophy the question of the meaning of being in general. We casually say that somethings are and others are not, but do we know what it means for something to be? The ontological distinction is the distinction between beings (entities) and the *being* of those beings. Everything is, but Heidegger calls the beings who make the question of their own being *Dasein.* Human beings are *daseins.* Human beings put forward the question of their own being and, as Heidegger says, have their own being, who they are, at stake when they pose such questions. Thinking about our moments of existential crisis, when the question "Who am I?" haunts us with poignant trenchancy, we know all too well what Heidegger is talking about. Unlike other animals and inanimate objects, human beings question themselves, their own being, their own existence.

Heidegger questions the possibility of completely bracketing all the acquired information we have about an experience before studying it. He points out that we always have some understanding or other of whatever we focus our attention on, or it wouldn't be possible to study it *qua* what it is. To pose the question "what is depression?", for instance, already implies we have some understanding of depression. Heidegger draws attention to this *hermeneutic circle,* following Schleiermacher and Dilthey into the discipline of interpretation, Hermeneutics. The point of phenomenology is to expose the meaning in terms of which entities are already understood.

With this starting point, Heidegger questions the primacy of Husserl's dichotomy between subjective acts and objective world. The subject doesn't find itself divided from the objects of the external world and other people. Human beings are, from the very beginning, already in-the-world and with-others. This being-in-the-world is what characterizes human existence. Primarily, we don't find ourselves perceiving external objects, rather objects show themselves in our experience and partly constitute what we are. Heidegger argues we are a clearing (*Lichtung*) or opening where things appear and show themselves in one way or another. Our most immediate relationship to objects is a circumspective concern. Unless something calls out for our conscious voluntary attention, we are usually immersed in our immediate surroundings. Heidegger's thought has clear clinical implications and the German philosopher himself tried to explore them in the famous *Zollikoner Seminare,* where he talked to psychiatrists about his thinking.

There are two related misconceptions about phenomenology that I find necessary to address:

The first worry is that since experience is always private, first-personal and subjective, and science and third-person enquiries are always public and objective, it's not possible to study experience directly and phenomenological psychology is impossible. The short reply is that the objection trades on a conceptual confusion and philosophical mistake. The objection is that since lived experience is subjective and psychology is objective, there can't be a direct study of experience in psychology but only cognition, language and memory though experimental trials. But this conflates two senses the words "objective" and "subjective" have: an epistemological sense and an ontological sense, as introduced by the philosopher John Searle (Searle, 2002). In the ontological sense, something is subjective if and only if it depends on the existence of a subject, self, or mind for its existence and objective only if it is independent of the mental states of sentient beings. In this sense, conscious experience is subjective. There's no experience without an experiencing subject. But this does not mean that consciousness is subjective in an epistemic sense. Something is epistemically subjective if and only if knowledge of that something depends solely on the personal preferences, value judgements and personal biases of the knower and epistemically objective if there's a method or process of finding out the truth about that something. Just because something is ontologically subjective, it doesn't follow that it is epistemically subjective. But is phenomenology also epistemically subjective? This brings us to the second worry.

The second related worry is that phenomenology seems liable to the same criticisms behaviourists directed against introspection and Wundt at the end of the twentieth century. Introspection is too unreliable and idiosyncratic to form the basis of a responsible practice in psychology (Zahavi, 2005). As John Watson puts it:

> 'Psychology as the behaviorist views it is a purely objective experimental branch of natural science. Its theoretical goal is the prediction and control of behavior. Introspection forms no essential part of its methods, nor is the scientific value of its data dependent upon the readiness with which they lend themselves to interpretation in terms of consciousness.'
> (Watson 1913, p. 158)

One problem with this objection is that it confuses due phenomenological reflection with naive introspection (Gallagher & Zahavi, 2012). Introspection is the untrained method the subject uses to find out whether she is having an experience of red or blue, the reaction times of her experiences, their bodily location, etc. Phenomenological and hermeneutical reflection, on the other hand, is always regulated by a method, and isn't just about the content of the experiences but about also about the form and structure of said experiences. When phenomenologists reflect on the experience of the divine, or depression, or euphoria, they don't need to be experiencing these things themselves, like it's the case with introspection, they can use reports of subjects, and try to find an essence/structure for these experiences that the persons themselves may not realize it is in their own experiences.

Is all of this relevant for psychotherapy and clinical practice? Existential-phenomenological therapists think so, as I will expound in the next section.

15.2.2 Existential psychotherapy

Drawing on phenomenological psychology, along with existential thinkers such as Kierkegaard and Nietzsche, some therapists adhere to the approach called existential psychotherapy (Binswanger, 1963; Yalom, 1980; van Deurzen-Smith, 2005; Spinelli, 2005; du Plock, 2016).

There are many forms of existential psychotherapy, from logotherapy, daseinanalysis to the British School (Cooper, 2005), but most share a few common tenets:

- Human being has no defined essence besides their existence – Unlike inanimate objects which have a fixed essence, e.g. chairs have to allow sitting, human beings are their existence and constantly mutating worldview.
- Existence is singular – Like Kierkegaard argued, concepts fail to match the flux of existence from which they are originated, because communication is only possible by abstracting from the particular concrete instances of experience so that we can communicate with others with similar, in some respects, but not identical experiences. Every experience and lived event is first-personal, so a completely third-person approach always leaves something out.
- Existence is temporal and finite – Human existence is temporally structured. We are thrown into the world and constantly project ourselves in future possibilities of our existence (Blattner, 2005). One of those possibilities is the impossibility of more possibilities – death. Every human being has to face the fact of his own mortality, either authentically or inauthentically. *Dasein* is being-towards-death.
- The authentic/inauthentic distinction – Among all our possibilities of life, some are our own and others are imposed, derived and designed by others or what Heidegger calls "The They". When we follow the stereotypes of our culture, live up to the expectations of others or immerse ourselves in the crowd, through seemingly juicy ambiguous empty talk, we're giving up our own most possibilities and our authentic project of living.
- An emphasis on freedom and responsibility – Human beings are, in Sartre's infamous words, condemned to be free. This comes with great responsibility for every decision is also a "cision" and denial of other possibilities. The word "decision" comes from the latin *de* "downwards/ on the basis of which / about" and *cider*

"cutting". When we make an important decision, we cut all other possibilities except one.

- An inquiry into the meaning and purpose of life and events – In our life's, events have a meaning and actions have a purpose. Our living is structured in a network of senses, references and signification. Therapy is a place to unfold an unveil this network and understand the world-view of the patient.

In my view, the clinical practice is the place where existential psychoterapists go from a third-personal psychological-philosophical discourse to a first-person case and back again to a third-person analytic discourse. The clinical encounter is not a space of pure philosophical reflection, such as that done by Heidegger in his analytic of *dasein*, nor can it remain in pure subjectivity, because we need conceptual categories with which we can analyze and understand our patients but is a halfway house between third and first person. This means going from the ontological – the domain of being – to the ontic – the domain of the human entity in front of me – back and forth.

Unlike standardized manualized treatment, where the clinical encounter is literally performed by following a manual and may leave little space to treat the patient as an individual, the kind of existential psychotherapy I practice fosters and focuses on the therapeutic alliance and the clinical space as the optimum ground for dwelling (*wohnen*) and lived experience to manifest itself as itself.

In this space of dwelling, we must not treat the patient in a detached objectivist calculating way, but in a caring way. Heidegger defines care (*sorge*) as the unifying feature of *dasein*'s existence and temporality. Care, in its two modes of occupation with everyday objects and preoccupation with persons, is a pre-intentional understanding and attunement to the world that what makes human beings who they are. In this respect, Heidegger tells us an illuminating and endearing fable about the centrality of care for human existence in *Sein und Zeit*:

> "Once when Care was crossing a river, she saw some clay; she thoughtfully took up a piece and began to shape it. While she was meditating on what she had made, Jupiter came by. Care asked him to give it spirit, and this he gladly granted. But when she wanted her name to be bestowed upon it, he forbade this, and demanded that it be given his name instead. While Care and Jupiter were disputing, Earth arose and desired that her own name be conferred on the creature, since she had furnished it with part of her body. They asked Saturn to be their arbiter, and he made the following

> decision, which seemed a just one: 'Since you, Jupiter, have given its spirit, you shall receive that spirit at its death; and since you, Earth, have given its body, you shall receive its body. But since Care first shaped this creature, she shall possess it as long as it lives. And because there is now a dispute among you as to its name, let it be called homo, for it is made out of humus (earth).'"
> (Heidegger, 1963, p. 408)

For Heidegger, this means that we always care in some way or another about what and who surrounds us, the world and what shows itself as itself to *dasein* matters to *dasein*. This care is more fundamental than body or spirit, as the fable makes clear. Also, interesting to note is that the judge and arbiter of men, is Saturn, the God of Time. Care is essentially related to time. When we project ourselves towards future possibilities, we are caring for those possibilities in one way or another. *Dasein* always cares about its being, whether it wins it or loses it.

In the next session, I will illustrate the contrast the existential first-person approach and the standard third-person approach in clinical practice.

15.3 Case study

The following clinical case will enlighten and make intelligible the clinical consequences of emphasizing a first-person discourse as opposed to a third person discourse and what is a good hermeneutical-existential clinical analysis in practice.

15.3.1 – Presentation of the case

Sandro, 55 years old, is a civil engineer in a public institution. He arrives at my private office, on the appointed time, self-conscious and with an excessively ceremonial "Do you allow me to come in?". He sits by the edge of the armchair and goes straight to business, with a noticeable quivering of the upper extremities.

"I'm feeling ghastly since my partner left me. I barely eat, can hardly sleep and I keep obsessing over the same thing daily. I feel a gloominess that keeps me from working in my day-to-day affairs tasks. The family physician allowed me to take a sick leave because I'm not in working conditions. As I live alone, I stay all day in bed and barely feed myself. I have lost 10 pounds. I just think that my life is meaningless since I'm so lonely. I don't have the courage to kill myself and I cry every day.".

When I questioned him about the end of that relationship, he replied:

"We met about a year ago through a gay dating app. I assumed that this was going to be a brief sexual encounter, but I ended up staying in his house all night. From then on, we were together whenever we could, after work hours and on the weekends. I practically moved into his house because it was closer to our offices. I thought I had met the one and I felt very happy. We had been on holiday together. I made the same old mistake and I pushed out my friends because I wanted to be only with him."

"Three months ago, I was hospitalized for surgery, previously scheduled, on an inguinal hernia. I was committed for 5 days. To my astonishment, he didn't show up for any visits. We spoke on the phone, but he was a lot colder. When I confronted him with this, he abruptly told me that I was pressing him and that he was sick of it and that everything is over between us. He didn't come back to see me again. I'm still trying to figure out what happened. I'm in a living hell." (2001)

Sandro had once consulted a psychiatrist in a private hospital. He described to me that episode:

"I decided to go to a psychiatrist because things couldn't go on like this. I choose a private hospital because I have special insurance (ADSE) and an appointment with the family doctor takes more than 5 months to get. The doctor was courteous, but I felt he didn't want me there very long. He looked at the clock a lot. Later I realized appointments were made to last 15 minutes. I spoke for a short while, and soon he politely interrupted me and gave me a prescription. I didn't follow that prescription and missed the next appointment 30 days later. The environment was very different from here; it embodied a very impersonal inter-relational style. The office was used by several medical specialists. I wasn't sitting in an armchair, like here next to you, but in front of a desk. The doctor wrote on his computer while I spoke and couldn't keep eye contact.".

Sandro accepted my request to contact his previous doctor, so he could forward me the clinical case, in accordance with the due deontological rules. Therefore, it was very easy for me to compare the doctor's notes (third-person discourse) with the description of the patient (first-person).

My colleague, very helpful, understood the situation and since he had little time made a copy paste of the official registry and sent it to me by email (the time allotted to each appointment was now 10 minutes, on administration's orders, so they could fit some more patients):

This is the official clinical registry:

"Patient, male sex, 55 years old, civil engineer, single. He has an anxio-depressive syndrome reactive to the breakdown of a homosexual

relationship. Previous history of surgical intervention due to an inguinal hernia. Without relevant previous psychiatric history, including his family.

Diagnostic: anxiety and depression mixed disorder.

Therapeutics: Elontril 150 id, zolpidem 10 id before bed time. Back in 30 days".

On his own part, Sandro describes the clinical encounter in these words:

"To tell the truth I don't have anything to criticize about the doctor. He was considerate and kind if we forget the time when he furrowed his eyebrows when I was speaking about my previous partner's withdrawal from my life. It seemed that he was doubting his hearing for a bit, but of course, I understood his reaction."

"The worst was that he didn't seem disposed to listen to me. I am sure he was under pressure because of the time, but it's not just that... He seemed more interested in hearing about my symptoms than in myself. As an engineer, I felt he was looking at me technically like I was just a broken machine. He told me I have depression, but I feel I am mourning my relationship, that, for me, it was meant to last a lifetime."

"I asked him why he was giving me an anti-depressive and he told that it was because my symptoms indicated unbalanced brain chemistry that lacked certain chemicals. To me, I have a mortal sadness. How could the substances disappear if before my hospital stay I felt so well? Have they diminished because of my break up? In that case, they are a consequence and not a cause of said depression. I don't know what to say or think! I feel very tired!"

At the end of the clinical interview, I proposed Sandro several possibilities:

a) He would be followed by me - both on a psychotherapeutically and pharmaceutical level;

b) He would keep his appointments with his psychiatrist and be followed by me psychotherapeutically;

c) I would take charge of the psychiatric consultation and suggest someone to be his psychotherapist;

Sandro chose (a) what allowed me to understand a bit more of his psychobiography. Sandro was born as the fruit of a brief relationship his parents had in their twenties. They separated when he was 6. He remembers suffering a lot with his father's absence and he wanted his father to come back. He wanted to have a family like every other family with his mother and father at home, a desire that has been with him since there. When he was 12, his father remarried and stopped talking and visiting him, maybe influenced by his new wife. With this new heartbreak he gets even more attached to his mother. She is described as an unconditional friend that supported him in everything but not very affectionate. In his words: "My mom didn't like hugging or kissing very much but she was always preoccupied with my wellbeing, she was very distrustful and thought I was very naive and could be easily fooled. She often commented – *you don't see the evil in people and you're going to suffer because you'll be fooled a lot, act smarter.*

When he was a teenager, he experimented with people from both sexes. In his twenties, he dedicated himself to the degree that was chosen by his mother that didn't want Sandro to "fall behind his father".

From 30 to 37, he had his first stable homosexual relationship, feeling completely in love, which ended because his partner traded him for someone 10 years younger. This was his first crisis, similar to the present one, although he was not yet ready to ask for clinical help. "I didn't think I was sick. I still don't. I was just heartbroken."

From 41 to 44, he had another relationship with a man, to which he enthusiastically dedicated himself again, with a very similar upshot. He was abandoned and felt very bad about it.

His mother is overly critical of his sexual orientation, which she considered shameful, and insisted in the possibility of changing his life, to marry and have a family, to know that when she died he wouldn't be alone. This opportunity arrived when Sandro reencountered an ex-girlfriend from his adolescence, with which he had never had sex. He invested in this relationship, showing off about the qualities of his wife, of how much in love he was and the, until then unknown, pleasure of PIV sex.

Contrary to what he expected, his mother wasn't thrilled about this relationship, nor with the engagement to his female partner, soon to be followed by marriage, because she understood that his girlfriend was a gold digger, since she had no means of subsistence of her own, two unemployed sons and loved to exhibit his engineer husband. Sandro didn't listen to his mom, which he attributed to jealousy, and lived two happy years with what he thought was his real family. When his wife demanded to marry with communion of goods and that he sold some of his expensive things to invest in an exquisite luxurious house, besides demanding he gave her the totality of his earnings because allegedly he didn't know how to manage it properly, Sandro realized everything and refused these demands. His fiancée broke off the engagement.

Funnily enough, when this relationship pattern was identified in therapy session, great initial enthusiasm followed by unforeseen and surprising abandonment, Sandro replied "It's curious I had never thought about it, I always lived my relationships as if they were completely unrelated".

Between these 4 relations, Sandro grew closer to his mother. They passed the weekends and vacations together. He also spent his time in his words, "a frenzy of sexual contacts", sometimes 3 per day, in gyms, public toilets and social dating apps.

15.3.2 – Discussion of the case

Is there a plausible hermeneutic-existential reading of Sandro's case? I have defended a hermeneutical critical approach, within an existential clinical setting (Amorim Rodrigues, 2016). This approach starts off hermeneutically from the person's action in the world (*welt*) to its understanding/interpretation from its worldview (Binswanger, 1963; Spinelli, 2005). This both unveils certain aspects of that world that are sometimes confused with its totality, and veils others, which remain obscure to the person herself.

In the next step of the hermeneutic-critical method, as I am proposing it, from the worldview of the patient we uncover its hidden assumptions. This means identifying the conditions of possibility of the intelligibility of the patient's discourse. Although the worldview of the patient already includes a partial awareness of its assumptions, very rarely have they been explicitly understood and reflected upon because they are precisely the point from which every understanding and meaning of what reveals itself to us in our world derives. In the most successful cases, we can perform a third step of this method and trace back the patient's worldview and assumptions to its psycho-biographical origin.

Taking our cue from Sandro's case, we observe a way of being-in-the-world where the being-towards-others predominates in detriment of being-for-oneself. Sandro's actions on his world are easily understood if taken to come from a worldview where the self appears incapable of taking care of itself to the point where it threatens, at a level of fantasy, his own survival. The non-existence of an adult presence and responsible leads to an affective state the oscillates between intense anguish, lived as a threat to his psychic survival, inactive despair, such as when he stays in bed crying all day, and acting out, when he searches for physical contact haphazardly.

In this case, we managed to trace back his belief in his own ineptitude to the subliminal message of his mother that abandonment, on the part of his father, is equivalent to death, when she said which meant "without me, you won't survive on your own". A valueless image of weakness was attributed to Sandro, by the words and looks of his mother. This assumption of his view of self and others permeates his way of being and allows its intelligibility.

This hermeneutical-critical approach has a psychodynamic character, in so far as it identifies conflict and defensive psychic processes, and existential, in so far as it starts from an analysis of being-in-the-world-with-others.

15.4 Conclusion

In this article, I have started by criticizing mainstream third-person approaches to the clinical lived encounter. Through an analysis of the latest issue of the *Journal of American Psychiatry* and the relevant literature, I have suggested that supposedly objective third-person approaches do not, at present, live up to rigor they proclaim and, in any case, need to be complemented by first-person approaches that can address problems outside their area of application.

Then I presented Husserl's phenomenological method and the improved Heideggerian hermeneutical approach. That approach is not completely subjective although it is about subjectivity. Existential therapists draw on these philosophers to breach the objectivity of the third person and address problems of the first-person individual existence with an analysis that goes back and forth between ontological and ontic perspectives.

Finally, through a complex case study, I applied a new hermeneutical method of my own, drawing extensively on the philosophers and existential psychotherapist previously cited. The result was a deeper understanding of the patient's worldview, its hidden assumptions and genealogy. This first-person understanding was not only critical to the clinical handling of the case but was also the main interest of the patient.

15.5 References

Baker, M. (2016). 1,500 scientists lift the lid on reproducibility. *Nature* 533, 452–454.

Baardseth, T. et al. (2013). Cognitive-behavioral therapy versus other therapies: redux. *Clinical Psychology Review,* 33(8):1253.

Begley, C. & Ellis, L. (2012) Drug development: Raise standards for preclinical cancer research. *Nature* 483, pp. 531-533.

Bennett, C. et al. (2009). Neural correlates of interspecies perspective taking in the post-mortem Atlantic Salmon: An argument for multiple comparisons correction. *Science* 39, pp. 538-547.

Binswanger, L. (1963) *Being-in-the-world.* London: Cordor books.

Blattner, W. (2005) *Heidegger's Temporal Idealism.* Cambridge: Cambridge University Press.

Bloch, S. (1999) *An Introduction to the Psychoterapies.* Oxford: Oxford University Press.

Boss, M. (1979) *Existential Foundations of Medicine and Psychology.* Northvale: Minutemen Press.

Broder, E. (2005). The Neuroscience of Psychotherapy: Building and Rebuilding the Human Brain. *The Canadian Child and Adolescent Psychiatry Review,* 14(4), pp. 121-125.

Button, K. et al. (2013) Power failure: why small sample size undermines the reliability of neuroscience. *Nature Reviews Neuroscience* 14, pp. 365-376.

Carter, C., Kofler, M., Forster, D., and McCulloch, M. (2015). A series of meta-analytic tests of the depletion effect: self-control does not seem to rely on a limited resource. *Journal of Experimental Psychology General* 144, pp. 796–815.

Clausen, J. & Neil L. (eds.), (2015) *Handbook of Neuroethics,* Netherlands: Springer.

Connolly, W. (2002) *Neuropolitics: Thinking, Culture, Speed.* Minnesota: University of Minnesota Press.

Cooper, M. (2003) *Existential Therapies.* Michigan: SAGE publications.

Doyen, S. et al. (2012). Behavioral priming: it's all in the mind, but whose mind?. PLoS One, 7, pp. 23-45.

du Plock, S. (2016) *Cases studies in Existential Therapy.* London: PCCS books.

Ebrahim, S. et al. (2015) Meta-analyses with industry involvement are massively published and report no caveats for antidepressants. *Journal of Clinical Epidemiology* 70, pp. 155 – 163.

Fallon, B. (2017) 'A Randomized Controlled Trial of Medication and Cognitive-Behavioral Therapy for Hypochondriasis'. *The American Journal of Psychiatry* 147(7) 123-125.

Frankl, V. (1946) *Man's Search for Meaning. An Introduction to Logotherapy.* Beacon Press: Boston.

Filippi, M. (2015) *Oxford textbook in neuroimaging.* Oxford University Press.

Fonzo et al., (2017a) Selective Effects of Psychotherapy on Frontopolar Cortical Function in PTSD. *American Journal Of Psychiatry* Volume 174(12), pp. 1175-1184.

Fonzo et al., (2017b) PTSD Psychotherapy Outcome Predicted by Brain Activation During Emotional Reactivity and Regulation. *American Journal Of Psychiatry* Volume 174(12), pp. 1163-1174.

Gallagher, S. & Zahavi, D. (2012) *The Phenomenological Mind.* Routledge: Routledge University Press.

Giorgi, Amedeo. (2009). *The Descriptive Phenomenological Method in Psychology: A Modified Husserlian Approach.* Pittsburgh: Duquesne University Press.

Giorgi, A. & Sousa, D. (2010) *Método Fenomenológico de Investigação em Psicologia.* Lisboa: Fim de Século.

Grawe, Karl (2007). *Neuropsychotherapy: How the Neurosciences Inform Effective Psychotherapy.* Routledge: Routledge University Press.

Glimcher, Paul W. and Ernst Fehr (eds), (2013) *Neuroeconomics.* Boston: Academic Press.

Heidegger, M. (1963). *Being and Time.* Oxford: Basil Blackwell.

Heidegger, M. (1998a) Letter on Humanism, translated by F. A Capuzzi and J. Glenn Gray, in D. F. Krell (ed.) *Martin Heidegger: Basic Writings, revised and expanded edition,* London: Routledge, pp. 217–65.

Heidegger, M. (1998b) Building Dwelling Thinking, translated by A. Hofstadter. In D. F. Krell (ed.) *Martin Heidegger: Basic Writings, revised and expanded edition,* London: Routledge, pp. 217–65.

Heidegger, M. (2001) *Zollikon Seminars: Protocols—Conversations—Letters,* translated by F. Mayr, Northwestern University Press, Illinois: Evanston.

Husserl, E. (1982) *Ideas Pertaining to a Pure Phenomenology and to a Phenomenological Philosophy—First Book: General Introduction to a Pure Phenomenology.* The Hague: Nijhoff

Hunsley, J., & Lee, M. (2007). Research-informed benchmarks for psychological treatments: Efficacy studies, effectiveness studies, and beyond. *Professional Psychology: Research and Practice.* 38(1), pp. 21-33.

Hollon, D., Thase, E., Markowitz, C. (2002). Treatment and prevention of depression. *Psychological Science in the Public Interest,* 3: pp. 39–77.

Ioannidis, J. (2005) Why Most Published Research Findings Are False. *PLoS Med* 2(8): pp.23-45.

Johnsen, TJ. &Friborg, O. (2016). The effects of cognitive behavioral therapy as an anti-depressive treatment is falling: A meta-analysis. *Psychol Bulletin* 141(4), pp.747-68.

Jones, C. et al. (2012) Cognitive behaviour therapy versus other psychosocial treatments for schizophrenia. *Cochrane Database System Review* 4, pp. 67-78.

Kirk, S. A., Gomory, T., & Cohen, D. (2013). *Mad Science: Psychiatric Coercion, Diagnosis, and Drugs.* London: Transaction Publishers.

Langdridge, D. (2007). *Phenomenological Psychology: Theory, Research and Method.* New York: Pearson Prentice Hall.

Makkreel, R. (1993) *Dilthey: Philosopher of the Human Sciences.* Harvard: Harvard University Press.

Mischel, W. (2008). Editorial: connecting clinical practice to scientific progress. Psychological Science in the Public Interest, 9(2): i–ii.

Moncrieff, J., Cohen, D., & Porter, S. (2013). The Psychoactive Effects of Psychiatric Medication: The Elephant in the Room. *Journal of Psychoactive Drugs,* 45(5), pp. 409–415.

Moynihan, Ray et al. (2002). 'Selling sickness: the pharmaceutical industry and disease mongering'. *British Medical Journal,* 324, pp. 886–891.

Newberg, A. (2010) *Principles of Neuroteology.* Michigan: Ashgate Press.

Nosek, B. (2015) Estimating the reproducibility of psychological science. *Science* 349.

Paisana, J. (1992) *Fenomenologia e Hermenêutica. A relação entre as filosofias de Husserl e Heidegger*. Lisboa: Editorial Presença.

Palmer, R. (1969) *Hermeneutics*. Northwestern: Northwestern University Press.

Renvoisé, P. & Morin, C. (2007). *Neuromarketing: Understanding the "Buy Buttons" in Your Customer's Brain*. London: London University Press.

Ricoeur, Paul. (1990) *Le Juste*. Paris: Édicions Esprit.

Searle, J. (1998) *The Rediscovery of Mind*. The MIT Press.

Shedler, J. (2015). Where is the Evidence for Evidence-Based Therapy. *The Journal of Psychological Therapies in Primary Care*, Vol. 4, pp. 47-59.

Simon, C. (2002) *A change of mind*. Washington: The Washington Post.

Spinelli, E. (2005) *The Interpreted World: An Introduction to Phenomenological Psychology*. London: SAGE.

Sundquist, J. et al, (2015) Mindfulness group therapy in primary care patients with depression, anxiety and stress and adjustment disorders: randomised controlled trial. *The British Journal of Psychiatry*, 206(2), pp. 128-35.

Tassone, Biagio G. (2012). *From Psychology to Phenomenology*. New York: Palgrave Macmillan.

Tolin, D. (2010) Is cognitive-behavioral therapy more effective than other therapies? A meta-analytic review. *Clinical Psychology Review*. 30(6), 710-20.

Van Deursen-Smith, E. (2005) *Existential Therapy*. London: Society for Existential Analysis.

Wardell, D. (2014) *Healing Touch: Enhancing Life through Energy Therapy*. New York: Palgrave Macmillan.

Watson, J. (1913) Psychology as the Behaviorist Views it. *Psychological Review* 20, pp.158-177.

Wilber, K. (2000). *Integral Psychology*. Boston: Shambhala.

Yalom, S. (1980) *Existential Psychoterapy*. New York: Basic Books.

Zahavi, D. (2003) *Husserl's Phenomenology*. Stanford: Stanford University Press.

Zahavi, D. (2005) *Subjectivity and Selfhood: Investigating the First-Person Perspective*. Cambridge: Cambridge University Press.

Chapter 16

"From D.I.R. to D.I.R.E.: the role of embodiment in the treatment of self-disorders"

Valeria Bizzari
Clinic for General Psychiatry, Centre of Psychosocial Medicine, Clinic University of Heidelberg, Heidelberg, Germany

In this paper, I try to combine Greenspan's "affective diathesis hypothesis" with a phenomenological account of subjectivity as a corporeal, kinesthetic entity. Greenspan's work elicited my interest for many different reasons. In my view, his account is not a mere attempt to affirm the importance of emotions in the development of subjective skills. On the contrary, the priority attributed to affective *interactions* and *lived experiences* makes his theory a rich source for the understanding of those pathologies which involve intersubjective deficits (such as the autism spectrum disorder) and gives reasons to think about a fruitful dialogue between his therapeutic proposal and phenomenology, a dialogue that may shed light on some important theoretical knots, such as the nature of self and the role of intercorporeality in the development of intelligence. After illustrating Greenspan's developmental perspective, I will point out these two theoretical themes, that can benefit from a comparison between the affective diathesis hypothesis and a phenomenological account. Then, I will propose a special kind of therapy integrating Greenspan and Wider's model with phenomenological insights.

16.1 "I feel therefore I am": Greenspan's "affective diathesis hypothesis" and the D.I.R. model

Imagine a child who pulls his father to the toy area and points to a particular toy. As the father gets it, the child immediately smiles. If we try to understand this apparently simple situation, we realize how many elements are required: in fact, the child described is involved in a complex chain of affective reciprocity that implies problem-solving interaction and a developed emotional life. In order to be able to indicate what he wants, the infant needs at least an *emotional desire*; then, he needs the ability to link this particular desire to an *action plan* (in this case, having the opportunity to play with his toy). These elements motivate his *interactive act* (pulling his father to the toy area), which implies not only that he is aware of the effects of specific gestures, but also that he conceives of his father as an intentional, reactive consciousness, embedded in a reciprocal affective relationship. Within this complex architecture, affect and emotion seem to work as a "team leader" which orchestrates the entire mental structure and covers the overall development of a child. In other words, *affect is responsible for motor planning and sequencing.*

This thesis contrasts with many tendencies which affirm, on the contrary, the priority of cognitive and representational skills (Chomsky, 1959, 1972). Nonetheless, many findings in neuroscience and developmental psychology (Meltzoff & Moore, 1977; Gallese, 2001) are increasingly embracing the view that emotions organize and orchestrate many of the mind's most important functions. In *The Growth of the Mind* (Greenspan, 1997) Greenspan claims that *emotional interactions influence intelligence.* According to his theory (the "affective diathesis hypothesis"), affective interactions emerge earlier than the sensorimotor schemes postulated by Piaget (1962), and they are used to understand and conceptualize our experiences in the world. Furthermore, even abstract thinking seems to be based on reflections about these personal affective experiences.

This happens through a multilayered process which comprehends different levels of interaffective engagement: the very first interaction happens during the first year of life and can be described as an affective interest in the world, which is due to a sensitive interaction between genetic proclivities and environmental experiences. In a second phase, this initial stage allows the infant to engage with a caregiver and not become self-absorbed.[1] The third developmental capacity is two-way purposeful communication, which involves affective reciprocity and does not imply inferential abilities. The fourth level, which occurs between 10 months and 18 months, involves the development of other capacities related to the infant's ability to engage in more complex affective

---

[1] This stage seems to correspond to the so-called "primary intersubjectivity" (Trevarthen, 1979), which is the ability to have immediate responsive conscious appreciation of the adult's communicative intentions.

interactions which involve problem-solving goals.[2] At this point, we register the arising of a bridge between affect and motor planning: the infant is able to connect his affect to his motor planning and sequencing (as illustrated by the episode described above). This particular phase is heavily emphasized by Greenspan, who considers this passage the real core of the affect diathesis hypothesis, which suggests that "affect invests not simply the capacity for complex interaction to give meaning to sounds, words, and behaviors, but also invests processing capacities, such as motor planning and visual spatial processing. It gives meaning and functional range to these capacities. Individuals can then use different processing capacities in a variety of ways—not just to solve an equation, but also for social, personal problem-solving" (Greenspan, 2001, p. 11). The toddler is learning a continuous flow of back and forth communication (the so-called "circle of communication"), and this allows the arising of a sense of self that is forming because the child is interacting and getting feedback from the environment.

At the fifth level, the affective system begins to imply ideas and complex thought[3]: for instance, the child can pretend play, an activity which requires both functional language and abstract thinking. The sixth stage involves the creation of logical bridges between emotions and ideas: the infant starts thinking at a symbolic level, connecting ideas of the immediate past and present.

In all of these stages, emotional and pre-reflective abilities are responsible for helping the child to go from simple interest in the world to social problem-solving, from sub-personal to symbolic knowledge. In particular, emotional interactions are considered responsible both for early social, cognitive and linguistic capacities, as well as for higher level intellectual and social skills. This thesis is very strong and even goes beyond Damasio's account, according to which emotions play a key role in judgment but are separated from overall intelligence (Damásio, 1994). According to Greenspan, emotional interactions are therefore the real core of behavioral and cognitive development.

---

[2] Trevarthen has labeled this stage "secondary intersubjectivity": a new form of cooperative intersubjectivity which involves person-person-object awareness (Trevarthen & Humbley, 1978).

[3]This stage seems to correspond to *tertiary intersubjectivity* (Fuchs, 2015), which is the ability to perceive others as intentional agents and to develop a self-other meta-perspective.

For these reasons, in 1997, S. Greenspan and S. Wieder developed a particular kind of therapy, called "D.I.R. model," aimed at the treatment of autistic patients, who lack emotional and cognitive skills.

We can summarize the D.I.R. model as follows:

- *Developmental:* The basic level which needs be taken into account by therapists involves the developmental capacities of the subject, which are essential for spontaneous and empathic relationships. This includes helping children develop the capacity to attend and remain calm and regulated, engage and affect basic gestures, engage in shared social problem-solving and intentional behavior involving a continuous flow of interactions in a row, and use ideas to communicate needs and think and play creatively;
- *Individual differences*: The D.I.R. model emphasizes the fact that each child has unique biologically-based manners to take in, regulate and respond to the environment and sensations. In fact, in the autism spectrum disorder some children are hyper sensitive to sound, others are less reactive to touch and so on;
- *Relationship-based:* This stage involves the learning relationships with caregivers and therapists, and, like phenomenology, sheds light on the intersubjective nature of the subject.

Greenspan and Wieder's therapy aims to facilitate mutual emotional interactions, promoting language development, motor skills and spatial-visual thinking.

We can thus affirm that according to the affective diathesis hypothesis:

1. Since birth, the very core of subjectivity is emotional and social;
2. Our cognitive and interpersonal skills depend upon the development of this emotional self.

These statements are not uncontroversial. On the contrary, they allow for a critical discussion of the real nature of subjectivity and intersubjectivity. Combining phenomenology with Greenspan's perspective, I will try to reflect upon these two important theoretical knots, and I will propose a modification of the D.I.R. therapeutic model, changing it from D.I.R. to D.I.R.E.

16.2. Theoretical insights on selfhood

16.2.1 Is our core self social?

There is a wide ongoing debate about the nature of selfhood: while the no-self view affirms that self does not exist or is an illusion (Hume, 2007; Metzinger, 2003), the narrative view considers it partly or wholly a narrative construction (Dennett, 1991; MacIntyre, 1981).Nonetheless, the most prominent view is the experiential phenomenological approach (Zahavi, 2005, 2014;): according to this approach, all experience has an initial personal character of phenomenal consciousness which implies a sense of self, the so-called 'minimal self' (Zahavi, 2005, 2017).

Emphasizing the pre-reflective, emotional domain of subjectivity, Greenspan also describes the core self as a "mineness" of subjectivity: in particular, he highlights the importance of the *lived emotional experience*. Furthermore, he underlines the intersubjective character of subjectivity, which seems to be engaged in the world from birth, and the importance of *affective relationality* for the arising of cognitive and motor abilities. In my opinion, a similar account of selfhood is coherent with those contemporary thinkers according to whom the core self is essentially interpersonal (Ratcliffe, 2017). In other words, our self is intimately connected to otherness to such an extent that we can affirm that it depends upon other selves.[4]

Despite the fact that Zahavi's description of the "minimal self" denies its intersubjective character, I believe that Greenspan is indeed coherent with his perspective, especially if we try to understand what Zahavi means by "intersubjective." In "Self-Awareness and Alterity" (Zahavi, 1999), in fact, he emphasizes the interdependency of the self and otherness, where otherness means an alterity that may be internal to the self, external (the world), or a concrete otherness. I believe that what Zahavi denies is the fact that there may be an immediate link between the core self and *concrete* otherness, but not the existence of an innate connection between the self and the first kind of alterity. In other words, Zahavi claims that relationality is present well before the concrete meeting with otherness. In perception, for instance, the implicit reference to an alterity is postulated as an a priori intersubjective structure (which he calls "*open intersubjectivity*," see Zahavi, 1999): when the subject perceives an object,

[4] The interdependence between selfhood and otherness is testified by several psychopathologies, especially in schizophrenia, a disorder where the inextricability between the self and its relation to intentionality and intersubjectivity dramatically emerges (Hipólito, Gonçalves, Pereira, 2018).

the absent profiles are co-intentioned *as if* they would be perceived from another perspective, another "here-and-now." Of course, intersubjectivity also involves other mechanisms: the effective *face-to-face encounter* and so-called "*common sense,*" that is, a sense of belonging to a community or a group that transcends the individual while permeating every moment of her life. Nonetheless, the ground of intersubjectivity is the capability itself to conceive of otherness, which is implicit in our self-consciousness prior to any concrete encounter and which does not need any simulation or conceptual abilities. According to this approach, the notion of self is necessarily linked to our interactions with others (Reddy, 2008). This basic interpersonal awareness is present from birth, as is the ability to engage with otherness and social emotions. Greenspan's approach confirms this hypothesis. As he affirms that the subject is interested and engaged in the world and the other from birth (stages 1 and 2), in the same way, Zahavi argues that the minimal self is not self-enclosed but the subject of intentional experiences[5].

Another interesting connection between Greenspan's account and the phenomenological approach is the emphasis on the motor and sensory skills of the subject. Nonetheless, while in phenomenology embodiment is considered very important for the definition of subjectivity as a dynamic intentional consciousness, in Greenspan's approach, corporeality is taken into account in terms of sensorial skills which emerge as a consequence of affect and interaction.

The goal of this paper is to integrate Greenspan's model with phenomenological insights. In other words, I would like to modify his *interaffective* proposal to create an *intercorporeal* proposal, without losing the emphasis on emotional components. I will do this by maintaining all of the original elements and adding the emphasis on embodiment. In this view, our core self is not just *social*, but also *embodied*.

---

[5]This is confirmed, in the field of developmental psychology, both by the "theory of innate intersubjectivity" (Trevarthen, 1979, 1998), according to which the infant is born with awareness receptive to subjective states in other persons; and by the fact that newborns are immediately involved in what Bateson (Bateson, 1975) calls "proto conversations," which are those interactions between mother and infant which involve the same patterns of adult communication. We are thus provided with a "natural sociability," a "mutual self-other-consciousness" (Trevarthen & Aitken, 2001) that plays a lead role in the development of the child's intelligence. The discovery of the mirror neurons system (see Rizzolatti & Craighero, 2009) has confirmed this intuition, supporting the intersubjective character of subjectivity to such an extent that several authors have labeled selfhood as a "dialogical self" (Hermans & Dimaggio, 2004; Muratori, 2007).

16.2.2 Is our core self embodied? A Husserlian account

One of the main achievements of the phenomenological method[6] is undoubtedly that of revealing the *intentional* bond that connects the subject and the world: in other words, thanks to the concept of intentionality, Husserl seems to succeed in the attempt to explain the experience of the world in all its *concreteness*[7].

Especially in recent decades, his main thesis and his description of subjectivity have started to be taken into consideration by several fields of study, many of them not even philosophical[8]. This growing interest is due, in part, to the development of a philosophical interpretation that is very far from the idea of an "absolute transcendental subjectivity" which has dominated for so many years; in part, however, the cause is the increasing interest of "empirical" disciplines in great existential questions, due to discoveries that have established the biological basis of many emotional aspects of our lives.

In fact, although several interpretations of Husserlian thought have labelled it as the main support for a theory of a “pure transcendental subject”, recent findings in the neurosciences and psychiatry have contributed to putting phenomenology and the empirical sciences in contact, and to renovating the image of the subject, offering new dignity to emotional and lived components.

---

[6]In philosophy, this discipline is the most committed to the attempt of discovering the fundamental structures of the subject, since it identifies the conditions of possibility and essential features of experiences of different kinds. I found it very useful and interesting comparing Greenspan's emphasis on pre-reflectiveness and emotions to this philosophical approach, in particular Husserlian phenomenology, which seems to underline the role of affective and pre-rational components in the development of subjectivity.

[7]At first sight, this statement may sound like a paradox: the very act that “puts in brackets” the elements that fall under the label of "facticity" is the one that reveals the true essence of things. For this reason, it is possible to argue that the Husserlian subject has a *constitutive* role towards the world, and everything is related to it. This thesis is neither an idealistic argument—which would consider the person confined in a solipsistic position that transcends reality—nor a realistic one; rather, it tends to exceed both orientations in favor of a *correlation* between the world and consciousness that, as such, is unique and concrete: in other words, a temporal, intersubjective and *embodied* consciousness.

[8]We can find this tendency in several areas, not only in phenomenology (see, for, instance, Jensen & Moran, 2013); but also, in the cognitive sciences (see Varela, Thompson, Rosch, 1991;); psychiatry and psychopathology (Fuchs, Schlimme, 2009); neuroscience (Gallese, 2005); and medicine (Toombs, 1998).

This tendency has helped to renovate the classical image of the subject and to provide renewed dignity and attention to the " lived ", emotional and "irrational" aspects of existence.

Cartesian dualism is challenged by positing phenomenology as a theoretical instrument able to affirm the reincarnation of the psyche into the body, of thought into feelings. In fact, the Husserlian subject is not a mere being characterized by the most complex levels of synthetic activity, but also a being completely rooted into the world and into the corporeity.

According to this approach, our body seems indispensable not only in the constitution of the self, but also for the establishment of intersubjectivity and the real world. What emerges is a vision of subjectivity as essentially embodied and situated in context.

In this view, the embodiment is not a mere possibility open to the subject, but his own *transcendental necessity:* in order to be an individual, the subject needs to exercise his essential functions (such as the constitution of the perceptive reality), functions which depend on his being embodied.

Like Greenspan, Husserl argues for a pre-reflective self immediately and intuitively aware of himself. In fact, in *Logical Investigations* (Husserl, 2001) he claims that, before the reflection and the perception of the intentional object, there is the *experience* of the intentional act. In other words, like in Greenspan's "affective diathesis hypothesis", the self seems to be primarily *experiential,* and pre-reflective components are described as antecedent any cognitive functions. Despite the fact that I do not 'perceive' the act (which will be thematized *later* through reflection), I am aware of it, thanks to a pre-reflective consciousness: "an *inner* awareness, not because it is a type of introspection, but because it belongs *intrinsically* to the very structure of the act itself" (Zahavi, 1998, p. 209). In fact, before analyzing my experiences, these are *lived* (I live them).

Activity and passivity, affection and reflection are inextricably intertwined. Self-affection, which appears as reducible to the self-manifestation of subjectivity, is linked to two main functions: temporality—which provides the *formal structure* to the pre-reflective self-awareness, enabling the experience of my own lived acts as a unity made up of retentions, protections and living present— and the *embodiment.* In fact, perception implies a lived body as an orientation center provided with *kinesthetic* sensations. For instance, when I see or touch the surface of an object, I also co-perceive it through the kinesthetic sensations of the movements of my fingers. This is referring to an implicit self-awareness of my own body and of my own motor possibilities.

While hylethic sensations refer to something external from themselves and the subject, kinesthetic sensations involve an immediate and corporeal self-awareness which is responsible for the distinction between the perceiver and what is perceived. Both of these experiences form a unitary experience, thanks to which I am aware of my movement *and* of the surrounding world. Nonetheless, hylethic experiences are subordinated to *kinesthetic* ones, which offer the adequate circumstances through which hylethic sensations can perceive the world. It seems to be clear how corporeity (which provides us with kinesthesis) is fundamental to making the subject aware of himself and of the surrounding world. In a similar manner, although with different terms, Greenspan argues for a link between affect and motor planning: more specifically, a sensory-affect motor pattern ("SAM") sets the agenda for mental and brain development. Of course, Greenspan's vision of corporeality is not the same as phenomenology: while the first conceives embodiment in terms of sensorial patterns and as a dynamic brain, Husserl's account focuses on a living consciousness. Nonetheless, the emphasis on the emotional components and the fact that Greenspan describes a *psychobiological* being (and not a mere biological entity), allows at least for a fruitful comparison of these two apparently distant approaches, which can also be read as attempts to reappraise the priority of the *lived, feeling* subject. On the other hand, I believe that phenomenology, and its analysis of corporeality, can be helpful in improving Greenspan's theory by adding a definition of embodiment as a "living body." Husserl himself speaks about consciousness as an "animated organism," a psychophysical being: his reflections can be helpful to clarify in which sense the subject is a lived, intentional entity that interacts with the environment (another element that is very important in Greenspan's account from the very first stage of development I described before).

Another important feature of selfhood is the fact that it is a *kinesthetic* entity. While Greenspan claims that our affective interactions work together with the motor system, giving birth to higher cognitive skills, Husserl—in *Cartesian Meditations,* paragraph 19, dedicated to the actuality and potentiality of the intentional life—emphasizes the concept of "*horizon*" and argues that perception has horizons which offer more perceptive possibilities. Husserl refers to this capability-consciousness with the term "I-can". Within this characterization, consciousness needs to be capable of *motility*: perception, for instance, is governed by a kinaesthetic activity that works according to a motivational "if-then" logic. While external things are presented through adumbrations, a similar consciousness is itself-given and can be defined as a system which comprehends all possibilities that are not currently actualized. The starting point of this system seems to be a sort of "primal motility," the possibility of movement itself: the possession of kinesthetic skills is the characteristic which constitutes the subject as an *animated organism.* The passive synthesis of association informs the subject about his possibilities in the world: the subject is therefore immediately aware of himself as an "I can" linked to the external realm in an interkinaesthetic manner. Every act of perception (*Noesis*) thus implies an epistemological horizon, which makes it possible to have complete knowledge of the object. In other words, perception is not a unique experience from a perspectival point of view, since it implies the possibility of adopting other perspectives. For this reason, the "I can" seems necessarily to be a moving body: otherwise, in fact, it would not be possible to postulate the existence of other horizons of perception and other points of view.

The central importance of kinesthesia emerges not only in the perception of objects but also in the perception of otherness: the first level of interpersonal understanding is, in fact, the co-subjectivity, that is, the ability to assume the existence of points of view different from our "here and now."

The "if-then" connections seem, then, to have a motivational role within the perceptual process: they provide not only a wider epistemological horizon, but the perceiving subject is also able to motivate his actions (thanks to the "legitimacy of reason") and to be, in turn, motivated in a pre-reflective and pre-theoretical sense. Within the perceptual act, the body has a central role because it is not a simple constitutive transcendental mechanism among others, but it is the element itself that shapes the whole experience of feeling.

Perception does not seem, therefore, to be an operation that can be resolved within the mere mind: the "constitution of the material thing" seems, rather, to be the result of an essentially corporeal experience.

Regarding the connection between pre-reflective, embodied components and higher functions, such as symbolic thinking and complex ideas (in other words, the 6th stage described by Greenspan), Husserl explains how the meaning of the material thing is immediately given because our perceptions are permeated by the functions of objects, by their values. The reality surrounding the subject is indeed "*vorhanden,*" ("at hand") and axiologically informed. According to this perspective, the world has a *practical* meaning before a theoretical connotation, and it is possible to distinguish between the "cultural" and "natural" world only *after* the reduction. Reason seems to enlarge its domain from a logic-theoretical framework to an emotional-practical one. This epistemological enlargement is not synonymous with a mere passage from a pure reflexive subjectivity to a mere bodily perceptual entity: on the contrary, the *dossic* or scientific consciousness—the one that reduces the world to its primary qualities, which are of *theoretical* interest) —-and the practical-emotional one (*Gemüt*) seem to be different fields which work together.

In other words, the cognitive process has *affective features*: movement, for example, is a resource for communication, a source of non-linguistic and kinesthetic concepts[9].

16.3. The role of the body in intersubjectivity. For an *intercorporeal* self

Being a *lived body* allows not only for self consciousness but also for the experience of the alterity: in fact, others are implied in my perceptual horizon before a concrete face-to-face encounter.

In general, in Husserlian thought we find two different ways of conceiving intersubjectivity:

- the intersubjective structure of experience, which is present before the concrete encounter with the alterity;
- the effective face-to-face encounter, which happens thanks to empathy.

Concerning the first kind of sociality, it appears clear when the subject perceives the surrounding world: during perception the thing is thematized, disclosing the other possible perceptions that we could have, that is, co-intentioning the absent profiles which would be perceived if I were another "here and now".

In fact, perception is never exhausted in the short-term but is an original reference structure for potential and anonymous co-perceivers. Since this implicit co-perceiver is characterized by perception capabilities, s/he is necessarily embodied: as I described in the previous paragraph, the subject can intend the world because s/he possesses kinesthetic abilities originally connected to his body. We can thus conclude that, in order to be perceived, the environment must be visible, tangible and so on to

[9]M. Sheets-Johnstone uses the example of walking, an essentially qualitative experience that can be phenomenologically described through terms that refer to the subject's kinesthesia. Husserl himself distinguishes between a merely perceptual kinaesthetic system and a practical kinaesthetic system which intervene in the world. Nonetheless, he claims that they are two intertwined modes of functioning (see Behnke, 2009).

everyone, not only to our individuality. The first openness in respect to otherness is, therefore, an eidetic variation[10] of myself[11].

However, otherness is not a mere change of myself, something possible, but an actual modification. The first true transcendence occurs only when the other appears within the field of perception of the subject, as a concrete psycho-physical being. Through the *Leib*, otherness is constituted as a new form of reality recognized as *analogous* thanks to the similarities between my body and that of others. Being embodied is essential: only through the *Leib* can I, in fact, enter into the field of perception of the other, which, in turn, recognizes me as *Leibanalogon*, a corporeality which finds its expression in an individuality.

This special kind of perceptual experience is called empathy and is the second kind of intersubjectivity: in the semester between 1928 and 1929 Husserl held his final class, "*Phänomenologie der Einfühlung in Vorlesungen und Űbungen*" on precisely this topic, evidence that it was of great interest to him and probably worthy of further study.

The procedure through which we experience an otherness is bodily and therefore takes place starting from the observation of the other's body and his or her movements: in other words, the appearance of the other's physical body within my perceptual field is a condition without which I would not have any experience of the other. The analogy is like the theoretical node on which the Husserlian theory of intersubjectivity is based: this passive association (*Paarung*) is indeed original, immediate, active and effective, and it emerges as early as the *Logical Investigations*, wherein Husserl argues that the other is not perceived through logical reasoning, but through a binding, a passive movement. Even in Ideas *II*, passive synthesis does not seem a simple preparation, but a way of perceptual organization. The bond with the other is inside the pure passivity: I recognize the other immediately as a *living animated body, expression of subjectivity*.

Embodiment in empathy seems to be a real *operational function*: the concrete experience of an alien subjectivity is, in fact, always an experience of the other as an incarnated being. The intersubjective relation seems to be a relationship between two living bodies. Looking at the bodies of others, which associatively invoke mine, there is a transposition (*Űbertragung)* through which I consider the other as having

---

[10]This is Husserl 's term for an intuitive act toward an essence or universal, in contrast to an empirical intuition or perception. He also called this act an essential intuition, eidetic intuition, or eidetic variation.

[11] This is coherent with the "open intersubjectivity" described by Zahavi (Zahavi, 2001).

bilateralism of kinesthesia in turn and also possessing a living body. However, the transposition is merely formal; it does not concern the contents of experience. Husserl's main purpose is, in fact, a theory of empathy through which I can have access to the other while keeping a primary differentiation between our flows of phenomenological experience. The physical body is, in fact, collectively available, but the psychic elements do not remain directly accessible. Thanks to the *Paarung* process, therefore, all consciousnesses are awakened, and thanks to de-centralization, they are put in a position of equality. Embodiment seems to be the necessary condition for all types of interpersonal experience.

Combining Greenspan's account with these phenomenological insights, we can thus affirm that selfhood is an *emotional, corporeal, kinaesthetic*[12] entity, an intentional dynamic consciousness which is innately intertwined with otherness.

In this view, we cannot define intersubjectivity either in terms of mental or abstract mechanisms, or as a simulating process: rather, intersubjectivity seems to include both sensory-motor experiences *and* contextual, pragmatic abilities. Greenspan's account and the multilayered, phenomenological description are suitable to emphasize this complexity, which dramatically emerges in several mental pathologies, such as the autism spectrum disorder[13], a disorder that seems to affect not only the cognitive abilities of the subject, but also the so-called *praktognosia* (i.e. the capability to relate to the world in a practical sense and not purely theoretically). Following a phenomenological perspective, we can claim that this disorder affects the social skills of the subject as early as the "open intersubjectivity", eroding inferential capabilities and especially the embodied interaffectivity. In a similar manner, Greenspan claims that the main problem in autism spectrum disorder can be found in the

---

[12]An introduction to the notion of kinaesthetic consciousness can be found in Claesges, 1964; Mickunas, 1974.

[13]Autism spectrum disorder is a syndrome which involves various symptoms, even though the most common are: difficulties in social interactions; restricted, repetitive or stereotyped behaviors and activities; defects in language development and other communication skills. This disorder was identified in the 1940s of the last century by Kanner and Asperger and raised the interest of many researchers concerning its real nature: while, on the one hand, numerous theories circumscribe it within a neural, merely cognitive disorder (Happè, 1994), it is often described as a behavioral deficit which prevents the subject from being interested in the world (Goldman, 2006). In contrast to both of these tendencies, Shaun Gallagher (Gallagher, 2004) has proposed an *interaction theory*, a vision of intersubjectivity as a multi-layered experience.

development of pre-reflective, intersubjective skills. According to Greenspan, while language, intelligence, personality and social skills are learned through meaningful relationships, emotions play an important role in integrating all human development aspects. In this view, the underlying sensory processes play an important function since they determine how children interpret and learn from their environment. In autism, we find a disruption in the ability of selfhood to be engaged with otherness, a deficit which prevents the arising of an affective chain of interactions and, accordingly, the development of cognitive abilities.

In other words, autism seems to be generated by a distortion of the bodily, kinesthetic and intersubjective self. Consequently, the disruption of affectivity is harmful to rationality as well: the result is represented by attempts to understand others by means of mechanical "mentalizing" strategies. Therefore, this syndrome represents an interesting case study to support an epistemological overturning between reason and emotions, emphasizing the centrality of the latter in intersubjective knowledge. Combining phenomenological remarks and Greenspan's approach, we can thus affirm that the subject is an embodied and affective consciousness.[14]

16.4. Therapeutic, phenomenological insights for the treatment of (inter)subjective, corporeal disorders

At this point, I think it might be instructive, and probably necessary, to ask ourselves what the practical consequences of such a vision of subjectivity are, and how we can deal with its disruptions. If the core of subjectivity does not have a theoretical, rational nature, but a pre-reflective, corporeal one, we can hypothesize treatments focused on these latter components.

In this view, phenomenology becomes a useful source not only because it is helpful in understanding the core of several subjective disorders, but also because it furnishes helpful tools for their treatment.

---

[14] Once again, we can find an interesting parallel in developmental psychology: in particular, according to Trevarthen and the motor theory of consciousness "Social isolation, socio-emotional and cognitive delay, and language disorder in children with autism are *secondary consequences* developed within socio-emotional systems as experience-dependent compensations for primary sensory-motor and affective integration errors and poorly regulated motor intentions " (Trevarthen & Delafield-Butt, 2013, p. 2).

The general direction of this kind of work is that of *somatics*, which consists in focusing on transformative body-oriented practices[15]: we can identify a Husserlian contribution to this topic in the draft of *Ideas III*, where the phenomenologist suggests a theoretical investigation whose specific focus is the living body.

Husserl stresses the uniqueness of the experience of the living body, and the shortcomings of a purely material analysis: on the contrary, the bodily and the psychic levels are inextricably intertwined and irreducible to each other. The study of the subject as a living body (*Leib*) resolves itself in a survey which comprehends both a first-person perspective analysis (a direct somatic perception), and a verifiable experience, combining the uniqueness and the concreteness of personal experience to the objectivity of science.

The relationship between the psyche and the living body is irreducible and necessary: you cannot investigate the one and ignore the other, since the subject is characterized by two main features — its essential transcendence and its intrinsic concreteness.[16] We can also describe this characterization as a layered ontology of human being, according to which "higher mental functions are irreducible to but one-sidedly founded in psychophysical existence; within the latter, the psyche is one-sidedly founded in the animate organism, so that everything pertaining to the psyche is provided with a location within the natural, spatiotemporal work ... And because of the psyche that animates it, this organism, in turn, is irreducible to but inseparable from the sheerly physical thing-stratum furnishing the self-sufficient basis upon which the entire hierarchical formation is founded" (Behnke, 2009, p. 14).

---

[15] We can find the roots of somatics in the late 19th and the beginning of the 20th century, but we can affirm that somatics became a professional field in the 1970s thanks to the work of Thomas Hanna (1928-1990), who emphasized the possibility of an integration of "first-person" and "third-person" perspectives for the study of the soma. According to this view, the soma is *the body experienced from within*: this is the reason why I believe that phenomenology and somatics can work together in the development of programs and studies centered on the living body.

[16] In other words, the transcendental ego does not perceive the world from an abstract place but is located in it through the body. This does not mean that the self should be reduced to its outward appearances and downgraded to the status of a "mere thing": the localization typical of a kinesthetic system does not involve a biological reductionism, but simply indicates the fact that the subject is embedded. Following this, the transcendental dimension is not compromised, but co-exists with the empirical one. We can thus claim that subjectivity is both transcendental *and* empirical, a characterization that is also called the "paradox of subjectivity" (see Crowell, 2003; Dodd, 1997).

For this reason, a somatologic, phenomenologically informed science seems to be necessary: the vision of subjectivity as a kinaesthetic bodily consciousness represents the founding dimension for a similar discipline which can be developed elaborating *transformative somatic practices* that are ways for addressing the body as a *living, moving whole.* One paradigmatic example of transformative somatic practices is the technique of "making a body," described by Behnke as "the ongoing (and typically anonymous) kinaesthetic organization of the body as a whole, even prior to carrying out specific acts of external perception or performing any specific practical gestures" (Behnke, 2009, p. 20). The aim of this technique is to strengthen the spontaneous and pre-reflective bodily pattern which is the ground for all subjective possibilities and for the link between the subject and the world.

Within this framework, I would like to integrate Greenspan and Wieder's model with the phenomenological emphasis on embodiment. The result is an affective, phenomenologically informed treatment whose goal is the strengthening of the pre-reflective, intersubjective and kinaesthetic bodily self. I think this could be very important for the treatment of those psychopathologies which involve not only an affective, emotional deficit, but also a detachment from the social milieu, such as autism spectrum disorder.

16.4.1 The D.I.R.E. model

My therapeutic proposal represents a manner in which phenomenology and psychology can fruitfully work together within a more scientific perspective. In fact, I intend to combine phenomenological-informed methods and notions with the D.I.R. model developed by S. Greenspan and S. Wieder in 1997 for the treatment of autistic patients[17]. The theoretical framework of this approach—the *affective diathesis hypothesis*— seems to be coherent with phenomenology and with a definition of subjectivity as a kinaesthetic corporeal entity. According to this approach, "The core psychological deficit in autism may, therefore, involve an inability to connect affect (i.e. intent) to motor planning and sequencing capacities and symbol formation" (Greenspan, 2001, p. 3).

[17] Greenspan, 1997, but also Greenspan & Wieder, 1998. Even though the D.I.R. model is aimed at treating autistic patients, I think that my modified version (D.I.R.E.) could be useful in the treatment of schizophrenia as well, not only because it takes into account corporeality, which seems to be the main disrupted dimension of this pathology, but also because autism, conceived as the loss of social attunement, could be read as a feature of schizophrenia (see Ballerini, 2002).

The possession of motor intentionality and affective abilities, as well as the establishment of *praktognosia* seem to be necessary requirements for motor and cognitive development: according to this logic, the interactions of the subject are central for his biological and psychic growth. In this process, the lived and moving body, which is intentionally directed toward the world, becomes a protagonist: recent studies have shown that motor skill improvements influence neural plasticity (Willingham,1998). In the early years of development, genetic properties and lived experiences influence each other, according to a vision very similar to contemporary phenomenological theories and the gestaltic, enactive approach (Fuchs & De Jaegher, 2009).

Like phenomenology, the D.I.R. hypothesis emphasizes the centrality of *intentionality* (the openness of the subject towards the world), *reciprocity*[18] (the fact that the subject constitutes herself in an intersubjective manner) and *emotions* (our engagement towards the world is emotionally and axiologically characterized). In this view, the main problem of autistic subjects does not seem to be a problem concerning their cognitive abilities, but rather a difficulty pertaining to the emotional and sensorial level: in other words, autistic children fail to connect with their *motility*, and therefore they are not able to deal with more complex emotional chains. Using a phenomenological vocabulary, we can affirm that such persons have difficulty in exercising their *body schema*: they do not recognize themselves as "I can." Bodily, affective components have priority over reflective, cognitive skills. Following this perspective, Greenspan then claims that "Children who have difficulty with motor planning become more deeply engaged when their caregivers learn how to position themselves in a way that makes the motor planning challenge simple" (Greenspan, 2001, p. 16). This is due to the fact that "The emotional interactions ... are not simply responsible for early social, cognitive, and language capacities, but for higher level intellectual and social capacities as well. A continuous flow of emotional interactions between children and caregivers is especially important for the development of the highest levels of human thinking involving self-reflection, making inferences, creating new problem-solving strategies,

---

[18]We can note the similarities with phenomenology: the uniqueness of the individual and the corporeal pre-reflective attunement towards the world are the central features of both of the accounts. The corporeal subjects are linked to the world in a unique kinaesthetic manner: for this reason, a good therapeutic approach cannot consist in a generalized program, but in specific, individual-based training.

having empathy and insight, and regulating mood and behavior" (Greenspan, 2001, pp.18-19).

Greenspan's and Wieder's therapy aims to facilitate mutual emotional interactions, promoting language development, motor skills and spatial-visual thinking. My proposal is to transform the D.I.R. model into the D.I.R.E. model, a therapeutic approach that also takes Embodiment into account. What I have tried to show in these pages, in fact, is the importance and the priority of corporeal awareness and pre-reflective capabilities for the development of cognitive faculties. This is the reason why I mean to propose the transformation of the D.I.R. model into the D.I.R.E one, adding a role for the Embodiment.

In fact, if we think of the subject as a kinaesthetic, corporeal, intersubjective self, *embodiment* is another element that therapy should take into account for the development of self-awareness and, accordingly, the improvement of intersubjective skills.

The general D.I.R. training is centered on *Floortime:* a specific technique which is aimed at both following the child's natural emotional interest and challenging him towards greater and greater mastery of this social, emotional capacity. Eliciting a spontaneous interaction between the autistic child and the adult, Floortime is helpful in improving motor and social skills.

On the other hand, the D.I.R.E. model, adding a role for Embodiment, makes it possible to enlarge the therapy by focusing more specifically on strengthening and recovering the kinaestetic, corporeal self, which seems to be the real core of subjectivity, and accordingly, of intersubjective, perceptual abilities.

This could be achieved in several manners, through motor, sensory, perceptual activities (or *transformative somatic practices)* such as:

- *Dance*: Dance could be used to improve self-awareness and kinaesthetic abilities. In fact, as argued by M. Sheets-Johnstone, movement and perception are ruled by "non-linguistic corporeal concepts" (Sheets-Johnstone, 2009, p. 151). This approach focuses in particular on the role of movement, which is considered the result of a combination of space, time and body strength. With this in mind, the living body itself is movement, and the subject can benefit from training his kinesthetic abilities. Furthermore, Dunkan and Earhart have shown, for example, that people suffering from Parkinson's disease can benefit from the Argentine Tango, characterized by slow movements which help them not only have more control of their own bodies, but also develop their intersubjective sensitivity;

- *Virtual Reality Simulations*: Virtual reality devices seem to be able to simulate sensorial, corporeal experiences through 360° videos. One example is furnished by Jones et al. (Jones et al., 2016), who noted that patients suffering from chronic pain feel better when they are busy playing with virtual reality devices, since their corporeal schema is focused on the 360-degree environment and intentionally directed beyond their biological bodies. Furthermore, VR simulations can elicit a sense of presence of otherness and can simulate sensorial activities and could be used not only for analyzing the subject's sense of (bodily) self, but also for training his social skills, enhancing embodied awareness and recreating the intentional link between the subject and the world (Cf. Bellani et al., 2011; Moore et al., 2005; Cheng & Ye, 2010);
- *The practice of "matching"*: This peculiar somatic technique is an exercise in self-awareness elaborated by Thomas Hanna and further developed by Elisabeth Behnke, who defines it as "a way of taking responsibility for my own somatic experience" (Behnke, 1998, p. 25). It consists of three main parts: awareness of something in my own body; an inner act of matching or aligning oneself with this; and an inner spontaneous change. At the center of this self-directed exercise are some features of my bodily experience, and being aware of these can improve perceptual capacities, proprioceptive experience and other- or world-centered perceptions. In matching, I literally grasp my pre-reflective embodiment; I become aware of me as an "I can." One example could be represented by learning to play a musical instrument, an activity in which bodily awareness can be really helpful because I need to practice and improve my motor skills. This kind of somatic learning "It's not a matter of adding concepts and effort on top what is already there. In this work we analyze how a body is in structure and function, evoking the person's awareness of it and then expanding the pattern to unveil and reveal new truth. In this way, we provide the option for a new pattern to emerge and grow. First, you have to match what is, the you evoke expansion of movement, then you progressively multiply this into new movement patterns" (Johson, 1980, p.12). The lived body, which is usually the one which makes the experience, becomes the object of the experience itself, an experience that, in its essence, is deeply dynamic and ongoingly

modifiable by correcting bodily patterns and replacing old habits with new ones. It is very important to underline the fact that in this kind of practice, as well as in phenomenology, the living body is considered a whole, as the real core of consciousness. Furthermore, Thomas Hanna has emphasized the fact that matching is not a solipsistic procedure, but an intercorporeal procedure, since the dynamic relationship between the subject and the world also involves the interconnection with otherness. Exercising our corporeal and pre-reflective awareness, therefore, allows us to be more connected with our own self, otherness and our possibilities of action in the world.

16.5 Conclusion

In this paper, I have compared Greenspan's "affective diathesis hypothesis" to a phenomenological account of subjectivity inspired by a very well-known interpretation of Husserl. Both of these approaches seem to put emphasis on affective and pre-reflective, lived components of selfhood to such an extent that we can define it as the result of interaffective chains of reciprocity and speak of a *social* core subjectivity. While in Greenspan's view the emotional domain is linked to a sensorial and motor system which drives the link between the subject and the world, in phenomenology, embodiment is an essential feature of subjectivity, necessary both for self-consciousness and intersubjectivity. I have thus attempted to integrate this description with Greenspan's hypothesis, claiming for the existence of an affective, embodied self.

Considering the subject a kinaesthetic, living body has undoubtedly important consequences not only in philosophy, but especially in more practical fields, such as medicine and psychopathology: firstly, in the relationship between clinician and patient, who needs to be recognized as a *consciousness* and not as mere organic thing; secondly, in the analysis and treatment of psychopathologies like autism spectrum disorder.

The D.I.R.E. model I proposed here represents not only a manner to emphasize the modernity and the concrete efficacy of phenomenological notions, but also an example of a transformative somatic practice which does not involve the manipulation of subjectivity, but the ongoing dialogue and intertwining among selfhood, otherness and the world.

We can thus claim that what is at stake is a form of *embodied ethics,* and that the practitioner "must accordingly respect what is going on in the interkinaestethic field, staying aware of boundaries, zones of permeability and resonance, and so on." (Behnke 2009, p. 23). In other words, as soon as we consider the other as a kinaesthetic, intersubjective, pre-reflective entity, we are respecting her as a person, as a dynamic consciousness.

16.6 References

Ballerini, A. (2002). *Patologia di un eremitaggio. Uno studio sull'autismo schizofrenico*, Torino: Bollati Boringhieri.

Bateson, M. C. (1975). Mother-infant exchanges: The epigenesis of conversational interaction. In: D. Aronson, R. W. Rieber (eds.), *Developmental psycholinguistics and communication disorders; Annals of the New York Academy of Sciences*, Vol. 263, New York, New York Academy of Sciences, 101-113.

Behnke, E. (1980). A Somatic Interview with Judith Aston. *Somatics 3:1.*

Behnke, E. (2009). The Human Science of Somatics and the Transcendental Phenomenology. *Zmogus ir Zodis* 2009 IV, 10-25.

Behnke, E. (1998). Matching. *Somatics* 6:4, 24-32.

Bellani, M., Fornasari,L., Chittaro, L., & Brambilla, P. (2011). Virtual reality in autism: state of the art. *Epidemiology and Psychiatric Sciences*. Vol. 20 no. 3, 235-238.

Bizzari, V., & Hipolito, I. (2016). Predictive Engagement and Motor Intentionality. *Esercizi Filosofici* 11, 2, 86-98.

Cheng, Y., and Ye, J. (2010). Exploring the social competence of students with autism spectrum conditions in a collaborative virtual learning environment: the pilot study. *Computers & Education*, vol. 54, no. 4, 1068-1077.

Chomsky, N. (1959). A review of BF Skinner's Verbal Behavior. *Language* 35, 1, 26-58.

Chomsky, N. (1972). *Language and Mind.* Hartcourt: Brace Jovanovich Enlarged edition.

Claesges, U., (1964). *Edmund Husserl Theorie del Raumkonstitution.* Den Haag, Martinus Nijoff.

Crowell, S. (2003). Facticity and Transcendental Philosophy in J. Malpas, *From Kant to Davidson. Philosophy and the Idea of Transcendental*, London.

Damasio, A. (1994). *Descartes' error: Emotion, reason, and the human brain.* New York, Putnam.

Dennett, D. (1991). *Consciousness Explained.* Boston: Little Brown and Company.

Dodd, J. (1997). *Idealism and Corporeity.* Kluwer Academic Publisher.

Duncan, R.P., & Earhart, G.M. (2014). Are the Effects of Community-Based Dance on Parkinson Disease Severity, Balance, and Functional Mobility Reduced with Time? A 2-year Prospective Pilot Study. J Altern *Journal of Complementary Medicine*, 20 (10), 757-763.

Fuchs, T., (2013). Depression, Intercorporeality and Interaffectivity, *Journal of Consciousness Studies* 20, No. 7-8.

Fuchs, T., De Jaegher, H. (2009). Enactive Intersubjectivity: Participatory Sense- Making and Mutual Incorporation. *Phenomenology and the Cognitive Sciences.* Vol. 8 no. 4, 465-486.

Fuchs, T., (2015). Pathologies of intersubjectivity in autism and schizophrenia. *Journal of Consciousness Studies,* 22 (1-2), 191-214.

Fuchs, T., Schlimme, J. (2009). Embodiment and Psychopathology: A Phenomenological Perspective. *Current Opinion in Psychiatry,* 12, 570-575.

Gallagher, S., (2000). Philosophical conceptions of the Self: Implications for Cognitive Science. *Trends in Cognitive Science* 4, 14-21.

Gallagher, S. (2004). Understanding Interpersonal Problems in Autism: Interaction Theory as An Alternative to Theory of Mind, in *Philosophy, Psychiatry and Psychology* Vol. 11, N. 3, The John Hopkins University Press.

Gallese, V. (2005). Embodied Simulation: from Neurons to Phenomenal Experience. *Phenomenology and the Cognitive Sciences,* 4, 28-48.

Gallese, V. (2001). The 'Shared manifold' hypothesis: From Mirror Neurons to Empathy. *Journal of Consciousness Studies* 8/5-7, 33-50.

Goldman, A. (2006). *Simulating minds: the philosophy, psychology, and neuroscience of mindreading,* New York: Oxford University Press.

Greenspan, S., (2001). The Affect Diathesis Hypothesis: The Role of Emotions in the Development of Intelligence and Social Skills. *Journal of Developmental and Learning Disorders* 5 (1), 1-45.

Greenspan, S. (1997). *The Growth of the Mind and the endangered origins of intelligence,* Reading, MA, Addison Wesley, Longman.

Greenspan, S. & Wieder, S. (1998). *The Child with Special Needs: Encouraging Intellectual and Emotional Growth.* Reading, MA: Perseus.

Jensen, R. T, & Moran, D. (2013). *The Phenomenology of Embodied Subjectivity,* Contributions to Phenomenology, Springer International Publishing, Switzerland.

Jones, T.; Moore, T., Choo, J. (2016). The Impact of Virtual Reality on Chronic Pain. *PLoSONE,* Vol. 11 no. 12, 1-10.

Johson, D. H. (1980). A Somatic Interview with Judith Aston. *Somatics 3:1.*

Hanna, T. (1988). *Somatics.* Reading, MA: Addison-Wesley.

Happè, F. (1994). An advanced test of theory of mind. *Journal of* Autism *and Developmental Disorders* 24, no. 2: 129-154.

Hermans, H., Dimaggio, G. (2004). *The dialogical self in psychotherapy,* Hove and New York, Brunner-Routledge.

Hipólito, I, Gonçalves, J, & Pereira, J. G. (2018). *Schizophrenia and Common Sense: explaining the relation between madness and social values.* Springer International Publishing. Studies in Brain and Mind. Vol 12.

Hume, D. (2007). *A Treatise of Human Nature.* New York, Oxford University Press.

Husserl, E. (1950). *Cartesianische Meditationem und Pariser Vorträge,* (ed. Strasser, S.) *Husserliana,* vol. I, Nijhoff, Den Haag.

Husserl, E. (1973). *Ding und Raum Vorlesungen 1907 (Husserliana XVI),* (ed. Claesges, U.). Den Haag, Nijhoff.

Husserl, E. (1950). *Husserliana.* Den Haag/Dordercht, Martinus Nijoff, Kluwer Academic Publishers.

Husserl, E. (1952). *Ideen zu einer reinen phänomenologischen philosophie,* Zweites Burch, hrsg. Von Marly Biemel, Martinus Nijhoff, Den Haag.

Husserl, E. (2001). *Logical Investigations,* transl. by J. N. Findlay, London and New York, Routledge (First published in German as *Logische Untersuchungen,* by M. Niemeyer, Halle 1900/1901).

Husserl, E. (1997). *Thing and Space: Lectures of 1907,* ed. U. Claesges, trans. R. Rojecewicz, Dordrecht: Kluwer.

MacIntyre, A. (1981). *After Virtue: A Study in Moral Theory.* Notre Dame, IN, University of Notre Dame Press.

Merleau-Ponty, M., (1945). *Phénoménologie de la perception,* Paris : Gallimard.

Meltzoff, A; Moore, K. M. (1977). Imitation of facial and manual gestures by human neonates. *Science* 198, 75-78.

Metzinger, T. (2003). *Being No One.* Cambridge, Mass, MIT Press.

Mickunas, A., (1974). The Primacy of Movement. *Main Currents in Modern Thought, vol.* 31:1, 8-12.

Moore, D., Cheng, Y., McGrath, P., and Powell, N. J. (2005). Collaborative virtual environment technology for people with autism. *Focus on autism and other developmental disabilities,* vol. 20, no. 4, 231-243.

Moulding R., Doron G., Bhar S.S., Nedeljkovic M., Mikulincer M. (eds), (2015). *The Self in Understanding and Treating Psychological Disorders.* Cambridge, Cambridge University Press.

Muratori, F. (2007). Autism as an effect of primary intersubjectivity disorder. *Giornale di Neuropsichiatria dell'Età Evolutiva,* vol.27, 425-454.

Piaget, J. (1962). The stages of intellectual development of the child. In: S. Harrison & J. McDermott (eds.), *Childhood Psyhcopathology,* New York, International Universities Press,157-166.

Ratcliffe, M. (2017). Selfhood, Schizophrenia and the Interpersonal Regulation of Experience. In: C. Durt, T. Fuchs, C. Tewes (eds.): *Embodiment, Enaction, and Culture: Investigating the Constitution of the Shared World,* Cambridge, Mass: MIT Press, 149-171.

Reddy, V. (2008). *How Infants Know Minds.* Cambridge, Mass, Harvard University Press.

Rizzolatti, G., Craighero, L. (2004). The mirror neuron system. *Annal Review of Neuroscience,* vol. 27, 169-192.

Sheets-Johnstone, M. (2009). *The Corporeal Turn.* Exeter: Imprintic Academic.

Trevarthen, C. (1979). Communication and cooperation in early infancy. A description of primary intersubjectivity. In: M. Bullowa, *Before Speech: the Beginning of Human Communication.* London, Cambridge University Press, 321-347.

Trevarthen, C. (1998). The concept and foundations of infant intersubjectivity. In: S. Bråten (eds.), *Intersubjective Communication and*

Emotion in Early Ontogeny, Cambridge, Cambridge University Press, 15-46.

Trevarthen, C., Aitken, K. J. (2001). Infant Inetrsubjectivity: Research, Theory, and Clinical Applications, in *Journal of Childhood Psychology,* Vol. 42, no. 1, 3-48.

Trevarthen, C., Dealfield-Butt, J. (2013). Autism as a developmental disorder in intentional movement and affective engagement, in *Frontiers in Integrative Neuroscience,* Vol. 7, 1-15.

Trevarthen, C. & Humbley, P. (1978). Secondary intersubjectivity: Confidence, confiding acts of meaning in the first year. In: A. Lock (eds.), *Action gesture and symbol,* London, Academic Press, 183-229.

Toombs, S.K. (1998). Illness and the paradigm of Lived Body. *Theoretical Medicine* 9 (2), 201-226.

Varela, F.J., Thompson, E., Rosch, E. (1991). *The embodied mind: Cognitive science and human experience,* Cambridge: The MIT Press.

Willingham, D.B (1998). A neuropsychological theory of motor skill learning. *Psychological Review,* n.105, 558–584.

Zahavi, D. (2014). Empathy and Other- Directed Intentionality, in *Topoi* Vol. 33, n. 1, Springer.

Zahavi, D., (2000). *Exploring the Self,* John Benjamins, Amsterdam.

Zahavi, D. (1997). Horizontal Intentionality and Transcendental Intersubjectivity in *Tijdschrift voor Filosofie,* Vol. 59, No. 2, 304-321.

Zahavi, D. (2001). *Husserl and Trascendental Intersubjectivity. A Response to the Linguistic- Pragmatic Critique,* trad. by E. Behnke, Athens: Ohio University Press.

Zahavi, D. (2005). *Subjectivity and Selfhood. Investigating the First-Person Perspective.* Cambridge, MA: MIT Press.

Zahavi, D. (1999). *Self-Awareness and Alterity. A Phenomenological Investigation.* Northwestern University Press.

Zahavi, D. (1998). Self Awareness and Affection. In N. Depraz and D. Zahavi (eds.): *Alterity and Facticity.* Kluwer Academic Publishers, Dordrecht, 205-228.

Zahavi, D. (2017). Thin, Thinner, Thinnest: Defining the Minimal Self. In: C. Durt, T. Fuchs, C. Tewes (eds.): *Embodiment, Enaction, and Culture: Investigating the Constitution of the Shared World,* Cambridge, Mass: MIT Press, 193-199.

Chapter 17

Psychoanalysis, social psychology and the therapeutic community

Robert Hinshelwood
Center for Psychoanalytic Studies, University of Essex

In this chapter, I shall reflect upon the origins of the therapeutic community movement in psychiatry, in Britain in particular. It was profoundly conditioned by the communalism widespread in Britain at war in the 1940s. People break down in a context, their family, work or neighbourhood. The history of the democratic therapeutic community is complex and now quite long – around 75 years. Its origins were in some new psychological ideas evolving in Germany and the US in the 1930s and tried out practically in psychiatry and psychoanalysis in Britain during the Second World War.

These developments were particularly associated with John Rickman and Tom Main. The general influence of psychoanalysis on culture and society of the time, and the communal spirit of a country at war were the values that provided ingredients for the enduring spirit of the therapeutic community. The position of therapeutic communities has varied somewhat between different times, cultures and countries as it has endured and spread. So, this Chapter may be largely relevant to Western cultures. The historical and current situation in Britain will be taken as typical of this general trend in the West and is especially relevant perhaps as the therapeutic community movement started here in Britain.

17.1 The new idea, T.F. Main and Northfield Hospital

The desperate plight of Britain in 1941 after the defeat at Dunkirk, led to many experimental efforts to gain advantage over the enemy. This thirst for novelty included military psychiatry even and was tolerated partly because the loss of men to mental disorders severely depleted the fighting forces.

The new idea that was tried out in various forms was that an individual is not an entity on his own. He/she is deeply enmeshed in a social context, or group. His/her behaviour and mental functioning are deeply dependent on the various social pressures and tensions that surround and impinge on him/her. This idea was novel to psychiatry at the time because, more usually, psychiatry has paid attention to an individual person's symptoms and analogous to a medical approach. It was an individual matter and would be amenable to physical and physiological correction, particularly with medication, applied to the individual patient.

This new view of an influential link with the social context was developed in the social psychology attributed to Kurt Lewin (1936), a psychologist in Berlin in the 1920s. Lewin eventually sought refuge in 1933 in the US, but his work strongly influenced Rickman and Main, and the post-war tradition of the Tavistock Clinic.

Lewin had been strongly influenced by gestalt psychology a psychological theory of perception in which what is perceived is a figure (the gestalt) that stands out from a background. Typical is an optical illusion such as a drawing that is immediately perceived as one figure, but can also be perceived, as a quite different figure – for instance, two profiles of faces can be alternatively seen as the outline of a symmetrical vase. From this tradition, Lewin had applied gestalt psychology to social groups. Thus, the individual in a group can be seen as a particular person (a figure) outlined against the rest of the group (the field) in which he is. And Lewin built up what came to be called social field theory.

Lewin made this distinction between the features of an individual in isolation and the individual in the context of the dynamic forces of his social field. He pointed out that this distinction occurs in science in general. Aristotle was interested in objects as entities with their own substances and characteristics. Whilst, in contrast, Galileo was interested in the entity in context; he did not investigate the substance of a stone, but its behaviour when in the field of gravity.

This social psychology impacted on psychiatry by pointing attention towards an individual's context and not just his symptoms. The focus thus shifted to the group dynamics (a term invented by Lewin) from which the individual's behaviour and experience emerge. It is not a network of interpersonal relations; the group is a whole entity of psychological forces which form a 'life-space' for each of the individuals.

John Rickman (1890-1951) was a key person in introducing this contextualising idea into psychoanalysis. He was a practicing psychoanalyst. He explored the psychoanalytic setting as a social context in which the analytic treatment happened, in other words, the transference and countertransference were the background of dynamics in a group of two, from which the individual patient emerged in the analytic setting, struggling with his own issues in the context of the psychoanalytic session. Rickman recognised how this connected with the object relations school of psychoanalysis developing in Britain. That school evolved from Klein's play technique for the psychoanalysis of children. She observed the child arranging toys together in a play which told a narrative of the relations between the objects the toys represented. She was not interested in the individual toys, but in the pattern of relations created in the child's play and showed his/her struggle with the influences affecting him/her in his/her context.

However, Rickman, a psychoanalyst, recruited Wilfred Bion (1897-1979), not yet a psychoanalyst, to a similar psychiatric project at Northfield Hospital in Birmingham that ran in 1943. They applied the therapeutic idea Rickman had worked out in psychoanalysis, but applied to psychiatry. They decided it was necessary to create a context within the rehabilitation service of a military mental health hospital that might resemble the conditions of a fighting unit of soldiers with a characteristic task, atmosphere and social structure (Bion and Rickman, 1943). Each individual was then seen against the background of the culture of the rehabilitation unit. The normal practice of individualised treatment was set aside when creating the need for soldiers to function again in their social context.

After the war, this idea of a patient in his context was carried into civilian psychiatry by several people including, Tom Main (1911-1990), who coined the term 'therapeutic community'. Main criticised the treatment of a patient in social isolation:

"Sooner or later the patient alone and unsupported must face the difficult task of returning to the society in which he became unstable... This task is no light one for a desocialised man however healthy he has become". (Main, 1946, pp. 7-8).

This became a movement to do something about the chronically institutionalised (and desocialised) state of patients generally in the mental health services in Britain (Martin, 1955, Barton, 1959).

Processes of de-institutionalisation were started up in a number of places: David Clark at Fulbourn (Clark, 1964, 1975), Elizabeth Schoenberg (Shoenberg, 1972), Mandelbrote and many others. They all involved engaging the seriously desocialised inmates in social activities. This movement married with other experiments that emphasised the social aspects of people who had a breakdown and emphasised the 'healthy' part of the patient as a well as the unhealthy. The best known was the work during the wartime at Mill Hill Hospital by Maxwell Jones (1909-1990, see Jones, 1952).

The experiments at Northfield led to these various ways of rethinking the institutions which provide mental health care, and in turn, they formed a whole social movement which cohered into the therapeutic community approach.

17.2 Psychoanalysis as social psychology

At the same time these fertile ideas led to innovations in other directions in the 1940s and beyond, which have fed back other influences into the developments of the therapeutic community. Turning briefly to these parallel developments after the war.

One of these was the development of group therapy emphasising the social forces and tensions of a group which are created by the individual members and which also push them into roles to play out in the group. This was started by Bion at the Tavistock Clinic, and became a form of group therapy on an out-patient basis. Connected to the development was a non-clinical project, sponsored by a substantial financial grant from the Rockefeller Foundation, to apply these ideas about unconscious group dynamics to large commercial and public service organisations. This became the Tavistock Institute, to which two psychoanalysts contributed; they were Elliott Jaques (1955) and Isabel Menzies (1959).

Jaques coined the term 'social defence system', describing how individuals will develop collective social practices that enable the individuals to avoid the worst of their anxiety. In working organisations, where there is an anxiety arising from the work, the work practices may develop in such a way to support the individuals' own psychological defences so that the anxiety becomes and remains unconscious.

This was exemplified empirically by Isabel Menzies (1959) who investigated a nursing service in a general hospital. The nursing organisation was failing to provide effective training for student nurses. She found it could be traced to the severe anxiety that nurses suffer in work arising from long hours in contact with ill, suffering, dying, mutilated and frightened patients. Many details of nursing practice grew up that enabled nurses to deny their emotional feelings of sympathy and responsibility which they might realistically or unrealistically feel. Such practices promoted various forms of depersonalisation of patients so that a nurse did not have any strong emotional responses to the persons she cared for. For example, there was a task list system in which various parts of the nursing work such as taking temperatures, blood pressure measurements, using the bed pan, and so on, were organised so that one nurse was responsible for the same task on each patient rather than responsible for all the tasks on one patient. As a result the suffering patient was not seen as a whole person, only as a part of one body amongst many. The purpose was, unconsciously to avoid a nurse developing a fuller emotional relationship with suffering patients.

Another typical practice Menzies found was to avoid making decisions by passing responsibility up the hierarchy; this avoided responsibility and thus avoided guilt for what happens to a patient. This had evolved as a set of *collective* practices in which the individual nurse worked. So, it is a social form of defensiveness, and separate from any internal psychological defences within the individual.

This collectively organised social form of support in the form of work practices has been confirmed by others in various forms of health provision and social care, including mental health (Hinshelwood and Skogstad, 2000, Valentine, 1994). The illustrations show how individuals are embedded in a social context, or field, at work which has been unconsciously elaborated to manage the unconscious issues of the individuals. This way of understanding organisational dynamics of psychiatry complements the therapeutic community. Whereas the social defence system concept can explain how strange ways of working grow up – like the institutionalisation in the old mental hospitals – the therapeutic community is a form of practice which tries to expose the general anxiety in the organisation with the aim of preventing unsatisfactory distortions to the practice of psychiatric care (Hinshelwood, 2001).

It seems that in these unconscious work practices whole groups of individuals allocate roles via unconscious pressures and enact certain dramas. In discussing Bion's approach to group therapy, Eric Trist described how it is important to make the group a safe enough place for these unconscious roles: "[H]e made it safe for the group to dramatise its unconscious situation... Experience of these episodes led me to think of group therapy as a theatre" (Trist, 1985, p. 30). This describes how a group may expose unconscious phantasies not normally allowed expression. It has been possible to turn this approach back to therapeutic communities understand how the members of a therapeutic community dramatize their disturbances, and anxieties and defences in their day-to-day activities (Hinshelwood, 1987).

The intensely felt anxieties and conflicts in mental health services cause tensions for staff and may generate dysfunctional kinds of practices and roles which thereby depersonalise and desocialise patients who are no longer fully persons. This parallels the nursing practices in a general hospital just described. This process is not limited to the large institutions but seems to have re-appeared in community mental health services (e.g. Foster and Roberts, 1998).

Like a psychiatric hospital, the therapeutic community may be seen as a kind of theatre in which the members dramatize unconscious situations as in a theatre (Hinshelwood, 2000, 2004). The therapeutic community, however, attempts to understand those processes and where they come from. The understanding of unconscious roles and relationships can lead us to very far-reaching explanations of social problems. The relations in traditional psychiatric hospitals become alienating, and Goffman (1961) showed their connection with a range of other 'total institutions', as he called them.

The withdrawal from emotional involvement described by Menzies (1959) amongst nurses in a general hospital is not so distant from the depersonalisation in those old mental hospitals that institutionalised patients. Their patients were very difficult; and they were difficult, in part, because of the difficult anxiety they cause the staff. Treating disturbed people will inevitably be disturbing for the staff, and such a contagion of disturbance is after all why the inmates have been ejected by their loved ones and friends back home. Staff may then need to keep emotional distance from disturbed people. It is all too understandable, even though, at the same time, it leads as Main remarked to the desocialisation of such people – on top of their psychiatric condition.

The attitude to a patient as a desocialised non-person impacts on certain patients. Typically for those in states of chronic schizophrenia it is detrimental because their condition includes a problem of identity anyway (Hinshelwood, 1999). So, to depersonalise them through the institutionalisation of psychiatric institutions enhances their inherent identity deficit. This convergence of emotional distancing with the lack of the person's firm identity creates a kind psychological side-effect of the culture of care, analogous to the side-effects of pharmacological medication.

The therapeutic community aim is to re-personalise those who are at risk of being depersonalised. For all the benefits that the scientific approach to psychiatry has brought, there is, at the same time, a risk of a hyper-objectivity of quantitative measurement and statistical analysis. Such objectivity can then be detrimental to the essential subjectivity of mental life and of the social health of the person. The therapeutic community remains a revolutionary idea however old it now is.

It is a general observation that a depersonalisation of human beings other than ourselves is remarkably prevalent. It seems to be prevalent in a variety of circumstances not just in mental health care. A more extreme version can be seen in those ordinary people who staffed the concentration camps in Nazi Germany. Somehow, they cut themselves off from the pain of noticing what they were doing to the human persons they worked and starved to death. Those guards went home after work to become family members again with their loved ones. Similarly, in war, fear and rage turns the enemies into non-humans. These examples are variants of the same process by which people with the best of caring intentions went to work in the old mental asylums and ended up overseeing regimes that desocialised people to become chronic inmates. How could ordinary and mostly well-meaning people cut off their humanity to devote themselves to such inhuman work? The psychoanalytic theory of anxiety and defence can go some way to answer that; and the degree to which it can provide an answer, is the degree to which it provides a basis for the therapeutic community re-humanising process.

We have turned, for a moment, from the therapeutic community itself, to a more general enquiry about unconscious social roles and tensions. In particular, the therapeutic community was born from the need to grasp the problem of psychiatric institutions relegating inmates to a non-person status. Psychiatry alone in struggling with the way we can reduce each other to non-human objects merely for extermination or at best for neglect and indifference. What is that inhuman process at the core of human interactions? I shall briefly take a wider scan of this human propensity.

17.3 Alienation and commodities

Karl Marx's notebooks, published as *The 1844 Economic and Philosophical Manuscripts* in 19**, dealt with the unfairness of a separation of a worker from the product of his labour. As he put it,

> *"[T]he object that labour produces, its product, stands opposed to it as something alien, as a power independent of the producer. The product of labour is labour embodied and made material in an object ... this realisation of labour appears as loss of reality for the worker, objectification as loss of and bondage to the object, and appropriation as estrangement, as alienation"*
> (Marx, 1844, p. 324).

If the product from work is taken away, then the producer is disconnected to some degree from what is felt – by him, and by others – to be him. It stands opposed, as something alien. The worker becomes a lesser person; as Marx said, 'The worker becomes poorer the more wealth he produces' (p. 323). He becomes poorer in a psychological sense. This is remarkably parallel to descriptions in the psychoanalytic literature (Hinshelwood, 1983). For example,

> In such fantasies products of the body and parts of the self are felt to be split off, projected into mother, and to be continuing their existence there (Klein, 1955, p. 142).

That is Melanie Klein describing the mechanism of splitting the ego, or identity, and subsequent projective identification of a part of that divided self into some other:

> [These mechanisms lead] to the weakening and impoverishment of the ego resulting from excessive splitting and projective identification' (Klein, 1946, p. 104).

For the idea of social alienation, it would seem that the internal psychological mechanism of splitting is necessary. The split-off part is then felt as an object. That indicates that the product of the worker 'stands opposed' to him, and it is projected, *'alien'* and *'independent'*, yet it retains a kind of identity alienated from the worker. These are two converging descriptions that come from quite unrelated fields – psychoanalysis and political theory (Hinshelwood, 1996).

Frank Braverman (1974) has developed this idea as the 'labour process'. Simply, an industrial enterprise engages a worker in making some physical product, shall we say nails, then that factory has costs that include the worker's daily wage to cover his/her living for a day, plus the maintenance and servicing costs of the factory and the enterprise itself. Then if the enterprise is successful, the worker makes enough nails to cover the proportion of the costs the enterprise runs up. But the intention is that the worker will make more nails than just covers these costs. Then the extra nails which the worker has made belong to the enterprise. The worker sells his time, not his products, so his material products (his 'embodied labour') are taken from him. Although the two processes – the psychological, and the economic – are quite different, they converge in their effects; and we can see that each build on the other. The point is that a person, or a 'part' of him can be used as a depersonalised commodity just as if it is an inanimate thing.

Returning to the psychiatric service, a person comes to be alienated as various aspects of his personhood are split off and lost to him. He has become weakened and impoverished. Not least is his personal autonomy which is lost, and the staff in his ward and on the psychiatric team take over his decision-making (see Hinshelwood, 1987) on his behalf and in effect 'become' that part of him. And thus, through this theory of alienation we can see what the therapeutic community stands for. It is the attempt to resist this reduction of the suffering person; he becomes a thing that does not have the essential characteristics of a person, such as his personal autonomy.

The anthropologist and psychiatrist Rob Barrett (1996) made a similar study of the way patients are admitted to a psychiatric ward. The psychiatric culture proceeds in three steps. At first, the patient is admitted and regarded as a disease with symptoms, rather than a person with volition and experiences. In Barrett's terms, he/she is not a 'moral' agent and does not have responsibility for how he/she is. He/she is not regarded as autonomous. In the second phase, the team engages in all the investigations of psychological and medical histories until they complete a compendium of facts. The patient is then a 'case', to be understood in terms of the neuro-biochemistry, psychological and social factors, all of a technical nature. The case is an agglomeration of facts.

Only in the third phase does something like a person begin to emerge again. When he is in progress of recovering, two things happen. First, he begins to be held responsible for the things he does and says. Then, second, members of staff begin to express personal views of liking or disliking the patient, and they can be said to be reconstructing a relationship with the person of the patient. He has achieved again the status of a moral being, who could be approved or disapproved of. Interestingly, Barrett observed that this last process, the more personal reactions of the staff to the person of the patient often occurred first of all off the ward and in the canteen during coffee and meal breaks. It is as if the person of the patient can be found best outside the professional arena in the ward itself. He becomes a person through his context of being with others.

17.4 The new public service ethos

I have described two processes – one social-economic, and one intra-psychic – by which persons come to be seen as non-persons. It is convenient to call such depersonalised entities by the term 'commodities', in line with the social alienation. The reduction of mental patients to this category probably pre-dates the development of the modern economic era. But the direction of economic development over the centuries has been in terms of the trade of impersonal commodities for impersonal money. In fact, the most recent development has been in the 1980s, in the

US and Europe, as a result of the reforms, sometimes called Reaganomics, which freed Western economies for the greater economic exploitation of commodities.

In addition, experiments, especially in Britain from the 1990s onwards, introduced commodity forms of economics into the public services as well as commercial enterprises. This commodity-oriented economics is, now often called 'neoliberalism' This new system for managing the public services has taken over the national health service in Britain, as well as social services and more recently education. It has had an impact on the mental health services where it piggy-backed, as it were, on the early 'commodification' of human beings that had taken the form of the institutionalisation of inmates. This climate of opinion has gathered impetus since then with the pressure for standardised and random controlled testing which can agglomerates patients into units, and indeed agglomerates the various research projects for meta-analyses. This approach to management has not necessarily been to the benefit of psychiatric patients already lacking subjective identity in themselves.

The therapeutic community, on the other hand, has existed to *preserve* the capacity of persons to remain functioning in their social relations whilst supported to be accountable for responsibilities they carry within the community. Therapeutic communities, therefore, exist in a very different environment now, compared with the time in the 1940s when the term 'therapeutic community' was first used. Today, pressures exist to justify the therapeutic community method as if it were a medical treatment operating on a material or bodily substance and tested according to prescribed methods of physical research. Without the results of that kind of research, therapeutic communities have suffered a decline as have other forms of psychotherapy.

Although we should readily accept that outcome studies are important for users and for the public purse, I have also tried to show previously that there are a number of issues that need careful consideration first and before applying the methods adopted by experimental psychology and medicine. We need to consider how to assess the personal world of relationships and subjective experience (Holmes, 2002, Hinshelwood, 2002), as opposed to measurable symptoms. At root is the contrast between instrumental methods aimed at psychological symptoms as if they were bodily ones on the one hand, and on the other, the relational methods aimed at the person and his personality within his human context. It is not that measurable effects should not be used, they should be, but something essential is left out in the case of each individual person, and indeed each individual therapeutic community. The challenge is how to bring back the personal, relating individual into his own case.

Moreover, that 'unit of care' approach does not easily fit the therapeutic community. Each therapeutic community is, like the individual person, a complex, self-organising system. It is composed of individual persons, and in turn enhances them through their wholesale participation (Hinshelwood, 2011). The therapeutic community's origin was in the exploitation of the idea that the individual is a function of his relations with his context, which impacts on him within a field of social forces and to which he contributes with his own influences, conflicts and tensions. The therapeutic community is allied with psychoanalysis as one of the group of psychiatric treatments which can be distinguished by being focused on relationships (Hinshelwood, 2011). Such relational treatments are each uniquely individual, depending on the specific persons involved; and they unhappily resist tick-box categories.

Put starkly, the therapeutic community method involves a considerable degree of bottom-up authority and responsibility, and a respect for the judgments of all members as 'moral' beings. This conflicts with the top-down arrangement of managerial authority and responsibility understandably familiar in medicine and most professionally managed organisations. The clash of cultures within psychiatric services is not always recognised by management – nor by therapeutic communities – with ensuing misunderstanding and suspicion. That incomprehension is not good for the mental health of either.

17.5 Summary and conclusions

Not only does the individual live within a social system which is a field of forces, but the therapeutic community too lives in a cultural environment which exerts forces on them. A management and economic environment which neglects the impact of the environment on the organic system risks inadvertently doing harm.

The ethos of the therapeutic community has been, from its inception, to confront the tendency in psychiatry for staff to withdraw emotionally from inmates for their own personal 'survival'. That withdrawal had led, over a period of a century and a half to psychiatry as a treatment of diseases rather than of persons. This may have been enhanced by the rise of scientific psychiatry which despite the advantages it brought, tended to look for aetiologies at the level of the body, instead of the person. This scientific tendency has blended with the cultural withdrawal from patients as persons.

The intention of the therapeutic community has been to preserve the view of patients as still-functioning persons, a view to be held in the minds of staff and of patients. The claim here is that a profound pillar of support has been the similar psychoanalytic project of prioritising the treatment relationship. I have suggested that the convergence of social field theory with object-relations psychoanalysis in the 1940s was the innovative foundation for what we must think of as the group of relational therapies (psychoanalytic psychotherapy, group therapy, therapeutic community). This complex of ideas and influences has lately come into conflict with a new movement in the management of public services that aims to 'sell' health as a commodity in line with the way that retail systems manage the sale of consumer goods.

Currently, the therapeutic community exists in a counter-cultural position in relation to the commodification of people and health in the 21st century. However, when has the therapeutic community movement not been struggling to represent something counter-cultural. One way or another, we have forever been challenging the conventional tide of opinion since the therapeutic community was conceived in the 1940s?

17.6 References

Barratt, R. (1996) *The Psychiatric Team and the Social Definition of Schizophrenia.* Cambridge: Cambridge University Press.

Barton, R. (1959) *Institutional Neurosis.* Bristol: Wright.

Bion, W. R. and Rickman, J. (1943[1961]) 'Intra-Group Tensions in Therapy'. In W. R. Bion (1961) *Experiences in Groups and Other Papers.* London: Routledge, pp. 11-26.

Braverman, F. (1974) *Labor and Monopoly Capital: The Degradation of Work in the Twentieth Century.* New York: Monthly Review Press.

Clark, David (1975) *Social therapy in Psychiatry.* London: Penguin.

Foster, A. and Roberts, Vega (eds.) (1998) *Managing Mental Health in the Community: Chaos and Containment in Community Care.* London: Routledge.

Goffman, Erving (1961) *Asylums: Essays on the Social Situation of Mental Patients and Other Inmates.* New York: Doubleday.

Harvey, D. (2005). *A Brief History of Neoliberalism.* Oxford: Oxford University Press.

Hinshelwood, R. D. (1983) Projective identification and Marx's concept of man. *International Review of Psychoanalysis 16*: 221-226.

Hinshelwood, R. D. (1987) *What Happens in Groups.* London: Free Association Books.

Hinshelwood, R.D. (1996) Convergences with psycho-analysis. In Parker, I. and Russell Spiers, R. (eds) *Psychology and Marxism* pp. 93-104. London: Pluto Press.

Hinshelwood, R.D. (1999) The difficult patient: the role of 'scientific' psychiatry in understanding patients with chronic schizophrenia or severe personality disorder. *British Journal of Psychiatry 174*: 187-190.

Hinshelwood, R.D. (2001) *Thinking about Institutions: Milieux and Madness* author (London: Jessica Kingsley).

Hinshelwood, R.D. (2002) Symptoms or relationships (Comment on Jeremy Holmes 'All you need is CBT'). *British Medical Journal* 324: 288-294.

Hinshelwood, R.D. (2004) *Suffering Insanity: Three Psychoanalytic Essays on Psychosis.* (London: Routledge).

Hinshelwood, R.D. (2011) Manual or matrix: How can we know our outcomes? *Therapeutic Communities* 31: 328-337.

Hinshelwood, R.D. and Skogstad, W. (2000) *Observing Organisations.* London: Routledge.

Holmes Jeremy (2002) All you need is cognitive therapy. *British Medical Journal* 324: 288-290.

Jaques, Elliot (1955) Social systems as a defence against persecutory and depressive anxiety. In: Klein, Heimann & Money-Kyrle eds *New Directions in Psychoanalysis.* London: Tavistock; pp. 478-498.

Jones, Maxwell (1952) *Social Psychiatry.* London: Tavistock.

Klein, M. (1946) Notes on some schizoid mechanisms, in *The Writings of Melanie Klein Vol.* 3: 1–24. London: Hogarth.

Klein, M. (1955) On identification. In *The Writings of Melanie Klein* Vol. 3: 141–175. London: Hogarth.

Layton, L. (2014) Grandiosity, neoliberalism, and neoconservatism. *Psychoanalytic Inquiry* 34:463-474.

Lewin, K. (1936). *Principles of Topological Psychology.* New York: McGraw-Hill.

Main, T.F. (1946) The hospital as a therapeutic institution. *Bull Menninger Clinic,* 10: 66-70. Republished (1989) in T.F. Main, *The Ailment and Other Psychoanalytic Essays,* London, Free Association, pp. 7-11.

Martin, D. (1955) Institutionalisation. *Lancet 2*: 1188-90.

Marx, Karl (1844) *Economic and philosophical manuscripts*: 279–400. In *Early Writings.* London: Penguin, (1975).

Menzies Lyth, Isabel 1959 The functioning of social systems as a defence against anxiety: a report on a study of the nursing service of a general hospital. *Human Relations* 13 95-121. Republished 1988 in Menzies *Containing Anxiety in Institutions.* London: Free Association.

Schoenberg, E (1972) *Hospital Looks at Itself: Essays from Claybury.* London: Bruno Cassirer.

Trist, E. (1985) Working with Bion in the1940s: The group decade in M. Pines (Ed.) *Bion and Group Psychotherapy.* London: Routledge and Keegan Paul, pp. 1-46.

Valentine, M. (1994). The Social Worker as 'Bad Object'. *The British Journal of Social Work* 24: 71–86.

Chapter 18

Ipsas aquas urere consuevit. Able to set even the waters aflame: psychiatric *praxis*, analytic thought, transference love

Giulio Caselli
Policlinico Gemelli, Rome, Italy

This paper reflects on psychiatric practice emphasizing the role of clinical practice based on relationship and transference, in opposition to a merely pharmacological and objectivizing approach. After a brief epistemological introduction, I will elucidate the central element of psychoanalysis: the phenomenon of transference love, and I will claim that it can be used in the psychiatric setting (as Gabbard's psychodynamic psychiatry shows). Then, I will focus on the analysis of the myth of Eros and Psyche as a symbol of the birth of the relational and reflective thought, a process that is necessary for communication to become a dynamic of care, and for the psychiatrist herself/himself to be considered an instrument of relation. In other words, I will argue for a renewed epistemology of psychiatry where emotions and love are primary tools for understanding and helping the patient.

18.1 Introduction: the need for a new epistemology in mental health

Since the beginning of the twenty first century, the dominant psychiatric paradigm has considered psychic phenomena as epiphenomena belonging to neural activities: in this way, symptoms and patients are reified and reduced to mere biological entities.[1] Nonetheless, we can affirm the existence of an antipsychiatry movement, which arose in the 1960s and that has been reasserted today through a critique that emphasizes the dehumanization of care and the need for a new path (Bracken et al., 2012; Cohen & Timimi, 2008; Fernando, 2014). In fact, so-called "biological psychiatry" conceives of mental health problems in terms of something that results from a "broken brain" or "chemical imbalance" and focuses on drug treatments (Baldessarini, 2014); while cognitive-behavioral psychology considers mental health problems as the result of maladaptive patterns of thought and action that can be modified by psychotherapy (Hofmann, Asmundson & Beck, 2013). As noted by Berrios and Markovà (Berrios & Markovà, 2015), these biological approaches are based on: Cartesian mechanism; a simplistic vision of man as a rational being; a theory of sign and language; and neuropsychological notions. In this view, mental disorders are treated as symptoms, signs, syndromes or diseases. In other words, what is at stake in a similar approach is a "medicalization" of mental pains, a process that implies that certain atypical behaviors are declared "abnormal" or "pathological," and that pathologies—and mental symptoms[2]— are localized in specific parts of the body (which has been reduced, in turn, to a body-brain). We can thus affirm that a biological, reductionist approach is based on a specific epistemological and ontological paradigm which involves a mere causal relationship between symptoms and brain structures and an objectifying gaze that the clinicians should adopt on the disturbed patients. In other words, the pathology is understood through a correlational mechanism between the brain's disruptions and mental symptoms, while the ill person is reduced to a malfunctioning objective and biological body. Nonetheless, speaking of mental disorders, are we dealing with symptoms as spatio-temporal objects located in the brain or with a different kind of complexity?

The first assumption we can discuss is the weak explanatory power of a similar approach: Berrios, who dedicated a consistent part of his work to

---

[1]As noted by Berrios and Markovà (Berrios & Markovà, 2015) there, are, of course, authors who have risen against the power of the neurosciences (for instance, Bennett & Hacker, 2003; Choudhury & Slaby, 2012), but it cannot be denied that the general trend has been to follow the official line.

[2] As underlined by Parnas et al.: "the psychiatrist confronts not a thing or body part or physiological process, but a person, i.e., another embodied consciousness embedded in the realm of meaning. The person does not manifest a series of independent referring symptoms/signs but rather certain unities of interpenetrating experiences, feelings, expressions, beliefs, and actions, all permeated by biographical details." (Parnas et al, 2012, p.6).

fighting against psychiatric biologism and reductionism, argues that to reduce mental symptoms to brain dysfunctions is a huge mistake, since this "functional specificity might ultimately be associated with a variety of mental symptoms, none of which necessarily preserves the original biological specificity [...] The converse could also be the case: namely that a mental symptom, whose content may seem to suggest a particular functional specificity (e.g. perception) may be related to a variety of core neurobiological signals. In other words, it could not be assumed that all patients complaining of, say, depersonalization or derealization would have the same neurobiological accompaniment" (Berrios & Markova, 2015, p. 15).

Accordingly, we can identify a significant epistemological problem: the dilemma between understanding and representing. In fact, if the object of the analysis is constituted by human behavior, emotions, relationships, and by the fact that many of our decisions and wishes cannot be determined by our conscious control, it seems clear that we cannot conceive of the human mind as a biological organ such as a kidney tubulus or a cardiac calcium channel—in other words, like an organic object of a radical interpretation and practice of the evidence-based medicine conceived of as a reductionist approach. The complexity of our mind excludes the possibility of analyzing it as a causal object and merely offering a solution to its mechanic malfunctioning.

A similar kind of cure has the effect of transforming and reducing the patient from an emotional, living and unique subject, inevitably linked to her/his wishes and feelings, to her/his symptoms, which are at the center of a statistic approach that destroys the peculiarity of the person, transforming it into an inconsistent slop where, for instance, x's anxiety—which is so intimate and close to her/his personal history—is considered in the same way as z's anxiety—which is also unique and intimate. Such a complex object of study needs a complex subject of analysis: Bleuer's psychiatry has taught us that the only instrument owned by the psychiatrists for performing their work is represented by the psychiatrists themselves as active and deep listeners, and not as simple observers—which, on the contrary, we can find in Kraepelin's attempt at objectifying psychiatry through the only available sense, sight, because we cannot auscultate, touch or move the patient's soul and inner world.

Furthermore, according to the epistemological paradigm of biological psychiatry, mental symptoms often have a mere referential function (they point to abnormalities of an anatomic-physiological substrate) and no intrinsic meaning. On the contrary, Berrios argues that symptoms are "hybrid objects" constituted by a layered process of different and interacting elements. Accordingly, we need a renewed epistemology, whose main important consequence would be a renewed kind of treatment. In fact, if a reductionist and biological epistemology involves a treatment merely based on drug prescription (since it is influenced by a mechanistic paradigm); a different kind of epistemology— which takes into account the complexity of the disorder and conceives of the patient as a living person— allows for a hermeneutic, dialogical therapy. For this reason, psychoanalytic thought can be helpful for the development of a similar kind of psychiatric practice.

We can hence argue for a shift from an objectifying approach—which conceives of the clinician in terms of an impartial and trained investigator who arrives at a diagnosis through a process of induction from evidence and verification—to a relational therapy. In fact, the ambiguity of care, as noted by Cargnello (Cargnello, 1999) can be schematized through an oscillation between being-with-someone[3] and having-something-in front of me—between a relational, dynamic account and an objectifying, distant gaze. The manners in which alterity is given to us can be identified in the other-as-object and the other-as-subject. Translated in terms of medical epistemology and hermeneutics, the therapeutic setting can be a relation between a distant object or a reciprocal, dynamic relationship with another subject. Buber's distinction between an "I-It" and an "I-Thou" relationship is very explicative: If we conceive of the therapeutic relationship as an "I-Thou" one, the ontology, epistemology and hermeneutic features of the medical encounter completely change with respect to the objectifying gaze of the "I-It" biological paradigm. The dialogical principle described by Buber involves an ideal "we," which is not a mere epistemological status where the "I" meets another subjectivity and knows it, but an ontological relationship which modifies the life-world of both, modifying (in the case of the therapeutic setting) both the world of the clinician and the world of the patient. Furthermore, there is a shift from causality to reciprocity (we cannot reduce anymore an activity to one subject or another, both are involved in a reciprocal relationship). There is a primacy of relationality and a dynamic movement that can be described very well by the phenomenon of transference love.

[3] There are many traditions (in philosophy, psychopathology and cognitive science) that argue for a relational vision of the subject. We can mention phenomenology (Husserl, Scheler, Merleau-Ponty), phenomenological psychopathology (Jaspers, Binswanger) and enactivism (especially the notion of participatory sense-making, see De Jaegher & Di Paolo, 2010). Even if my approach is coherent with these, what I would like to emphasize here is the relationality at stake in the psychoanalytic tradition, a perspective which can find interesting parallelisms with Buber's philosophy in particular.

18.2. Toward a relational praxis: the role of psychoanalysis in psychiatry

What is truly needed, therefore, is a change in epistemology, a new direction able to clarify which kind of *structures* and *objects* are at stake in psychiatry. For these reasons, I will claim that the *object* of psychiatry is not a "statistical person"[4] but a living, relational and unique subjectivity who needs a kind of caring whose *structure* is relational as well.

In other words, what is truly needed is a psychiatric approach that starts from its fundamental epistemological building block: the relation, i.e. the encounter of the subject with the other. This other is the bearer of a symptom, of a narrative, and of an undefeated subjectivity, as well as of a code that can be shared, while also being unfamiliar (*unheimlich*) and therefore disturbing. The relationship at stake in the therapeutic setting can be conceived of as an interest in the other, a turning towards it, a fascination or even an attraction: an *eros* that creates a bridge between the two people involved and that can be defined as the "transference love,"[5] to use the words of classical psychoanalysis.

For these reasons, it seems that psychoanalysis can play a fundamental role in the definition of a renewed epistemology of psychiatry or, at least, could offer fruitful tools for understanding the objects of psychiatry as relational ones and practicing a kind of cure which is primarily relational and dynamic.

In fact, as argued by Gabbard et al., "psychoanalytic treatments, by which we mean psychoanalysis and psychoanalytic therapies of varying frequency and duration, must remain closely integrated with psychiatry's repertoire of treatments, for the welfare of both psychoanalysis and psychiatry" (Gabbard, Gunderson & Fonagy, 2002, p. 505). In other words,

---

[4]About the statistical person, Lacan observes that "It is necessary to get rid of the average man, who does not exist. It is nothing but a statistical fiction. There are individuals, and that's it...No one resembles from one point of view to the other, nobody has the same phobias, the same anguishes, the same way of telling" (Lacan, 1975).

[5]The idea of transference love is an important clinical concept which has received little attention in recent years. This notion can be traced back to Freud, who, in 1915, wrote that the phenomenon of love between patient and therapist was of dynamic clinical importance. If this love occurs within the context of the transference, then we can term this phenomenon "transference love." The concept has been developed by authors such as Glover and Eichhoff and has been absorbed into different kinds of discussion about transference psychosis, erotic or eroticized love, etc. (see also Kohut, 1971). The main aim of this paper is to emphasize the fact that *transference love is an essential ingredient to a successful treatment.*

psychoanalysis can be fully integrated into psychiatry, not only in the treatment of specific conditions, such as depression and anxiety disorders, but also in more enduring challenges, since it is able to affect long-rage vulnerability by altering the way the patient deals with stressors.

Nevertheless, I am not arguing for a dichotomy according to which psychoanalysis can face psychological problems, while biological, pharmacological therapies should be applied to disorders of the brain. Kandel underlined the fact that psychotherapy acts at the synaptic level in the brain and seems to be an "environmental experience" which influences gene expressions in the patients (see Singh & Singh, 2004). Accordingly, psychiatry should integrate psychoanalysis *and* pharmacotherapy. The role of psychoanalysis is fundamental to understanding the etiology and the treatment of mental disorders and seems to offer an interesting paradigm of the therapeutic relationship. In the next pages, I will focus especially on the notion of "transference love", and I will argue that it could be considered as the vehicle of cure because it provides the impulse necessary for understanding and translating the language of the patient.

In particular, I will take into account the myth of Eros and Psyche in order to explain what the dynamics of transference means. This is already alluded to in the title of this paper, "Able to Set Even the Waters Aflame," as Apuleius states in describing the power of Eros. More specifically, what I would like to emphasize is the fact that *Eros and Psyche* could be read as a paradigm of the clinical encounter based on transference love, a kind of therapy that could be helpful not only in psychoanalysis, but also in a form of psychiatry that is based on the key concept of "eros" in the therapeutic bind.

18.3. "Essentially, the cure is effected by love"[6]

18.3.1 Transference love in psychoanalysis

Freud (Freud, 1971) was convinced that erotic attraction was an active component of the psychoanalytic treatment. In his view, the analyst who

[6]The title of this section consists of a statement that Freud wrote to Jung in a private letter, where he talked about a dynamic power: *libido*, able to grasp and translate the unconscious. In this sense, the erotic bond and attraction seem to be elements which are actively involved in the therapeutic process. What is striking here is the dynamic character of phenomena, like transitional states implied both in an objectual regressive relationship and in a new and real relationship at stake in the psychiatric setting. In this sense, transference love is a dynamic and perspectival power able to repair the injuries of attachment through an internalization of care.

listens with care and concern to the patient's thoughts and seeks to understand her/him, solicits a kind of sexual dependency in the patient. On the other hand, the patient might have a similar effect on the analyst, who could not reciprocate that love, but can use it to build a model of a cure. This lack of reciprocity was considered as the element which differentiated transference love from extra-analytic love, that is the love existing outside of the therapeutic setting.

Freud did not focus his work on transference, but this notion has remained at the center of psychoanalytic thought, and many thinkers developed the idea that transference can be reciprocated to varying degrees (see, for instance, Joseph, 1993). In any case, the most important thing is that, regardless of the analyst's internal response to the patient's love, s/he *must analyze* the transference love. In this view, the reciprocity which is usually implied in "non-therapeutic" love may be felt but cannot be articulated in the same way (see Gabbard, 2004). The analyst should treat the love as *material to be understood* in all of its deep and unconscious meanings. This process transforms romantic space into a space of analysis, that is a space where emotions and passions are used to build symbolic links between personal meanings and what is happening between patient and analyst. The conflicts and themes that arise in the therapeutic relationship are useful for illuminating the nature of problems in the patient's relationships outside of therapy.[7]

In this phase, the therapist should have a profound respect for the patient's autonomy, s/he cannot impose her/his own bias but, on the contrary, s/he promotes individuation. Furthermore, transference allows for a greater impact in the therapeutic alliance, creating a supporting, empathic environment which is really helpful to make the patient open toward therapist's interpretations. In this way, transference increases reflective functioning and helps the patient learn about her/his internal objects.

We can thus affirm that:

---

[7] A study by Høglend (Høglend, 2006) reported a randomized trial of dynamic psychotherapy designed to determine the impact of transference interpretations (1-3 per session) in a once-weekly psychotherapy for the duration of 1 year. The study was applied on one hundred patients and the results have shown that they benefited more from the therapy using transference interpretation than from alternative treatments.

- From an *epistemological* perspective, transference love has an *explanatory* power, since it allows the therapist to interpret the patient's feelings and build logical hypotheses;
- From a *practical* point of view, transference love is crucial in creating an *attitude of care.*

In other words, transference seems to be the pivot upon which the entire structure of psychoanalytic treatment rests. Accordingly, psychoanalysis can still have a role in psychiatry, both in its epistemological structure and in its practical stances. Gabbard's psychodynamic psychiatry and its use of transference love represent an interesting example.

18.3.2 Transference love in psychodynamic psychiatry: Gabbard's perspective

Transference phenomena are the center of a psychodynamic psychiatry[8], as Gabbard's reflection (Gabbard, 2002, 2003) emphasizes, specifying how transfer originates from premature relationships of attachment developed during childhood and influenced by the therapist's behavior. Accordingly, all the relationships developed within the clinical setting consist of a real bond *and* of a transferal phenomenon. In the context of psychodynamic psychiatry, we can observe two kinds of transference: a repetitive dimension, in which the patient fears and expects that the psychiatrist behaves like her parents; and a perspectival-individuating dimension, in which the patient strongly wishes for a healing experience which was missing in her childhood.

These features continuously oscillate between the patient's first-person perspective, the psychiatrist's impressions and the relational field existing between them. Jung himself underlined the fact that one of his main important epistemological theses was the libido's split into a negative and positive direction, and he also argued that there is not an intimate relationship without transferal phenomena which favor or disturb such a relationship. In other words, according to Jung, transference is "the alpha" and "the omega" of the analytic method and content.

As I have described before, these phenomena are very well known in psychoanalytic work, but what I would like to emphasize here is their role

[8] Nowadays, "psychodynamic" is broadly used as a synonymous for "psychoanalytical". In a nutshell, dynamic psychiatrists use principles from psychoanalytical theory and technique. Another key component is that the clinician treats the *person* and not just the illness, taking personality into account (Gabbard, 2012).

in the field of psychiatry. In fact, even the "pharmacological", objectifying psychiatric approach can have a transferal character: consuming or refusing the "object-medication" seems to be deeply linked to those feelings of caring involved in the prescription-ingestion movement.

In this view, the act of refusing a pill can be linked to problems caused by submission experiences, the over-control of a powerful, sometimes parental person while accepting to take a pill can express a total release of responsiveness in respect of the patient's condition.

In the same way, transferal dynamics also involve the psychiatrist, who could prescribe a significant number of pills because of a contra-transferal[9] desperation caused by the failure of her/his cure, or, on the contrary, could refuse to prescribe psycho-pharmacological therapies because of a narcissism linked to a potential lack of efficacy of psychotherapy on that specific patient. The patient's request for attention and caring is linked to the manner in which the psychiatrist conceives of the relationship, creating a peculiar de-centered, dialectically shaped subject.

As Gabbard noted, the analytic reflection makes it possible to transform the erotic transferal (and contra-transferal) field into an "analytic space," a place where it is possible to create personal meanings and play with them, where fantasy and reality live in a dialectic relationship. Transference love makes it possible to deal with an illusion which is felt simultaneously as real and unreal: it is real because "it involves a unique current relationship, and unreal, or displaced, in the sense that it has elements of past object relationships that have been internalized and then reactivated in the analytic dyad" (Gabbard, 2004, p.35).[10]

Gabbard speaks of "psychoanalytic love" (Gabbard, 2003) in terms of certainty with regard to the patient's possibilities of growing into the analytic process and in terms of trust with regard to the analytic relationship itself as caring and looking for the patient's truth, being

---

[9]Anna Reich describes the phenomenon of contra-transference as "the effects of the analyst's unconscious needs and conflicts on his understanding or technique. The patient represents for the analyst an object of the past to whom past feelings and wishes are projected." (Reich, 1951).

[10] This psychic movement is described very well by the myth of Eros and Psyche, which will be at the center of my analysis in the next section. The stall of impulses imposed by the relationship makes it possible to establish symbolic links integrating and interiorizing different kinds of emotions and affects, instead of simply acting out.

emotionally involved in a process where the feeling of being loved is probably a necessary component.

To conclude, we can claim that psychoanalytic thought itself is "a manifestation of love" (Lear, 1990, p. 28). Nonetheless, the psychiatrist's attitude is directed at *reflection, contemplation and analysis,* and not to action. If we take into account that in analytic psychiatry love is a matter of knowledge, research and exploration, we can very well argue that the myth[11] of *Eros and Psyche* can explain these psychic movements in an extremely deep and intuitive manner.

18.4. *Eros and Psyche* as a paradigm of the therapeutic encounter

18.4.1 The myth

Preserved as a form of narration in the tale "*Asinus aureus*" ("*The Golden Ass*") by Lucius Apuleius, *Eros and Psyche* is one of the most famous stories in classical mythology.

Psyche is a stunningly beautiful girl and people traveled from all over the world to admire her. Aphrodite, goddess of beauty, became jealous of her and asked for her son, Eros, the god of love, to make Psyche fall in love with the most detestable creature in the whole world. Nonetheless, upon seeing her, Eros immediately fell in love with her. Instead of bringing her a monster, he spirited her to his magnificent palace and made her his wife. However, because Psyche was a mortal, Eros could not allow her to know either who he was or what he looked like. He visited her only at night and made her agree never to look upon him.

Psyche lived a happy life with Eros, although one of mystery. Psyche's sisters, upon seeing the splendid palace where Psyche lived, became envious and devised a plot to ruin her. They poisoned her mind, assuring Psyche that if her husband was so secretive about his identity and appearance, there must be something dreadfully wrong with him.

For this reason, confused and conflicted, Psyche turned on a lamp one night as her husband lied next to her. Oil fell from her lamp and severely

---

[11]The use of this myth has a twofold aim: on the one hand, it represents the paradigm of the clinical encounter, which needs to resolve itself in an attitude of emotionally *and* reflectively characterized care; on the other hand, the myth itself seems to have an explanatory power, and to be an epistemological tool able to overcome the limits of biological epistemology and hermeneutics. In other words, a more humanized treatment not only requires a change in the definition of mental symptoms and treatments, but also a change in the way we speak about them.

burned Eros' shoulder. He awoke and finding that his wife had betrayed him, fled.

In anguish at her faithlessness and at having hurt him and then lost him, Psyche vowed to show Eros how much she loved him by spending the rest of her life searching for him. In the meanwhile, Aphrodite decided to punish Psyche severely and told her that she must perform certain tasks. These tasks were clearly impossible, but Psyche hoped to complete them and find her lost love.

First, Aphrodite took some tiny seeds of wheat, poppy, and millet, mixed them and dropped them in a single pile. She gave Psyche until nightfall to separate the seeds. But Psyche succeeded and survived the night easily. Another task consisted in retrieving a golden fleece from the river. Psyche was helped by a reed and completed the mission. Then, Aphrodite ordered Psyche to fill a flask from the mouth of the River Styx. When Psyche reached the head of the river, she realized that this task was impossible because of the rocks. This time, an eagle helped her and filled the flask. Aphrodite still did not give in. In the last trial, Psyche must descend into Hades to capture Persephone's beauty and bring it to Aphrodite: a jar of ointment that brings with it a divine, dangerous beauty that deals with death. Psyche did everything well and overcomes all difficulties: she went down and up, she was capable of plunging back into the world of the shadows, of depression and the boundaries of death. However, she was curious and opened the jar. She entered a cloud of sleep, and all seemed to be lost. Finally, Eros asked for her immortality: Zeus made Psyche immortal and allowed her to taste ambrosia, the drink of the Gods. Even Aphrodite was happy because, now that Psyche was living in the sky with her husband, men on earth had forgotten all about her and were again worshiping the true goddess of beauty.

18.4.2 A psychoanalytical interpretation: from romantic to reflective space

The occurrence of a powerful relational thrust within the psychiatrist/patient pair is well described and symbolized by some aspects of the myth of *Eros and Psyche.*

There are many papers in psychoanalysis that reflect the importance of this myth of the transformational factors capable of carrying the psyche towards integration and awareness of its various instances. The work and efforts of Psyche in the myth are paradigmatic to understanding how the development of an emotional *and* reflective consciousness is crucial for resolving personal complexes and causing the best integration of selfhood.

According to Schwartz-Salant (Schwartz-Salant, 1997) the tale of Eros and Psyche explains what it means to get into a space in which relationships are the essences of the psychic affair. Hillman (Hillman, 1979) also notes that, in Apuleius' myth, Psyche's struggles represent the *reflexive* instinct through which the affection is transformed into an experiential content: through reflection the psyche brings relationality into eros, transforming its original nature.

In other words, this myth can be used to reflect on *how the relationship and the transference love can be the foundation of a psychiatric care as care of the psyche.* This kind of care—conceived of as a transformation of eros in two instances, the reflective *and* the emotional— is very well represented in the myth.

In the Palace of Eros, we can observe the difference between aphroditic-incestual love and the psychic-relational love. In this initial phase, the reflection is intolerable (from Latin is *re-flecto*, "I lean back") because it is a depressive and uncomfortable movement. Aphrodite instead expresses the fundamental anti-psychic component of this kind of love.

The labors of *Psyche* are also labors of the *psyche*—the labor of the myth is the labor of the mind. Much has been written about the various tests of Psyche in order to understand mental and psychodynamic instances; however, I would like to focus on the particular aspects represented by some of the trials:

- In the first task, the effort of the separation of the various seeds from an undifferentiated chaos can be read as the labor of the psyche before the psychotic conditions of dissociation and fragmentation. Conditions represented symbolically by the room full of grains of different types upon which is set the test to differentiate and group them;
- In the second task, Psyche is obliged to collect the Golden Fleece of rams from the forest and must wait for the night. These animals have an instinctual impulse; they are aggressive, relentless and dangerous. In the patient-psychiatrist relationship the clinician often feels like the Golden Fleece is ahead of her/him and that s/he can just catch it. Clinical interaction—for example, with borderline patients— teaches something about the explosive aggression of the rams from the myth;
- Then, Psyche meets Pan, through Syrinx, the nymph of watercourses. The nymph sings a wistful song that puts Psyche in a depressive condition, since she puts Psyche into communication

> with her internal sweet and reflective voice. Now Psyche can sustain herself in depression, she can talk with her depression and she can get the depressive message, she can have a relationship with her own body, with her emotional soul and she cannot dissociate these aspects. This movement does not take place under the devouring impulse of a big, yellow sun, but in the cool twilight, in the dusk, when the idea has reached sufficient maturity and impulses are at rest. From this movement comes a thoughtful creativity and not an impulsive or accelerated one. This movement of Psyche reflects the movement of the psyche: in this sense, analytical psychiatry pays particular attention to the movements that go towards *individuation*, becoming the "direction of the cure" (Recalcati, 2016).

Finally, after this journey, Eros and Psyche can meet again. He stings her with an arrow of love in order to wake her, but the wound is small and does not cause her pain because, in a mature conscience of depression, the wound is just a small pain. Psyche produces an important transformational process, turning venereal beauty into a beauty conscious of the underworld.

The true interest of the relationship is not partial or complexual but is a stand in front of the irreducible force of Eros and the continuous moving of Psyche. A desire not reduced to the violence of fusional embrace in the castle of Aphrodite, but as a constant dynamic that brings the characters into deepness, makes them curious, sometimes makes them lost, and always enables them to find each other again in this dynamic dyad.

This myth can be read as a reflection on the meeting between psychiatrist and patient. In this view, objectifying psychiatry seems similar to the castle of Eros in the dark, concerned only in silencing the symptom in a containing way, without the self-reflecting component capable of producing a transformation in the therapeutic relationship. Nonetheless, Psyche is very curious, and thanks to her curiosity can be identified as the subject of the myth. Curiosity is a vital part of the psyche and enables the game of mutual awakening between the two, an awakening that creates the balance of the Self[12]. The relational Eros, therefore, allows the patient

[12] The Self (Selbst) in Carl Gustav Jung's vision of analytical psychology is defined as the psychic totality with respect to which the self, our conscious part, is only a small part.

to become curious about her own narrative, with respect to deepening in her own internal world, to her emotions. A catalyst for this reaction can be precisely the interested listening of the psychiatrist.

In the descent into the underworld, Psyche is warned not to perform acts of piety or care towards people she meets there because she would be deceived and she would remain trapped there. It is very important to emphasize this step because it is very hard for her to avoid being sentimental towards what is destined to die, end and suffer. A similar attitude in therapy might be for the psychiatrist to surrender to piety towards the patient—if the psychiatrist indulges in this sentiment, aiming at keeping alive something dead by following the *furor sanandi*. Instead, the psychiatrist cannot try to cure something that cannot be cured, but only seen, understood and integrated into the internal world of the patient. In this sense, in order to penetrate the waters of Hades and remain undamaged, it is necessary to work with conscious attention *transforming the romantic space into a reflective space*. What seems to be important from the clinical point of view is the fact that the psychic activity, in order to be developed, actually needs the desire of a relational eros. Therefore, the evolutionary process passes through the erotic transference: in the movement of Psyche towards Eros, the relationship is *therapeutic*. In this view, transference love is central because: it allows the therapist to interpret the patient's feelings; it promotes individuality; it creates a caring and empathic environment. Its twofold nature—real and unreal, emotional and reflective—is represented by the dynamic movement between Eros and Psyche. The myth can be considered, as Neumann does (Neumann,1989), an archetype of a therapeutic relationship based on transference love: from an unconscious, elementary, impersonal attraction, the mythical story turns into a complex relationship (from an "I-it" relationship, we witness the arising of an "I-Thou" bond, in Buber's terms (see Buber, 1923).

Jung believes that the task of the analytic activity is to establish a hierarchical relationship between self and myself, between the whole and the part, able to satisfy the conditions for a resumption of the evolutionary movement that he calls "individuation. In Jung's vision, "The self is not only the center, but also the whole circumference which embraces both conscious and unconscious; it is the center of this totality" being "the archetype of wholeness and the regulating center of the psyche as a transpersonal power that transcends the ego. As an archetype, it cannot be seen directly, but by ongoing individuating maturation and analytic observation, can be experienced objectively by its cohesive wholeness making factor" (Jung, 1981).

I would like to conclude this section by referring to James Hillman's thought (Hillman, 1979) that wherever a so-called "therapeutic relationship" is present, a transference occurs, and wherever a transference phenomenon arises, we are inside a therapeutic relationship. In this sense, a psychiatric practice should always take into account the relational component.

18.5. Conclusion: What kind of love in the therapeutic setting?

This paper has a twofold aim: accounting for transference love and its importance in the therapeutic relationship and arguing that psychoanalysis can still have a role in psychiatry's epistemology and practice.

In the first part, I argued for a shift from an objectifying approach to a relational therapy. Then, I described the role of transference love in the clinical setting, and I claimed that it can allow for a deep understanding of patient's thought *and* create an attitude of care which is necessary for the therapeutic relationship. In other words, transference love seems to have both *epistemological* and *practical* relevance. As also showed by Gabbard's psychodynamic psychiatry, transference can be fully integrated into psychiatry, working together with other approaches.

In other words, the kind of love which is possible for the psychiatrist is not only a relational exchange: transference is not a repetition of an ancient objectual relationship, which can be more or less traumatic, but is also a new kind of love, an expression of the transformative power of relationality, able to create a space where emotions and reflection co-exist. The myth of Eros and Psyche seems to represent well the rise of an emotional and reflective kind of love, thanks to which psyche's struggles can be defeated. We can mention that, in the beautiful text *Grundformen und Erkenntnis menschlichen Daseins* (Binswanger, 1942), Binswanger claims that we really know someone through love, and especially in the understanding of an ill person, we need love. In this view—as in Buber's—the dimension of "we" precedes the "I" and the "Thou," and we exist only in an authentic encounter. V. E. Gebsattel also claims that there is no therapeutic relationship without a close bond between patient and clinician (von Gebsattel, 1964).

To conclude, a renewed epistemological psychiatric paradigm needs to take into account the relational character of the patient and the expressive power of symptoms as its *object* and should have a relational *structure* where love is a primary component (and the psychiatrists themselves relational tools). As Loewald observed: "In our work we can truly be said that in our best moments of dispassionate and objective analyzing, we love our object, the patient, more than at any other time and are compassionate with his whole being" (Loewald, 1970, p. 297).

18.6 References

Apuleio, L. (2002). *La Favola di Amore e Psiche,* Roma: Grandi Tascabili Economici Newton.

Baldessarini, R. J. (2014). The impact of psychopharmacology on contemporary psychiatry. *Canadian Journal of Psychiatry,* 59(8),401-405.

Bennett, M. R., & Hacker, P. M. S. (2003). *Philosophical foundations of neuroscience.* Oxford, England: Blackwell.

Berrios, G. (2010). *On evidence-based medicine,* www.psicoevidencias.es

Berrios, G. (2016). *Per una nuova epistemologia,* Roma: Giovanni Fioriti Editore.

Berrios, G., Markovà, I. (2015). Toward a New Epistemology of Psychiatry. *Re-Visioning Psychiatry. Cultural Phenomenology, Critical Neuroscience and Global Mental Health* (eds.) L. Kirmayer, R. Lemelson, C. Cummings, Cambridge, Cambridge University Press, 41-64.

Binswanger, L. (1942). *Grundformen und Erkenntnis menschlichen Daseins,* Zürich, Niehans.

Bracken, P., Thomas, P., Timimi, S., Asen, E., Behr, G., Beuster, C., Yeomans, D. (2012). Psychiatry beyond the current paradigm. *British Journal of Psychiatry,* 201(6), 430 4. http://dx.doi.org/10.1192/bjpbp.112.109447

Buber, M. (1923). L'Io e il Tu. *Il Principio dialogico* (1958), Ed. di Comunità, Milano.

Cargnello, D. (1999). Ambiguità della psichiatria. *Comprendre,* 9, 7-48.

Choudhury, S., & Slaby, J. (Eds.). (2012). *Critical neuroscience.* Oxford, England: Wiley Blackwell.

Cohen, C. I., & Timimi, S. (2008). *Liberatory psychiatry: Philosophy, politics, and mental health.* Cambridge, England: Cambridge University Press.

De Jaegher, H., Di Paolo, E, Gallagher, S. (2010). Can Social Interaction Constitute Social Cognition. *Trends in Cognitive Sciences* 14 (10), 441-447.

Eickhoff F.W. (1987). A short annotation to Sigmund Freud 's observation on transference love. *International Review of Psychoanalysis* 14:103-109.

Fernando, S. (2014). *Mental health worldwide: Culture, globalization and development.* New York, NY: Palgrave Macmillan. http://dx.doi.org/10.1057/9781137329608

Freud S (1971). Observation on Transference Love, in *Standard Edition,* London, Hogarth Press, 16:431–463.

Gabbard, G.O. (2002). *Psichiatria Psicodinamica,* Milano: Raffaello Cortina Editore.

Gabbard, G.O. (2004). *Love and Hate in the Analytic Setting,* Oxford: Rowman and Littlefield Publishers.

Gabbard, G. (2012). Current Psychodynamic Approaches in Psychiatry. *Oxford Medicine Online,* Oxford University Press, *DOI* 0.1093/med/9780199696758.001.0001.

Gabbard, G., Gunderson, G., Fonagy, P. (2002). The Place of Psychoanlytic Treatments Within Psychiatry. *Archive of General Psychiatry* 59, 505-510.

Gebsattel, V.E. von (1954). *Prolegomena einer medizinischen Anthropologie,* Berlin-Göttingen-Heidelberg, Springer.

Glover E. (1951). *The Technique of Psychoanalysis,* New York, International University Press.

Hillman, J. (1988). *Saggi sul Puer,* Milano: Raffaello Cortina Editore.

Hofmann, S. G., Asmundson, G. J., & Beck, A. T. (2013). The science of cognitive therapy. *Behavior Therapy,* 44(2), 199 212. http://dx.doi.org/10.1016/j.beth.2009.01.007

Høglend P, Amlo S, Marble A, Bøgwald K-P, Sørbye Ø, Sjaastad MC, Heyerdahl O. (2006). Analysis of the patient-therapist relationship in dynamic psychotherapy: an experimental study of transference interpretations. *American Journal of Psychiatry* 163:1739–1746.

Joseph, B. (1993). On transference love: some current observations. *On Freud's "Observations on Transference Love",* eds. Person, E.; Hagelin, A., Fonagy, P., New Haven: Yale university Press, 102-113.

Jung, CG. (1981). *Psicologia e alchimia,* Torino: Editore Boringhieri.

Kohut H. (1971). The analysis of the self. *The Psychoanalytic Study of the Child,* New York, International Universities Press.

Lacan, J. (1975). *Escritos 2.* Mexico DF: Siglo XXI Editores.

Loewald, E.P. (1970). Psychoanalytic theory and psychoanalytic process. *Papers on Psychoanalysis,* 277-301

Lopez Pedraza, R. (2005). Su *Amore e Psiche. Una favola per l'anima,* Bergamo: Moretti & Vitali Editori.

Neumann, E. (1989). *Amore e Psiche. Un'interpretazione nella psicologia del profondo,* Roma: Casa Editrice Astrolabio - Ubaldini Editore.

Paris, G. (1997). *La rinascita di Afrodite,* Bergamo: Moretti & Vitali Editori.

Parnas, J., Sass, L. Zahavi, D. (2012). Rediscovering Psychopathology: the Epistemology and Phenomenology of the Psychiatric Object. *Schizophrenia Bulletin,* doi:10.1093/schbul/sbs153

Recalcati, M. (2016). Jacques Lacan. Volume II. *La Clinica Psicanalitica: Struttura e Soggetto,* Milano: Raffaello Cortina Editore.

Reich A. (1951). On counter-transference. *International Journal of Psychoanalysis* 32: 25-31.

Schwartz-Salant, N. (1997). *Borderline. Visione e Terapia,* Milano: La Biblioteca di Vivarium.

Singh A, Singh S. (2004). Resolution of the secolarization of ideologies and approaches in psychiatry. *Mens Sana Monographs* 2004; 2(2): 5-32.

List of Acronyms

AD	Antidepressant
AD	Anxiety Disorders
ADHD	Attention Deficit Hyperactivity Disorder
AIQ	Achievement Intellectual Quotient
ANS	Autonomous Nervous System
ASD	Autistic Spectrum Disorder
ATC	Association of Therapeutic Communities
AVHs	Auditory Verbal Hallucinations
BS	Basic Symptoms
BSI	Brief Symptom Inventory
BPD	Borderline Personality Disorder
BPRS	Brief Psychiatric Rating Scale
CBT	Cognitive-behavioral therapy
CGI-C	Clinical Global Impression - change scale
CGI-S	Clinical Global Impression - severity scale
CGOM	Client-generated Outcome Measure
CGOMS	Client-Generated Outcome Measures
CHD	Coronary Heart Disease
CHERRIES	Checklist for Reporting Results of Internet E-Surveys
CHR	Clinical High Risk
CORE IMS	Clinical Outcomes in Routine Evaluation Information Management Systems

CORE-net	Clinical Outcomes in Routine Evaluation - Net
CORE-OM	Clinical Outcomes in Routine Evaluation – Outcome Measure
CRM	Centro de Recuperaçao de Menores
DD	Double Diagnosis
DLD	Delayed Language Development
DIR	Developmental, Individual difference, Relationship-based
DIRE	Developmental, Individual difference, Relationship -based, Embodied
DSM	Diagnostic and Statistical Manual of Mental Disorders
ECT	Electroconvulsive Therapy
FDA	US Food and Drug Administration
FEP	First Episode of Psychosis
FSIQ	Full Scale Intellectual Quotient
HOC	Higher-Order Cognition
HPA	Hypothalamic-Pituitary-Adrenal Axis or System
HRSD	Hamilton Rating Scale for Depression
ICD	International Classification of Diseases and Related Health Problems
IDD	Intellectual Developmental Disorder
IIF	Interpersonal Interpretative Function
IIF-A	Interpersonal Interpretative Function – Affective-Oriented Processes
IIF-C	Interpersonal Interpretative Function – Cognitive-Oriented Processes
IIHSCJ	Irmãs Hospitaleiras do Sagrado Coração de Jesus
IPPS	Individualized Patient-Progress System
LLE	Living-Learning Experience

MBI	Maslach Burnout Inventory
MRI	Magnetic Resonance Imaging
NHS	National Health Service (in the United Kingdom)
NICE	NICE National Institute for Health and Care Excellence
OT	Occupational Therapy
PANSS	Positive and Negative Syndrome Scale
PAS-ADD	Psychiatric Assessment Schedules for Adults with Developmental Disabilities
PASTOR	Positive Appraisal Style Theory of Resilience
PD	Personality Disorder
p factor	psychopathology factor
PM	Psychomotricity
PQ	Personal Questionnaire
PSYHLOPS	Psychological Outcomes Profiles
QI	Intelligence Quotient or Level
RCPsych	Royal College of Psychiatrists (London)
RCT	Randomised Controlled Trial
rTMS	Repetitive transcranial magnetic stimulation
SCL-90-R	Symptom Checklist - 90 – Revised
SIPS	Structured Interview for Prodromal Syndromes
SLI	Specific Language Impairment
SPSS	Statistical Package for the Social Sciences
ST	Speech Therapy
TC, TCs	Therapeutic Community, Therapeutic Communities
TRD	Treatment-resistant depression
UHR	Ultra High Risk approach

VIQ	Verbal Intellectual Quotient
WHO	World Health Organization
WISC-II	Weschler Intelligence Scale for Children

Index

A

B

C

D

E

F

G

H

I

J

K

L

M

N

O

P

www.ingramcontent.com/pod-product-compliance
Lightning Source LLC
LaVergne TN
LVHW050804220926
848485LV00013B/1426/J
* 9 7 8 1 6 2 2 7 3 4 3 3 7 *